ENCYCLOPAEDIA OF
MATHEMATICS

ENCYCLOPAEDIA OF
MATHEMATICS

Surender Sachit

ANMOL PUBLICATIONS PVT. LTD.
NEW DELHI - 110 002 (INDIA)

ANMOL PUBLICATIONS PVT. LTD.
H.O.: 4374/4B, Ansari Road, Darya Ganj,
New Delhi-110 002 (India)
Ph.: 23278000, 23261597

B.O.: No. 1015, Ist Main Road, BSK IIIrd Stage
IIIrd Phase, IIIrd Block
Bangalore - 560 085 (India)
Visit us at: www.anmolpublications.com

Encyclopaedia of Mathematics

First Published, 2007

PRINTED IN INDIA

Printed at Mehra Offset Press, Delhi.

Preface

The Encyclopaedia of Mathematics covers all aspects of Mathematics. This Encyclopaedia covers a range of various Mathematics disciplines such as Logic & Model Theory, Computability Theory & Recursion Theory, Set Theory, Category Theory , Group Theory , Ring Theory, Field Theory, Module Theory, Galois Theory, Number Theory, Combinatorics , Algebraic Geometry, Real Analysis & Measure Theory > Calculus, Complex Analysis, Tensor & Vector Analysis, Differential & Integral Equations , Numerical Analysis, Functional Analysis & The Theory of Functions, Euclidean Geometry, NonEuclidean Geometry, Absolute Geometry, Metric Geometry, Projective Geometry, Affine Geometry, Discrete Geometry & Graph Theory , Differential Geometry , PointSet or General Topology, Algebraic Topology, Probability Theory, Statistics, Game Theory and Systems & Control Theory.

This new encyclopaedia integrates all of the diverse approaches that have gone into the making of the modern field of Mathematics. It is unique among available reference works in that it covers the most recent topics of research. Encyclopaedia of Mathematics is intended for all those interested in this field. It will be of immense value to students of all kinds, especially students of different fields of mathematics.

It will also be a major source of reference for professionals and students, who require an authoritative and up-to-date guide to assist them in their work. In preparation of this book, the author has freely consulted large number of books and journals so no authenticity is claimed. Author is especially thankful to Anmol Publications Pvt. Ltd., New Delhi for shaping this book in its final form. Suggestions for further improvement of this book are not only welcome but also greatly appreciated.

Author

A The metric prefix symbol a- (metric prefix: atto-) denotes a multiple of ten to the negative eighteenth power (10^{-18} = 0.000 000 000 000 000 001) as defined by Le Système International d'Unités (SI).

AAS Reference to solving a triangle given the measure of two angles and the length of a non-included side.

Abscissa Horizontal axis.

Absolute maximum The all-time, one-and-only, single, absolute and total maximum value of a function over a specified domain of the function. (Although it is the unique maximum value, it could occur at more than one point, as when you have two mountain peaks of exactly the same height.) The absolute maximum is sometimes also called the global maximum.

Absolute minimum On a graph , it is a point that is having a really bad day.

Absolute risk Probability of an event over a period of time; expressed as a cumulative incidence like 10-year risk of 10% (meaning 10% of individuals in the group of interest will develop the condition in the next 10 year period). It shows the actual likelihood of contracting the disease and provides more realistic and comprehensible risk than relative risk/odds ratio.

Absolute value The absolute value is defined for real and complex numbers. For real numbers, the absolute value coincides with the number itself if the latter is either positive or zero. The absolute value of a negative number is obtained by multiplying the number by –1, i.e. by changing its sign. The absolute value if a number r is denoted |r|. Therefore, |r|=r for r≥0 and |r|=–r for negative r. In other words, |r| is the distance form number r to the origin. In this form, the definition applies to complex numbers that are identified with points in the plane.

Absolute value of 2 Both 2, i.e., |2| = 2 and |–2| = 2.

Absolute value of a complex number The absolute value of a complex number a + bi is the square root of a square plus b square.

Abundant A number is abundant if sigma(n) > 2n, or, equivalently, if index(n) > 2. Another way of saying this is that sum of the proper divisors of n exceeds n.

Any multiple of an abundant number or of a perfect number is abundant. An abundant number which is neither a multiple of another abundant number nor a multiple of a perfect number is called primitive abundant.

Acceleration Acceleration is the rate of change of the velocity. Since the rate of change of a function is its derivative, the acceleration is the derivative of the velocity function. Since the velocity function is itself the derivative of the function giving your position, the acceleration function is the second derivative of the function giving your position.(In maths, a= $dv/dt = d^2s/dt^2$, where s is the position function).

Acceptance testing The process whereby actual users test a completed information system, the end result of which is the users' acceptance of it.

Access interaction An access interaction is a shorthand for a bidirectional interaction with the system being modeled. It is used to obtain information about objects, usually regarding relationships and related objects.

Ace 1. The noun and adjective ace means one (1). An ace in parlor games is a playing card, die, or domino that represents the number one. An ace in golf is a hole made in one stroke, also called a hole-in-one. An ace in racket sports, volleyball, or handball is a point made on a serve without contact by the opponent.

2. The noun and adjective ace means of first (1st) or high rank or quality.

3. In military aviation, the noun ace means an aviator who has destroyed five (5) or more enemy aircraft. More generally, ace means one who excels at something.

Action An action is a process in states or a procedure in transitions that allow states or transitions to do some processes.

Action stubs That part of a decision table that lists the actions that result for a given set of conditions.

Activation The time period during which an object performs an operation.

Actor An external entity that interacts with the system (similar to an external entity in data flow diagramming).

Acuity Acuteness; keenness, as of thought or vision.

Acute angles An acute angle is an angle measuring between 0 and 90 degrees.

Example: The following angles are all acute angles.

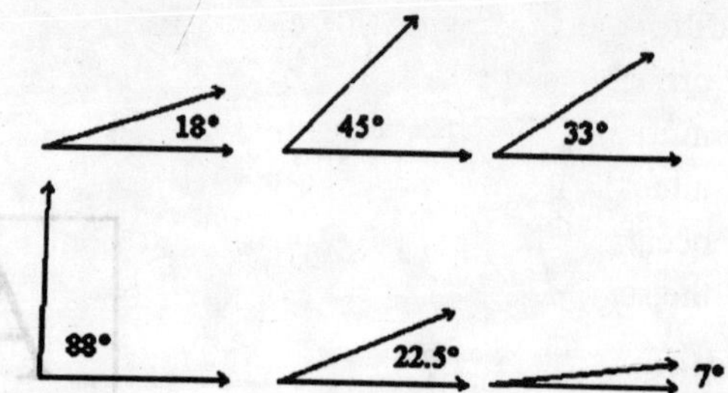

Acute triangle A triangle having three acute angles.

Examples:

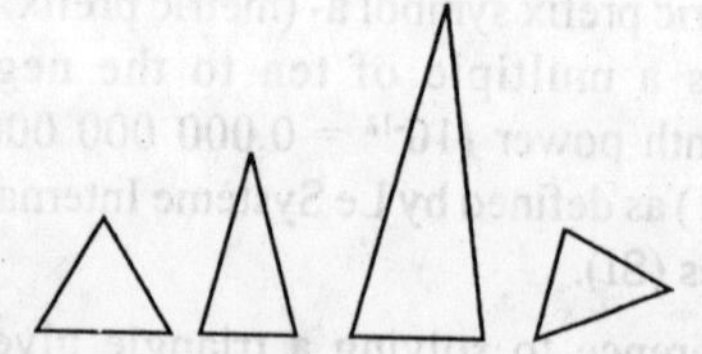

Adaptation A change in behavior to meet situational demands.

Adaptive maintenance Changes made to a system to evolve its functionality to changing business needs or technologies.

Add interaction An add interaction adds an object that already exists in the model to an object class.

Addition rule The probability of any of one of several mutually exclusive events occurring is equal to the sum of their individual probabilities. A typical example is the probability of a baby to be homozygous or heterozygous for a Mendelian recessive disorder when both parents are carriers. This equals to 1/4 + 1/2 = 3/4. A baby can be either homozygous or heterozygous but not both of them at the same time; thus, these are mutually exclusive events.

Addition The operation, or process, of calculating the sum of two numbers or quantities. Additive inverse of a matrix For each matrix A, there is a matrix -A(the additive inverse of A), such that A+(-A) = Z.

Additive inverse Two numbers are additive inverses of each other if their sum is 0 (e.g., since $-4 + 4 = 0$, then -4 and 4 are additive inverses of each other).

Additive law of probability The rule giving the probability of the occurrence of one or more mutually exclusive events. Given a set of mutually exclusive events, the probability of the occurrence of one event or another is equal to the sum of their separate probabilities.

Adjacency Matrix Adjacency matrix of an (undirected) graph is a matrix whose order is the number of nodes, and whose entries record which nodes are connected to each other by a link or edge of the graph.

If two nodes I and J are connected by an edge, then $A_{i,j}=A_{j,i}=1$. All other entries of the matrix are 0. Thus, an adjacency matrix is a zero-one matrix. The usual convention is that a node is not connected to itself, and hence the diagonal of the matrix is zero.

The product $A^2=A\times A$ is a matrix which records the number of paths between nodes I and J. If it is possible to reach one node from another, it must be possible in a path of no more than n–1 links. Hence, the reachability matrix, which records whether it is possible to get from node I to node J in one or more steps, can be determined by taking the logical sum of the matrices I, A, A^2, ..., A^{n-1}.

(Loops:) If edges are allowed which begin and end at the same node, (sometimes called loops) then the adjacency matrix can have some 1's on the diagonal.

If edges are assigned a direction, then it is possible that there is an edge from node I to node J, but not in the reverse direrection. In this case of a directed graph, an adjacency matrix can be defined as well. The only difference is that the adjacency matrix now need not be symmetric. The reachability matrix can still be defined in the same way.

If multiple edges are allowed between two nodes, (which might be distinguished by color in a drawing), then the adjacency matrix can record this information by storing in $A_{i,j}$ the number of edges between the two nodes. The reachability matrix can still be defined in the same way, and still counts the number of different routes correctly.

Adjacent angles Two angles that share a ray, thereby being directly next to each other.

Adjacent side The side of a right triangle that is next to the reference angle. (Not hypotenuse).

Adjacent values Actual data points that are no more extreme than the inner fences.

Adjoint matrix The word adjoint has a number of related meanings. In linear algebra, it refers to the conjugate transpose and is most commonly denoted A^H. The analogous concept applied to an operator instead of a matrix, sometimes also known as the Hermitian conjugate (Griffiths 1987, p. 22), is most commonly denoted using dagger notation $A^\dagger$ (Arfken 1985). The adjoint operator is very common in both Sturm-Liouville theory and quantum mechanics. For example, Dirac denotes the adjoint of the bra vector

$< P|\alpha$ as $\alpha^\dagger|P >$, or $\bar{\alpha}|P >$.

Given a second-order ordinary differential equation

$$\tilde{\mathcal{L}} u(x) \equiv p_0 \frac{d^2 u}{d x^2} + p_1 \frac{d u}{d x} + p_2 u \quad (1)$$

with differential operator

$$\tilde{\mathcal{L}} = p_0 \frac{d^2}{d x^2} + p_1 \frac{d}{d x} + p_2, \quad (2)$$

Where $p_i \equiv p_i(x)$ and $u \equiv u(x)$, the adjoint operator $\tilde{\mathcal{L}}^\dagger$ is *defined* by

$$\tilde{L}^{\dagger} u \equiv \frac{d^2}{d x^2}(p_0 u) - \frac{d}{d x}(p_1 u) + p_2 u \quad (3)$$

$$\equiv p_0 \frac{d^2 u}{d x^2} + (2p_0' - p_1)\frac{du}{dx} + (p_0'' - p_1' + p_2)u \quad (4)$$

Writing the two linearly independent solutions as $y_1(x)$ and $y_2(x)$, the adjoint operator can then also be written

$$\tilde{L}^{\dagger} u = \int (y_2 \tilde{L} y_1 - y_1 \tilde{L} y_2)\, d x = \left[\frac{p_1}{p_0}(y_1' y_2 - y_1 y_2')\right] \quad (5)$$

In general, given two adjoint operators $\tilde{A}$ and $\tilde{B}$,

$$(\tilde{A}\tilde{B})^{\dagger} = \tilde{B}^{\dagger}\tilde{A}^{\dagger}, \quad (6)$$

which can be generalized to

$$(\tilde{A}\tilde{B}\cdots\tilde{Z})^{\dagger} = \tilde{Z}^{\dagger}\cdots\tilde{B}^{\dagger}\tilde{A}^{\dagger}.$$

Adjusted correlation (r_{adj}) A correction to the computed correlation coefficient to adjust for the number of predictors relative to the sample size.

Adjusted means Means that have been adjusted for differences on a covariate.

Adjusted odds ratio In a multiple logistic regression model where the response variable is the presence or absence of a disease, an odds ratio for a binomial exposure variable is an adjusted odds ratio for the levels of all other risk factors included in a multivariable model. It is also possible to calculate the adjusted odds ratio for a continuous exposure variable. An adjusted odds ratio results from the comparison of two strata similar at all variables except exposure (or the marker of interest). It can be calculated when stratified data are available as contingency tables by Mantel-Haenszel test.

Adult learners Students age 25 or older.

Affine cipher Affine ciphers use linear functions to scramble the letters of secret messages.

Affine space The best known example of the affine space is the plane although a straight line is another example. One important feature of affine spaces is the ability to fix a point and coordinate axes through it so that every point in the space can be represented as a tuple of its coordinates. Every (ordered) pair of points A and B in an affine spaces is associated with a vector AB, A being the beginning while B the end point of AB. For any three points A, B, C the following holds AB+BC+CA=0.

Affirmative The adjective and noun affirmative is a logical value. Negative and affirmative are sometimes used as a binary logic pair.

Algebraic expressions Expressions that are made up of variables, numbers, grouping sym- bols, operation signs, and exponents.

Algebraic number An algebraic number is a number that can be expressed as a root of a polynomial equation with integer coefficients. Examples include 0, 1, i, $2^{½}$, φ, –3.456, 7+8i, $9\cdot e^{i\cdot\pi/4}$, etc.

Algebraic vector An ordered pair of numbers representing the terminal point of a standard vector.

Algorithm A fixed process which if caried out systematically produces a desired result. Thus the Euclidean algorithm is a set of rules which when applied to two integers produces their common divisor.

All The pronoun and adjective all means the whole of or every member of.

All subsets regression The result of a stepwise multiple regression when the program chooses that subset of variables that has the highest correlation with the criterion.

Almost The adverb and adjective almost means slightly less than the specified amount.

Alpha (α) Statisticians use the Greek letter alpha (α) to indicate the probability of rejecting the statistical hypothesis tested when in fact, that hypothesis is true. Before conducting any statistical test, it is important to establish a value for alpha. For most psychologists, and for many other scientists, it is customary to set alpha at 0.05.

This is the equivalent of asserting that you will reject the hypothesis tested if the obtained statistic is among those that would occur only 5 out of 100 times that random samples are drawn from a population in which the hypothesis is true. If your obtained statistic leads you to reject the hypothesis tested, it's not because you believe that the obtained statistic could not have occurred by chance.

It's that you are asserting that the odds of obtaining that statistic by chance only are sufficiently low (one out of twenty) that it reasonable to conclude that your results are not due to chance. Could you be in error? Of course you could, but at least you know the probability of such an error. It is exactly equal to the value you have previously established for alpha.

Alpha testing User testing of a completed information system using simulated data.

Alternate exterior angles For any pair of parallel lines 1 and 2, that are both intersected by a third

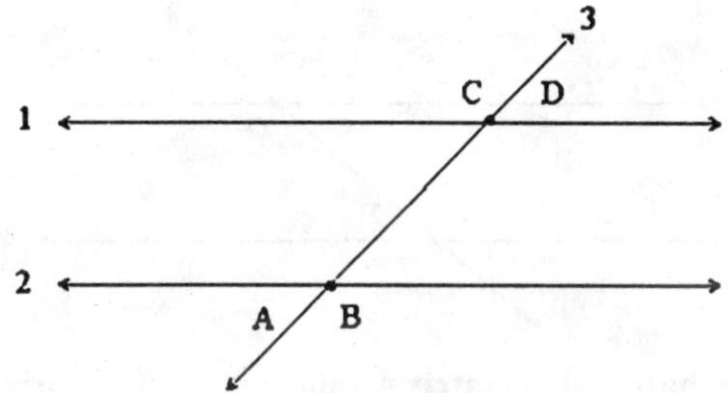

line, such as line 3 in the diagram below, angle A and angle D are called alternate exterior

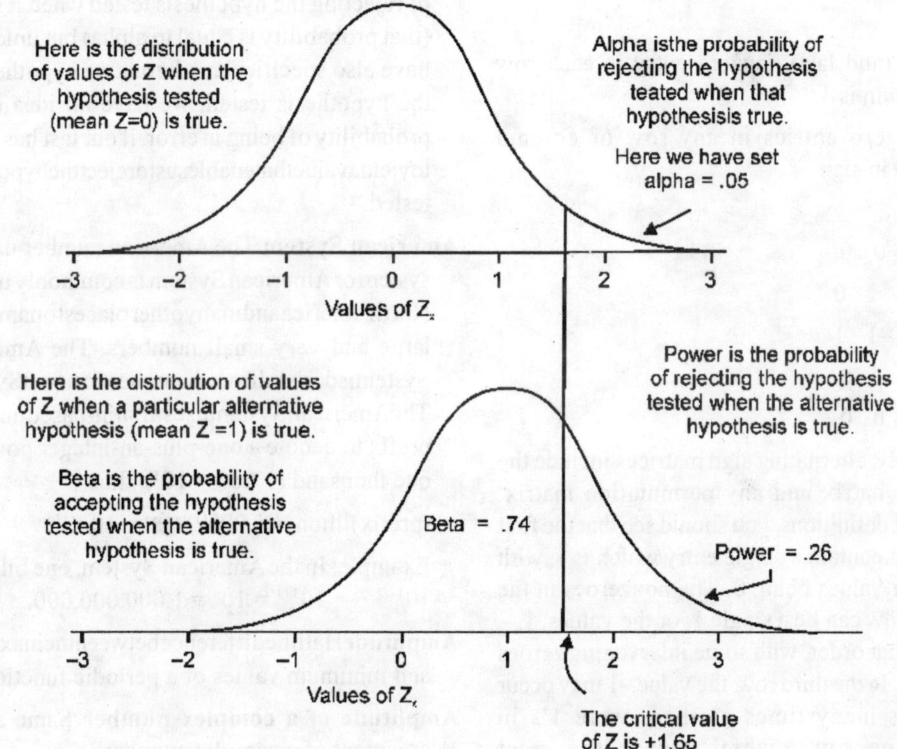

angles. Alternate exterior angles have the same degree measurement. Angle B and angle C are also alternate exterior angles.

Alternate interior angles For any pair of parallel lines 1 and 2, that are both intersected by a third line, such as line 3 in the diagram below, angle A and angle D are called alternate interior angles. Alternate interior angles have the same degree measurement. Angle B and angle C are also alternate interior angles.

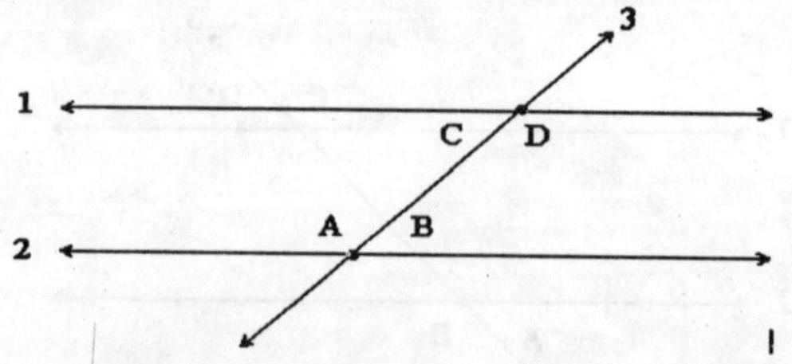

Alternating sign matrix An alternating sign matrix is an integer matrix with the properties that the entries are only 0, +1, or –1;

the sum of the entries in each row and column is 1;

the first (and last) nonzero entry in each row and column is 1;

the nonzero entries in any row or column alternate in sign.

Example:

0 1 0 0 0

1 –1 0 1 0

0 1 0 –1 1

0 0 0 1 0

0 0 1 0 0

Obviously, alternating sign matrices include the identity matrix and any permutation matrix. From the definitions, you should see that the first row must contain a single entry which is 1, with the other values being 0. The nonzeroes in the second row can be a single 1, or the values, 1, –1, 1, in that order, with some intervening zeroes possible. In the third row, the value –1 may occur up to as many times as there were 1's in preceding rows, which means the most interesting row could be 1, –1, 1, –1, 1, –1, 1. Thus the number of possible nonzeroes grows until the central row of the matrix is reached. Since the same restrictions apply from the bottom reading up, the number of possible nonzeroes must now decrease. Similar reasoning controls the nonzero population of the columns.

IfweletA$_n$denotethenumberofdistinctalternating sign matrices of order n, then it has only recently been proved that

A_n = Product (0 <= I <= N–1) (3*I+1)! / (N+I)!

Givingthesequence1,2,7,42,429,7436,218348, 10850216, .

Alternativehypothesis Thetestofagivenstatistical hypothesisentailsanassessmentofwhetherornot our sample (or samples) have yielded a statistic that is among those cases that would only occur alpha proportion of the time if the hypothesis tested is true.

In these circumstances we know the probability of rejecting the hypothesis tested when it is true (that probability is equal to alpha) but unless we have also specified an alternative hypothesis to the hypothesis tested, we have no idea of the probability of being in error, if our test has failed toyieldavaluethatenablesustorejectthehypothesis tested.

American System The American number naming system or American System is commonly used in NorthAmericaandmanyotherplacestonamevery large and very small numbers. The American SystemisderivedfromtheearlierChuquetSystem. TheAmericanSystemusesaLatin-basednumeric prefix to denote a one-plus-an-integer power of one thousand (1,000 = 10^3), thus:

(prefix)illion = $1{,}000^{(1+\text{prefix})} = 10^{3\times(1+\text{prefix})}$

Example: In the American System, one billion = $10^{3\times(1+2)} = 10^{3\times3} = 10^9 = 1{,}000{,}000{,}000$.

Amplitude Halfthedifferencebetweenthemaximum and minimum values of a periodic function.

Amplitude of a complex number Same as the argument of a complex number.

Analysis of covariance (Ancova) An analysis of variance in which the data are adjusted (or controlled) for the effects of one or more other variables.

Analysis of molecular variance (AMOVA) A statistical (analysis of variance) method for analysis of molecular genetic data. It is used for partitioning diversity within and among populations using nucleotide sequence or other molecular data. AMOVA produces estimates of variance components and F-statistic analogs (designated as phi-statistics). The significance of the variance components and phi-statistics is tested using a permutational approach, eliminating the normality assumption that is inappropriate for molecular data.

Analytic view Definition of probability in terms of an analysis of possible outcomes.

Angle 1. The shape made by two straight lines meeting in a point.

2. The space between those lines.

3. The measurement of that space in degrees or radians.

Angle bisector An angle bisector is a ray that divides an angle into two equal angles.

Example: The blue ray on the right is the angle bisector of the angle on the left.

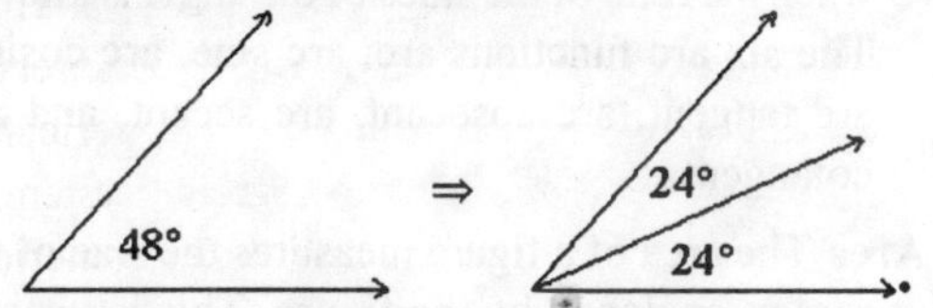

The red ray on the right is the angle bisector of the angle on the left.

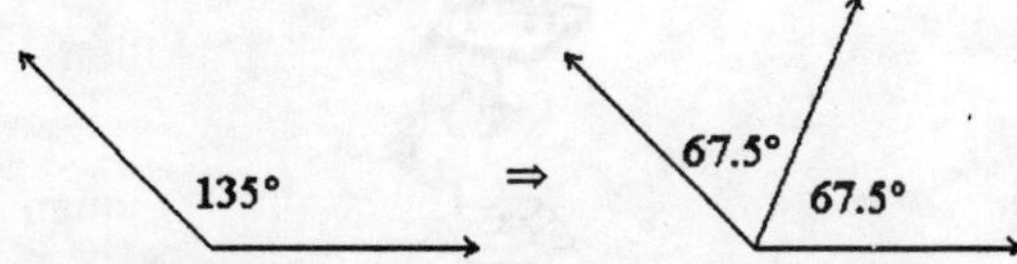

Angle measure The measure in degrees or radians of the radial distance between two rays that meet at a point.

Angle of turn The angle created when a line of work is turned. The angle of turn is formed between where the original line would have gone (if it had not turned) and the line going in the new direction.

Angle or depression An angle that measures between 0° and 90° formed by a horizontal ray and a ray from the observer to the object observed below.

Angular velocity Defined in terms of angle of rotation and time.

Annual The adjective annual means recurring once (1) each year.

ANOVA (analysis of variance) test for significant differences between multiple means by comparing variances. It concerns a normally distributed response (outcome) variable and a single categorical explanatory (predictor) variable, which represents treatments or groups. ANOVA is a special case of multiple regression where indicator variables (or orthogonal polynomials) are used to describe the discrete levels of factor variables. The term analysis of variance refers not to the model but to the method of determining which effects are statistically significant. Major assumptions of ANOVA are the homogeneity of variances (it is assumed that the variances in the different groups of the design are similar) and normal distribution of the data within each treatment group. Under the null hypothesis (that there are no mean differences between groups or treatments in the population), the variance estimated from the within-group (treatment) random variability (residual sum of squares = RSS) should be about the same as the variance estimated from between-groups (treatments) variability (explained sum of squares = ESS). If the null hypothesis is true, mean ESS / mean RSS (variance ratio) should be equal to 1. This is known as the F test or variance ratio test. The ANOVA approach is based on the partitioning of sums of squares and degrees of freedom associated with the response variable. ANOVA interpretations of main effects and interactions are not so obvious in other regression models. An accumulated ANOVA table reports the results from fitting a succession of regression models to data from a factorial experiment. Each main effect is added on to the constant term followed by the interaction(s). At each level an F test result is also reported showing the extra effect of adding each variable so it can be worked out which model would fit best. In a two-way ANOVA with equal replications, the order of terms added to the model does not matter, whereas, this is not the case if there are unequal replications. When the assumptions of ANOVA are not met, its non-parametric equivalent Kruskal-Wallis test may be used .

Antiderivative This is the opposite of the derivative.The antiderivative of a function f(x) is another function F(x) whose derivative is f(x). Also called the indefinite integral of f(x). The 'antiderivative' terminology is traditionally usually used just before the introduction of indefinite integrals , and then never used again, having been forever replaced with the term 'indefinite integral'.

Antidifferentiation The process of taking an antiderivative.

Antisymmetric matrix A square matrix A is antisymmetric if it is equal to the negative of its transpose:

$A = -A'$.

Every matrix A can be decomposed into the sum of an antisymmetric and a symmetric matrix:

$A = B + C = (1/2) * ((A - A') + (A + A'))$

Simple facts about an antisymmetric matrix A:

A has a zero diagonal;

A has pure imaginary eigenvalues;

A is normal, hence unitarily diagonalizable.

I – A is not singular;

The matrix (I + A) * Inverse (I – A) is orthogonal;

An antisymmetric matrix is also called skew symmetric.

In complex arithmetic, the corresponding object is a skew Hermitian matrix.

A-Orthogonal Vectors Two vectors u and v are said to be A-orthogonal if

$(u, A * v) = 0$.

Here A should be a positive definite symmetric matrix, which in turn guarantees that the expression (u, A * v) may be regarded as an inner product of the vectors u and v, with the usual properties.

This concept is useful in the analysis of the conjugate gradient method.

Apéry's constant Apéry's Constant is an irrational number approximately 1.202 056 903 160.

Application independence The separation of data and the definition of data from the applications that use these data.

Application server A middle-tier software and hardware combination that lies between the Web server and the corporate network and systems.

Application software Software designed to process data and support users in an organization. Examples of application software include spreadsheets, word processors, and database management systems.

Arc A curved line whose points are equal distances from a single point.

Arc function (also called inverse function) The function used to find the degrees of an angle when the ratio of the sides of the angle is known. The six arc functions are: arc sine, arc cosine, arc tangent, arc cosecant, arc secant, and arc cotangent.

Area The area of a figure measures the size of the region enclosed by the figure. This is usually

expressed in terms of some square unit. A few examples of the units used are square meters, square centimeters, square inches, or square kilometers.

Area of a circle The area of a circle is Pi $\times r^2$ or Pi $\times r \times r$, where r is the length of its radius. Pi is a number that is approximately 3.14159.

Example: What is the area of a circle having a radius of 4.2 cm, to the nearest tenth of a square cm? Using an approximation of 3.14159 for Pi, and the fact that the area of a circle is Pi $\times r^2$, the area of this circle is Pi $\times 4.2^2 \cong 3.14159 \times 4.2^2 = 55.41\ldots$ square cm, which is 55.4 square cm when rounded to the nearest tenth.

Area of a parallelogram The area of a parallelogram is $b \times h$, where b is the length of the base of the parallelogram, and h is the corresponding height. To picture this, consider the parallelogram below:

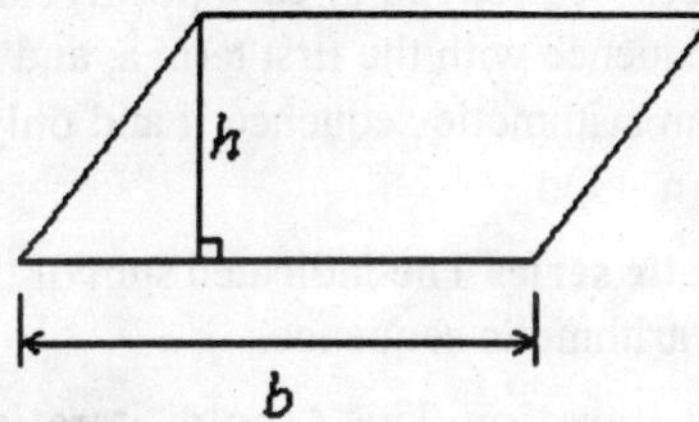

We can picture cutting off a triangle from one side and pasting it onto the other side to form a rectangle with side-lengths b and h. This rectangle has area $b \times h$.

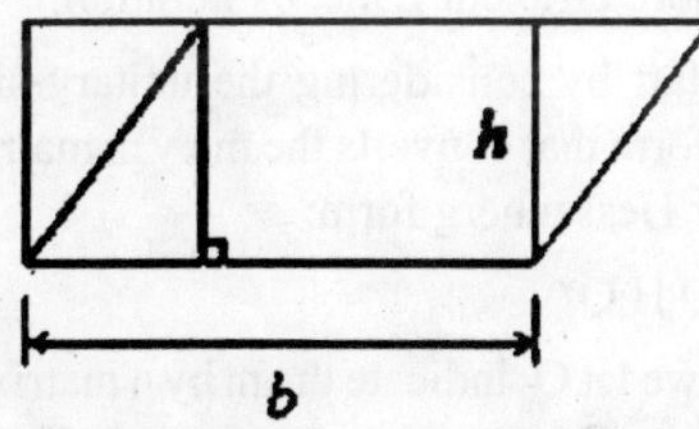

Example: What is the area of a parallelogram having a base of 20 and a corresponding height of 7? The area is the product of a base and its corresponding height, which is $20 \times 7 = 140$.

Area of a rectangle The area of a rectangle is the product of its width and length.

Example: What is the area of a rectangle having a length of 6 and a width of 2.2? The area is the product of these two side-lengths, which is $6 \times 2.2 = 13.2$.

Area of a square If l is the side-length of a square, the area of the square is l^2 or $l \times l$.

Example: What is the area of a square having side-length 3.4? The area is the square of the side-length, which is $3.4 \times 3.4 = 11.56$.

Area of a trapezoid If a and b are the lengths of the two parallel bases of a trapezoid, and h is its height, the area of the trapezoid is

$1/2 \times h \times (a + b)$.

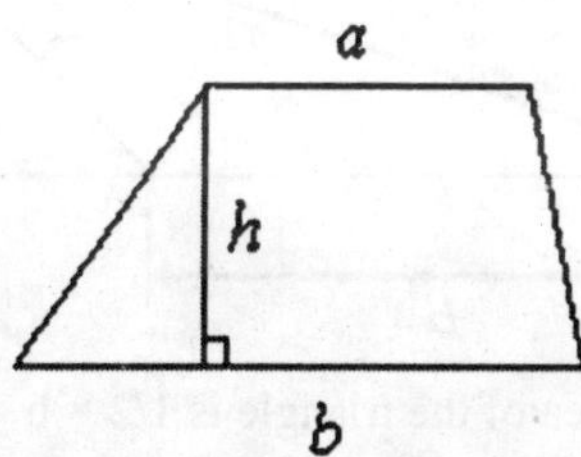

To picture this, consider two identical trapezoids, and turn one around and paste it to the other along one side as pictured below:

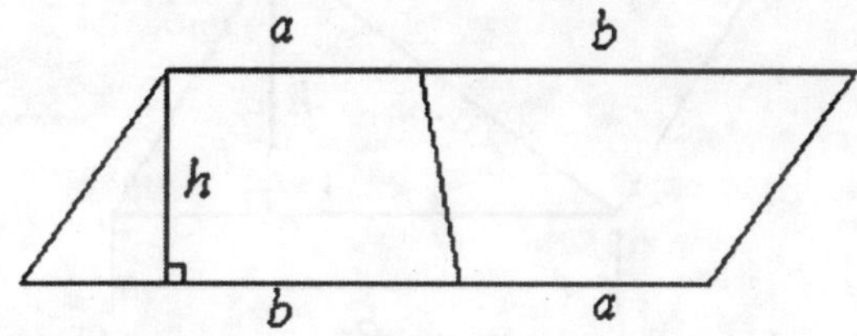

The figure formed is a parallelogram having an area of $h \times (a + b)$, which is twice the area of one of the trapezoids.

Example: What is the area of a trapezoid having bases 12 and 8 and a height of 5?

Using the formula for the area of a trapezoid, we see that the area is

1/2 × 5 × (12 + 8) = 1/2 × 5 × 20 = 1/2 × 100 = 50.

Area of a triangle Consider a triangle with base length b and height h.

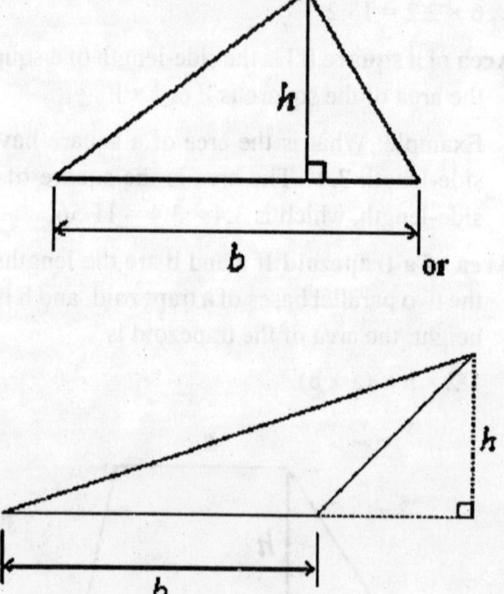

The area of the triangle is 1/2 × b × h.

To picture this, we could take a second triangle identical to the first, then rotate it and paste it to the first triangle as pictured below:

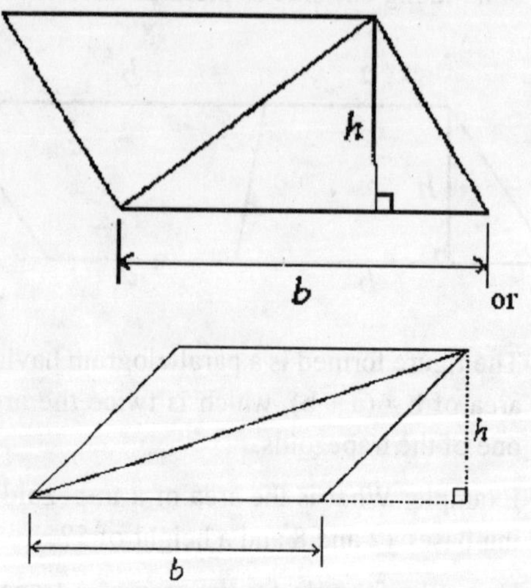

The figure formed is a parallelogram with base length b and height h, and has area b × ×h.

This area is twice that of the triangle, so the triangle has area 1/2 × b × h.

Example: What is the area of the triangle below having a base of length 5.2 and a height of 4.2?

The area of a triangle is half the product of its base and height, which is 1/2 ×5.2 × 4.2 = 2.6 × 4.2 = 10.92..

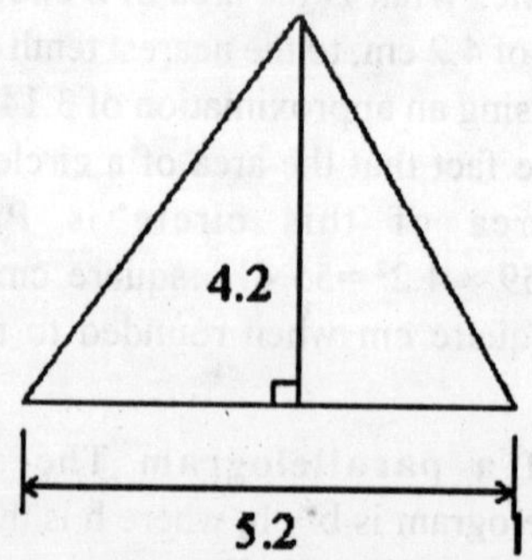

Arithmetic means Terms of an arithmetic sequence that are between two given terms.

Arithmetic sequence For each positive integer n, the sequence with the first term a, and nth term a, is an arithmetic sequence if and only if an = a1 + (n – 1)d.

Arithmetic series The indicated sum of the terms of an arithmetic sequence.

Arnoldi iteration The Arnoldi iteration is an iterative method that seeks to approximate some of the eigenvalues of a very large matrix.

If the matrix is symmetric or Hermitian, then the Arnoldi iteration is usually reformulated into the more efficient Lanczos iteration.

We start by considering the unitary similarity transform that converts the m by m matrix A into upper Hessenberg form:

$A = Q H Q^*$

Now we let Q_n indicate the m by n matrix formed from the first n columns of Q; similarly, H_n is the n+1 by n+1 submatrix of H. Then we write

$A Q_n = Q_{n+1} H_n$

In particular, we can look at the subsystem involving the n-th column of Q_n:

$A^*q_n = h_{1,n}\, q_1 + \ldots$

$+ h_{n,n}\, q_n$

$+ h_{n+1,n}\, q_{n+1}$

This can be interpreted as a recursive relationship for q_{n+1} in terms of the previous vectors. The Arnoldi iteration allows us to build up the matrices Q_n and H_n. For reasons not completely understood, even for n much smaller than m, the eigenvalues of H_n will tend to some of the eigenvalues of A, and typically the extreme ones.

The new vector q_{n+1} is a Krylov vector; actually, because of orthogonalization, it represents the new component introduced into the Krylov subspace by the product A^*q_n.

Thus, a typical procedure is to compute H_n and apply a standard eigenvalue procedure for Hessenberg matrices to extract its eigenvalues.

An efficient form for the Arnoldi iteration is:

```
Let b be an arbitrary initial vector;
Set q(1) = b / ||b||;
for n = 1, ...
v = A * q(n)
for j = 1 to n
h(j,n) = conjugate (q(j)) * v
v = v – q(j) * h(j,n)
h(n+1,n) = ||v||
q(n+1) = v / h(n+1,n)
end for
end for
```

Array The set of Y values associated with a given X, or the set of X values associated with a given Y in a bivariate distribution of X and Y.

ASA Reference to solving a triangle given the measure of two angles and the length of the included side.

Association A statistically significant correlation between an environmental exposure or a biochemical/genetic marker and a disease or condition. An association may be an artefact (random error-chance, bias, confounding) or a real one. In population genetics, an association may be due to population stratification, linkage disequilibrium, or direct causation. A significant association should be presented together with a measure of the strength of association (odds ratio, relative risk or relative hazard and its 95% confidence interval) and when appropriate a measure of potential impact (attributable risk, prevented fraction, attributable fraction/etiologic fraction).

Association role The end of an association where it connects to a class.

Associative entity An entity type that associates the instances of one or more entity types and contains attributes that are peculiar to the relationship between those entity instances.

Associative property This property applies both to multiplication and addition and states that you can group several numbers that are being added or multiplied (not both) in any way and yield the same value. In mathematical terms, for all real numbers a, b, and c, (a+b)+c=a+(b+c) or (ab)c=a(bc).

Assumptions Certain conditions of the data that are required to be met for validity of a statistical test. ANOVA generally assumes, normal distribution of the data within each treatment group, homogeneity of the variances in the treatment groups, and independence of the observations. In regression analysis, main assumptions are the normal distribution of the response variable, constant variance across fitted values, independence of error terms, and the consistency of underlying hazard rate over time (proportionality assumption) in Cox Proportional Hazard Models.

ASTC An acronym representing which trigonometric functions are positive in the I, II, III, and IV quadrants respectively.

Asymmetric relationships Log-linear models where at least one variable is treated as an independent variable and at least one variable is treated as a dependent variable.

Asymptote An asymptote for the graph of a function is a line sitting in the x-y plane that the graph of the function approaches, getting closer and closer as we travel along the line. Functions that have had one too many may weave back and forth across an asymptote, but still, the further out you go, the closer they get.

Asymptotically unbiased In point estimation, the property that the bias approaches zero as the sample size (N) increases. Therefore, estimators with this property improve as N increases.

Atomic event Another name for elementary event.

Atto The metric prefix atto- (prefix symbol: a-) denotes a multiple of ten to the negative eighteenth power (10^{-18} = 10 raised to power –18 = 0.000 000 000 000 000 001) as defined by Le Système International d'Unités (SI).

Attributable fraction (etiologic fraction) It shows what proportion of disease in the exposed individuals is due to the exposure.

Attributable risk (AR) Also called excess risk or risk difference. A measure of potential impact of an association. It quantifies the additional risk of disease following exposure over and above that experienced by individuals who are not exposed. It shows how much of the disease is eliminated if no one had the risk factor (unrealistic). The information contained in AR combines the relative risk and the risk factor prevalence. The larger the AR, the greater the effect of the risk factor on the exposed group.

Attribute A named property or characteristic of an entity that is of interest to the organization.

Audit trail A record of the sequence of data entries and the date of those entries.

Augmented matrix The coefficient matrix for a system of linear equations with a final column included for the constants of the equations.

Average deviation The average deviation is one of several indices of variability that statisticians use to characterize the dispersion among the measures in a given population.

To calculate the average deviation of a set of scores it is first necessary to compute their mean and then specify the distance between between each score and that mean without regard to whether the score is above or below the mean. The average deviation is defined as the mean of these absolute values.

Algebriacally the average deviation is specified as follows:

54 77 67 68 46 64 62 56 38
Height (inches)

$$\text{Average Deviation} = \frac{\Sigma\,|X - \mu|}{N}$$

$$\text{For these data} \quad \frac{\Sigma\,|X - \mu|}{N} =$$

$$= \frac{|54-59.11|+|77-59.11|+\ldots+|56-59.11|+|38-59.11|}{9}$$

$$= \frac{5.11 \quad 17.89 + \ldots + 3.11 \quad 21.11}{9}$$

$$= 9.43$$

Average It is better to avoid this sometimes vague term. It usually refers to the (arithmetic) mean, but it can also signify the median, the mode, the geometric mean, and weighted means, among other things.

Average expected payoff An estimate of the amount that will be gained in a game of chance, calculated by multiplying the probability of winning by the number of points won each time.

Axiomatic completeness An axiomatic theory is complete if every syntactically correct statement in the theory can be proven either right or wrong.

Axioms of probability There are three axioms of probability: 1. Probability is always more than zero

2. The chance that something happens is 1,or 100%

3. If two events cannot both occur at the same time, the chance that either one occurs is the sum of the chances that each occurs.

Axis of symmetry A line in the plane of a graph such that the part of the graph on one side of the line is a reflection of the part on the other side.

Azygous The adjective azygous means occurring singly (1) and not as a pair (2).

B

B The symbol B represents a logarithmic power measurement in bels.

Backward elimination A stepwise regression procedure in which we start with all predictors in the model and then eliminate those that either do not contribute significantly or fail to meet some other predetermined standard.

Baker's dozen The compound noun baker's dozen means thirteen (13).

Balanced design An experimental design in which the same number of observations is taken for each combination of the experimental factors.

Balancing The conservation of inputs and outputs to a data flow diagram process when that process is decomposed to a lower level.

Band matrix A band matrix is a matrix whose entries are all zero except for the diagonal and a few of the immediately adjacent diagonals.

If the band matrix is large enough, then many significant efficiencies can be achieved in storage and matrix operations. Because so many elements are zero, the band matrix can be stored more compactly using band matrix storage. And because so many elements are zero, many algorithms can be speeded up to execute more quickly, including matrix multiplication and Gauss elimination.

Special cases of band matrices include those which are also symmetric or positive definite.

Band matrices with very tight bandwidths are of interest, and include matrices which are diagonal, bidiagonal, tridiagonal, and in some cases pentadiagonal.

Here is an example of a band matrix:

$$\begin{matrix} 11 & 12 & 0 & 0 & 0 & 0 \\ 21 & 22 & 23 & 0 & 0 & 0 \\ 31 & 32 & 33 & 34 & 0 & 0 \\ 0 & 42 & 43 & 44 & 45 & 0 \\ 0 & 0 & 53 & 54 & 55 & 56 \\ 0 & 0 & 0 & 64 & 65 & 66 \end{matrix}$$

This matrix has an upper bandwidth of 1, and lower bandwidth of 2, and an overall bandwidth of 4.

In some cases, a band matrix can be LU-factored without pivoting. This is always true if the matrix is positive definite. If a banded matrix A with lower and upper bandwidths ml and mu can be LU factored without pivoting, then its factorization can be written as

$$A = L * U$$

The usual P factor is missing because there is no pivoting, but more is true. The factor L, which is unit lower triangular, is in fact also a band matrix, with lower bandwith ml, and U, the upper triangular factor, is also band matrix with upper bandwith mu.

If the matrix A is positive definite symmetric banded, with a half-bandwidth of m, then the banded LU factorization becomes:

A = L * U

with L and U having m as their lower and upper bandwidths respectively. Moreover, if we factor out the non-unit diagonal of U by writing

U = D * V

where V is unit upper triangular, we may write

A = L * U = L * D * V = L * D2 * D2 * V = R' * R

where D2 is the obvious square root of D (which must have positive entries because A is positive definite), and R is an upper triangular banded matrix with upper bandwidth m. Another way of looking at this is to say that the Cholesky factor of a banded positive definite symmetric matrix is itself banded with the same (half) bandwidth as the original matrix.

LAPACK and LINPACK include special routines for a variety of band matrices. These routines can compute the LU factorization, determinant, inverse, or solution of a linear system.

LAPACK and EISPACK have routines for computing the eigenvalues of a symmetric banded matrix.

Bandwidth The bandwidth of a band matrix is, roughly speaking, the number of diagonals that contain nonzero entries.

More precisely, define ML, the lower bandwidth of a matrix A to be the maximum value of (I – J), and MU to be the maximum value of (I – J), for all nonzero matrix entries $A_{i,j}$. Then the bandwidth M is defined by:

M = ML + 1 + MU.

This definition always treats the (main) diagonal as nonzero, and is not misled by a matrix which has only two nonzero diagonals, which are actually widely separated. All the territory between the diagonals must be included when measuring bandwidth.

Bar graph A graphical display representing data in different categories or groups. The length of a rectangle or bar is used to represent the numerical amount.

Barely The adverb barely means only slightly more than a required minimum.

Base The noun and adjective base means a number notation system based on the hyphenated suffix number. Thus, base_ten (base_10) is the common decimal positional number notation system.

Base eight The compound noun and adjective base eight (base_8) means a number notation system based on the number eight (8). The modern octal or base eight positional number notation system uses the numerals 0, 1, 2, 3, 4, 5, 6, and 7 to represent the integers zero through seven.

Base five The compound noun and adjective base five (base_5) means a number notation system based on the number five (5).

Base sixteen The compound noun and adjective base sixteen (base_16) means a number notation system based on the number sixteen (16). The modern sexdecimal or base sixteen positional number notation system uses the numerals 0, 1, 2, 3, 4, 5, 6, 7, 8, 9, A, B, C, D, E, and F to represent the integers zero through fifteen.

Base ten The compound noun and adjective base ten (base_10) a number notation system based on the number ten (10). The common decimal or base ten positional number notation system uses the numerals 0, 1, 2, 3, 4, 5, 6, 7, 8, and 9 to represent the integers zero through nine.

Base twelve The compound noun and adjective base twelve (base_12) means a number notation system based on the number twelve (12).

Base twenty The compound noun and aajective base twenty (base_20) means a number notation system based on the number twenty (20).

Base two The compound noun and adjective base two (base_2) means a number notation system based on the number two (2). The modern binary or base two positional number notation system uses the numerals 0 and 1 to represent the integers zero and one.

Baseline modules Software modules that have been tested, documented, and approved to be included in the most recently created version of a system.

Baseline project plan (BPP) The major outcome and deliverable from the project initiation and planning phase that contains an estimate of the project's scope, benefits, costs, risks, and resource requirements.

Basic linear algebra subprograms (BLAS) The BLAS or Basic Linear Algebra Subprograms, are a set of routines offering vector and matrix utilities. They are extensively used as part of LINPACK and LAPACK, to simplify algorithms, and to make them run more quickly.

The Level 1 BLAS provide basic vector operations such as the dot product, vector norm, and scaling. Level 2 BLAS provide operations involving a matrix and a vector, and Level 3 BLAS provide matrix-matrix operations.

There are also sets of sparse BLAS and parallel BLAS available.

Here are the Level 1 BLAS routines for real, single precision vectors:

ISAMAX returns the index of maximum absolute value in vector SX;

SASUM computes the sum of absolute values of vector SX;

SAXPY adds a scalar multiple of one vector to another;

SCOPY copies vector SX into SY;

SDOT computes the dot product of two vectors;

SMACH estimates the largest and smallest machine values, and roundoff value;

SNRM2 computes the Euclidean norm of a vector SX;

SROT applies a Givens rotation;

SROTG generates a Givens Rotation;

SSCAL scales a vector by a constant;

SSWAP interchanges two vectors SX and SY.

Basis A basis for a linear space X of dimension N is a set of N vectors, {v(i) | 1 <= i <= N} from which all the elements of X can be constructed by linear combinations.

Naturally, we require that each of the vectors v(i) be an element of the space X. Moreover, it is not enough that these vectors span the space; we also require that they be linearly independent, that is, there should be no redundant vectors in the set.

The columns of the identity matrix form a basis for the linear space of vectors of dimension N. A square matrix of order N is not defective exactly when its eigenvectors form a basis for the linear space of vectors of dimension N.

Given a particular basis, the representation of a vector x is the unique set of coefficients c(i) so that

x = Sum (1 <= I <= N) c(i) * v(i)

The coefficients must be unique, otherwise you can prove that the basis is not linearly independent.

If the basis vectors are pairwise orthogonal, then the basis is called an orthogonal basis. If the basis vectors have unit length in the Euclidean norm, the basis is a normal basis. If both properties apply, it is an orthonormal basis. The columns of an orthogonal matrix are an orthonormal basis for the linear space of vectors of dimension N.

Bayes' method in genetic counselling This method uses available additional information to modify risks calculated purely by Mendelian probabilities. It combines prior and conditional probabilities to give joint and posterior probabilities of unknown events.

Bayesian inference An inference method radically different from the classical frequentist approach which takes into account the prior probability for an event. Established as a new method by Reverend Thomas Bayes.

Bearing An angle measured clockwise from due north to a vector.

Behavior Represents how an object acts and reacts.

Behrens Fisher problem An old name given to the problem of how to compare two independent means when we can not assume homogeneity of variance.

Bel The noun and adjective bel (symbol: B) is a logarithmic comparison of power levels representing a ratio of ten (10). Two (2) bels equal a power ratio of one hundred (100), and three (3) bels equal a power ratio of one thousand (1,000).

Bell curve See normal distribution .

Bernoulli distribution Models the behaviour of data taking just two distinct values (0 and 1).

Bernoulli trial A event with one of two mutually exclusive outcomes—such as pass/fail.

Beta (β) Statisticians use the Greek letter beta (β) to indicate the probability of failing to reject the hypothesis tested when that hypothesis is false and a specific alternative hypothesis is true. For a given test, the value of beta is determined by the previously elected value of alpha, certain features of the statistic that is being calculated (particularly the sample size) and the specific alternative hypothesis that is being entertained. While it is possible to carry out a statistical test without entertaining a specific alternative hypothesis, neither beta nor power can be calculated if there is no specific alternative hypothesis. It is relevant to note here that power (the probability that the test will reject the hypothesis tested when a specific alternative hypothesis is true) is always equal to one minus beta. (i.e. Power = 1 – beta)

Beta testing User testing of a completed information system using real data in the real user environment.

Between-subjects designs Designs in which different subjects serve under the different treatment levels.

Biannual The adjective biannual means recurring twice (2) per year.

Bias Can refer either to a sample not being representative of the population, or to the

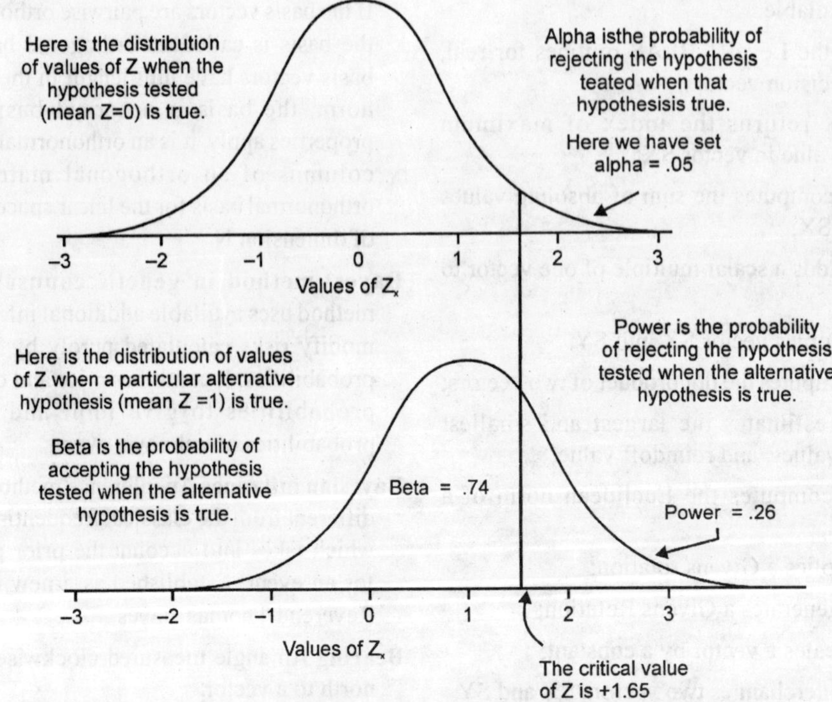

difference between the expected value of an estimator and the true value.

Biased sample A sample that is not random.

Biathlon The noun biathlon means an athletic competition consisting of two (2) separate events.

Bicentenary 1. The noun and adjective bicentenary means a group of two hundred (200).

2. The noun and adjective bicentenary means a period of two hundred (200) years.

3. The noun and adjective bicentenary means a two hundred (200) year anniversary.

4. The adjective and noun bicentenary means two hundredth (200th) in rank or order.

Bicentennial 1. The adjective and noun bicentennial means recurring every two hundred (200) years.

2. The adjective and noun bicentennial means lasting two hundred (200) years.

3. The noun and adjective bicentennial means a two hundred (200) year anniversary.

Bicentennium The noun and adjective bicentennium means a period of two hundred (200) years.

Bicentiquadragesimal The adjective bicentiquadragesimal means based on the number two hundred forty (240).

Bicentisexquinquagesimal The adjective bicentisexquinquagesimal means based on the number two hundred fifty-six ($2^8 = 256$).

Bidiagonal matrix A bidiagonal matrix has only two nonzero diagonals. The matrix is called upper bidiagonal if these are the main diagonal and the principal superdiagonal. The matrix is called lower bidiagonal if these are the main diagonal and the principal subdiagonal.

A simple example of an upper bidiagonal matrix is:

1 2 0 0 0

0 3 4 0 0

0 0 5 6 0

0 0 0 7 8

0 0 0 0 9

The Jordan Canonical Form is an example of an upper bidiagonal matrix.

A bidiagonal matrix is automatically a:

tridiagonal matrix;

triangular matrix;

band matrix.

with all the rights and privileges appertaining thereunto.

If a tridiagonal matrix can be LU-factored without pivoting, then the L factor is lower bidiagonal, and the U factor is upper bidiagonal.

Bidirectional interaction A bidirectional interaction is used when there are two one way interactions that are so closely related that they can be treated as being the same interaction.

Biennary The noun and adjective biennary means a group of two (2) or a period of two (2) years.

Biennial 1. The adjective and noun biennial means recurring every two (2) years.

2. The adjective and noun biennial means lasting two (2) years.

3. The noun and adjective biennial means a two (2) year anniversary.

Bilateral The adjective bilateral means having two (2) sides.

Billiard The noun and adjective billiard means multiplied by ten to the fifteenth power (10^{15} = 10 raised to power +15 = 1,000,000,000,000,000) in the modified Chuquet number naming system.

Billion The noun and adjective billion means multiplied by ten to the ninth power (10^9 = 10 raised to power +09 = 1,000,000,000) in the American number naming system, but means multiplied by ten to the twelfth power (10^{12} = 10 raised to power +12 = 1,000,000,000,000) in the Chuquet number naming system.

Billions and billions The expression billions and billions means an astronomically large number. This expression was popularized by American astronomer Carl Sagan (1934-1996).

Bimillenary 1. The noun and adjective bimillenary means a group of two thousand (2000).

2. The noun and adjective bimillenary means a period of two thousand (2000) years.

3. The noun and adjective bimillenary means a two thousand (2000) year anniversary.

4. The adjective and noun bimillenary means two thousandth (2000th) in rank or order.

Bimillennial 1. The adjective and noun bimillennial means recurring every two thousand (2000) years.

2. The adjective and noun bimillennial means lasting two thousand (2000) years.

3. The noun and adjective bimillennial means a two thousand (2000) year anniversary.

Bimillennium The noun and adjective bimillennium means a period of two thousand (2000) years.

Bimodal Having two modes, which are the most frequently occurring number in a list.

Binal The adjective binal means based on the number two (2).

Binary The adjective binary means having two (2) parts. A numeral system based on the number two (2) is properly called binal. The more common term binary is also appropriate since a binal digit has two (2) possible binary states: zero (0) or one (1), FALSE or TRUE, low or high, etc.

Binary relationship A relationship between instances of two entity types.

Binomial A polynomial with two terms.

Binomial distribution The binomial distribution gives the probability of obtaining exactly r successes in n independent trials, where there are two possible outcomes one of which is conventionally called success

Bis The prefix bis- means two (2).

Bisect The verb bisect means to divide into two (2) usually equal parts.

Bisecting line A line which divides an object or shape into two equal parts.

Biserial correlation (r_b) The correlation between a continuous variable and a dichotomous variable, where we assume an underlying normality to the dichotomous variable. Rarely used.

Bit In computer terminology, the noun bit means a single 1. Binary unit having two

2. Possible states. Bit is a contraction of binary digit.

Bivariate normal model A regression model in which both X and Y are subject to random error and have a multivariate normal distribution.

Biyearly 1. The adjective biyearly means recurring every two (2) years.

2. The adjective biyearly can also mean recurring twice (2) per year.

Block matrix A block matrix is a matrix which is described as being built up of smaller matrices. For example, a tridiagonal block matrix might look like this:

2 4 | 3 9 | 0 0

6 | 0 3 | 0 0

1 0 | 2 4 | 3 9

5 5 | 4 6 | 0 3

0 0 | 1 0 | 2 4

0 0 | 5 5 | 4 6

but for certain purposes, it might help us to see this matrix as really being a tridiagonal matrix, whose elements are themselves little matrices:

a | b | 0

c | a | b

0 | c | a

An algorithm suitable for a tridiagonal matrix can often be extended, in a natural manner, to handle a block tridiagonal matrix. Similar extensions can be made in some cases for other types of block matrices. A block banded matrix can be factored by a variant of banded Gauss elimination, for instance.

Blocking When the available experimental units are not homogeneous, grouping them into blocks of homogeneous units (stratification) will reduce the experimental error variance. This is called blocking where differences between experimental units other than those caused by treatment factors are taken into account. This is like comparing age-matched groups (blocks) of a control group with the corresponding blocks in the patients group in an investigation of the side effects of a drug as age itself may cause differences in the experiment. Block effects soak up the extra, uninteresting and already known variability in a model. Blocking is preferable to randomisation when the factors that might affect the outcome are known.

Blocks Homogeneous grouping of experimental units (subjects) in experimental design. Groups of experimental units that are relatively homogeneous in that the responses on two units within the same block are likely to be more similar than the responses on two units in different blocks. Dividing the experimental units into blocks is a way of improving the accuracy between treatments. Blocking will minimise variation between subjects that is not attributable to the factors under study. Blocking is similar to matching in two-sample tests or stratification to control for confounding.

Bonferroni correction This is a multiple comparison technique used to adjust the error level.

Bonferroni inequality An inequality on which the Bonferroni test is based. It states that the probability of the occurrence of one or more events can never exceed the sum of their individual probabilities.

Bonferroni test A multiple comparison procedure in which the familywise error rate is divided by the number of comparisons. It is based on the Bonferroni inequality.

Boost A set of prime powers which provide support for each in other in a cycle. For example, 43^2 supports 631, 631 supports 79, and 79^2 supports 43, so 43^2 79^2 631 is a boost

Bootstrap An application of resampling statistics. It is a data-based simulation method used to estimate variance and bias of an estimator and provide confidence intervals for parameters where it would be difficult to do so in the usual way.

Border banded matrix A border banded matrix is a 2 by 2 block matrix comprising a (large) leading block which is a square banded matrix, two dense rectangular side strips, and a (small) trailing block which is a square dense matrix.

For example, a toy border banded matrix might look like this:

2 –1 0 0 0 0 0 | 1 2

–1 2 –1 0 0 0 0 | 2 5

0 –1 2 –1 0 0 0 | 7 8

0 0 –1 2 –1 0 0 | 3 3

0 0 0 –1 2 –1 0 | 4 2

0 0 0 0 –1 2 –1 | 3 1

0 0 0 0 0 –1 2 | 7 8

3 7 8 3 2 3 1 | 5 2

1 2 4 7 9 2 4 | 3 6

which we can regard as being made up of the blocks:

A11 | A12

A21 | A22

where, as we specified, A11 is a square banded matrix, A22 is a square dense matrix, and A21 and A12 are rectangular strips.

It is desirable to take advantage of the banded structure of A11. We can specify an algorithm for solving a linear system A * x = b that can be written in terms of operations involving the submatrices A11, A12, A21 and A22, which will achieve this goal, at the expense of a little extra work. One problem with this technique is that it will fail if certain combinations of the matrices A11 and A22 are singular, which can happen even when A is not singular.

The algorithm for solving A * x = b rewrites the system as:

A11 * X1 + A12 * X2 = B1

A21 * X1 + A22 * X2 = B2

The first equation can be solved for X1 in terms of X2:

$X1 = -A11^{-1} * A12 * X2 + A11^{-1} * B1$

allowing us to rewrite the second equation for X2:

$(A22 - A21 * A11^{-1} * A12)\ X2 = B2 - A21 * A11^{-1} * B1$

which can be solved as:

$X2 = (A22 - A21 * A11^{-1} * A12)^{-1} * (B2 - A21 * A11^{-1} * B1)$

The actual algorithm doesn't compute the inverse, of course, but rather factors the matrices A11 and $A22 - A21 * A11^{-1} * A12$.

Cartesian Basis Vectors

The Cartesian basis vectors are simply the N columns of the identity matrix, regarded as individual column vectors.

These vectors form the standard basis for the set of vectors R^N. The vector corresponding to the I-th column of the identity matrix is often symbolized by e_i.

Thus, if we are working in a space with dimension N of 4, the basis vectors e_1 through e_4 would be:

1 0 0 0

0 1 0 0

0 0 1 0

0 0 0 1

Facts about the Cartesian basis vectors:

The vectors have unit Euclidean norm, are pairwise orthogonal, and form a basis for the space of N dimensional vectors;

$A * e_j$ yields a vector which is the J-th column of the matrix A;

the outer product of e_i and e_j a matrix A which is all zero except for the single entry $A_{i,j}=1$.

The Cauchy-Schwarz Inequality The Cauchy-Schwarz Inequality is a relationship between a vector inner product and a vector norm derived from that inner product. In particular, if the norm ||*|| is defined by an inner product (*,*) as follows:

|| x || = sqrt (x, x),

then the Cauchy-Schwarz inequality guarantees that for any vectors x and y it is the case that:

| (x, y) | <= || x || * || y ||.

The Cayley-Hamilton Theorem The Cayley-Hamilton Theorem guarantees that every (square) matrix satisfies its own characteristic equation.

For example, if A is the matrix:

2 3

1 4

then the characteristic equation is

$lambda^2 - 6 * lambda + 5 = 0$.

which is not true for all values lambda, but just a few special values known as eigenvalues. The Cayley-Hamilton theorem guarantees that the matrix version of the characteristic equation, with A taking the place of lambda, is guaranteed to be true:

$A^2 - 6 * A + 5 * I = 0$.

Centrosymmetric Matrix A centrosymmetric matrix is one which is symmetric about its center; that is,

$A_{i,j} = A_{m+1-i,n+1-j}$

Example: 1 10 8 11 5

13 2 9 4 12

6 7 3 7 6

12 4 9 2 13

5 11 8 10 1

A centrosymmetric matrix A satisfies the following equation involving the Exchange matrix J: J*A*J = A

Both The plural pronoun and adjective both means two (2) of two (2).

Boundary condition An extreme value in an inequality problem. Solutions will be less than or greater than that value.

Boundary The line that marks the inside and outside of a system, and that sets off one system from other systems in the organization.

Box and whisker plot A graphical representation of the dispersion of a sample.

Boxplot A graphical representation of the dispersion of a sample.

Braids Theory. Braids Theory was invented by Emil Artin and is a part of the Knot Theory. Braids are collections of lines whose ends are attached to two parallel straight lines. There is an interesting puzzle related to the Braids Theory.

Break even analysis A type of cost-benefit analysis to identify at what point (if ever) benefits equal costs.

Build routines Guidelines that list the instructions to construct an executable system from the baseline source code.

Business case A written report that outlines the justification for an information system. The report highlights economic benefits and costs and the technical and organizational feasibility of the proposed system.

Business process reengineering (BPR) The search for, and implementation of, radical change in business processes to achieve breakthrough improvements in products and services.

Byte In computer terminology, the noun byte means a set of eight (8) binary units having two to the eighth power ($2^8 = 256$) possible states. A byte can be used to represent a single alphanumeric character. A byte is equal to eight (8) bits, four (4) crumbs, or two (2) nibbles.

C The traditional metric prefix symbol c- (metric prefix: centi-) denotes a multiple of one one-hundredth (1/100 = 0.010).

Calculated (or computed or derived) field A field that can be derived from other database fields.

Candidate key An attribute (or combination of attributes) that uniquely identifies each instance of an entity type.

Capacity The maximum amount that can be contained by an object. Often refers to measurement of liquid.

Cardinal In common usage, a cardinal number is a number used in counting (a counting number), such as 1, 2, 3, ...

In formal set theory, a cardinal number (also called the cardinality) is a type of number defined in such a way that any method of counting sets using it gives the same result. (This is not true for the ordinal numbers.) In fact, the cardinal numbers are obtained by collecting all ordinal numbers which are obtainable by counting a given set. A set has $\aleph_0$ members if it can be put into a one-to-one correspondence with the finite ordinal numbers. The cardinality of a set is also frequently referred to as the power of a set.

In Georg Cantor's original notation, the symbol for a set A annotated with a single overbar $\overline{A}$ indicated A stripped of any structure besides order, hence it represented the order type of the set. A double overbar $\overline{\overline{A}}$ then indicated stripping the order from the set and thus indicated the cardinal number of the set. However, in modern notation, the symbol $|A|$ is used to denote the cardinal number of set.

Cantor, the father of modern set theory, noticed that while the ordinal numbers $\omega_{+1,}$... were bigger than omega in the sense of order, they were not bigger in the sense of equipollence. This led him to study what would come to be called cardinal numbers. He called the ordinals $\omega, \omega_{+1,...}$ that are equipollent to the integers the second number class (as opposed to the finite ordinals, which he called the first number class). Cantor showed

1. The second number class is bigger than the first.

2. There is no class bigger than the first number class and smaller than the second.

3. The class of real numbers is bigger than the first number class.

One of the first serious mathematical definitions of cardinal was the one devised by Gottlob Frege and Bertrand Russell, who defined a cardinal number $|A|$ as the set of all sets equipollent to A. Unfortunately, the objects produced by this definition are not sets in the sense of Zermelo-

Fraenkel set theory, but rather proper classes in the terminology of von Neumann.

Tarski (1924) proposed to instead define a cardinal number by stating that every set A is associated with a cardinal number $|A|$, and two sets A and B have the same cardinal number if they are equipollent. The problem is that this definition requires a special axiom to guarantee that cardinals exist.

A. P. Morse and Dana Scott defined cardinal number by letting A be any set, then calling $|A|$ the set of all sets equipollent to A and of least possible rank.

It is possible to associate cardinality with a specific set, but the process required either the axiom of foundation or the axiom of choice. However, these are two of the more controversial Zermelo-Fraenkel axioms. With the axiom of choice, the cardinals can be enumerated through the ordinals. In fact, the two can be put into one-to-one correspondence. The axiom of choice implies that every set can be well ordered and can therefore be associated with an ordinal number.

This leads to the definition of cardinal number for a set A as the least ordinal number b such that A and b are equipollent. In this model, the cardinal numbers are just the initial ordinals. This definition obviously depends on the axiom of choice, because if the axiom of choice is not true, then there are sets that cannot be well ordered. Cantor believed that every set could be well ordered and used this correspondence to define the $\aleph_s$ For any ordinal number α $\aleph_\alpha = \omega_\alpha$ An inaccessible cardinal cannot be expressed in terms of a smaller number of smaller cardinals.

Cardinality The number of instances of entity B that can (or must) be associated with each instance of entity A.

Cardioid A special case of epicycloids. A plane curve traced by a point on a circle rolling on the outside of a circle of equal radius.

Carryover effect Any effect of a drug that lasts beyond the period of treatment. This is a worry in drug trials with crossover design and the reason for the washout period between treatments.

Cartesian coordinates Most often in the plane, but also in the higher dimension spaces, the Cartesian coordinates are defined by a number (2 in the plane) of perpendicular number lines - coordinate axes. A point is then defined by its projections on the axes. In the plane, points are defined by 2 such projections that are written as an ordered pair of real numbers, (x, y).

Cartesian plane That's a plane upon which we have Cartesian coordinates.

Cascade The primes or prime powers resulting from factoring sigma of some prime power, then factoring sigma of each prime or prime power in the resulting factorization, and so on.

Case-control study A design preferred over cohort studies for relatively rare diseases in which cases with a disease or exposure are compared with controls randomly selected from the same study base. This design yields odds ratio as opposed to relative risk from cohort studies.

Catalan's constant Catalan's Constant is a real number approximately 0.915 965 594 177.

Categorical (nominal) variable A variable that can be assigned to categories. A non-numerical (qualitative) variable measured on a (discrete) nominal scale such as gender, drug treatments, disease subtypes; or on an ordinal scale such as low, median or high dosage. A variable may alternatively be quantitative (continuous or discrete).

Categorical data Data representing counts or the number of observations in each category.

Cauchy sequence A sequence $x_0, x_1, \ldots$ of elements of a metric space is said to be a Cauchy sequence if differences $|x_{n+m} - x_n|$ are uniformly small in m (i.e. do not depend on m) and tend to 0 as n grows.

Causal relationship It does not matter how small it is, a P value does not signify causality. To establish a causal relationship, the following non-statistical evidence is required: consistency (reproducibility), biological plausibility, dose-response, temporality (when applicable) and strength of the relationship (as measured by odds ratio/relative risk/hazard ratio).

Cell The combination of a particular row and column. The set of observations obtained under identical treatment conditions.

Censored data Data which have been categorized into two or more groups on the basis of a cutoff score on some criterion variable. Often a consideration in logistic regression.

Censored observation Observations that survived to a certain point in time before dropping out from the study for a reason other than having the outcome of interest (lost to follow up or not enough time to have the event). Thus, censoring is simply an incomplete observation that has ended before time-to-event. These observations are still useful in survival analysis.

Cent The noun cent (symbol: ¢) means divided by one hundred (1/100 = 0.010). In currency, a cent is one one-hundredth of a dollar ($0.01), euro (€0.01), or other monetary unit. The plural of cent is normally cents, but the preferred plural of the euro cent is cent.

Centenarian The noun and adjective centenarian means one hundred (100) years of age or more.

Centenary 1. The noun and adjective centenary means a group of one hundred (100).

2. The noun and adjective centenary means a period of one hundred (100) years.

3. The noun and adjective centenary means a one hundred (100) year anniversary.

4. The adjective and noun centenary means one hundredth (100th) in rank or order.

Centennial 1. The adjective and noun centennial means recurring every one hundred (100) years.

2. The adjective and noun centennial means lasting one hundred (100) years.

3. The noun and adjective centennial means a one hundred (100) year anniversary.

Centennium The noun and adjective centennium means a period of one hundred (100) years.

Center length The distance of the straight line or piece of material between the arcs of an offset.

Center mark back The distance used in marking and cutting a miter from the center line.

Center of a circle The point in a circle from which the distance to any point on the circle is a constant.

Center of an ellipse The intersection of the axes of symmetry.

Centering The process of converting data to deviation scores by subtracting the mean from each observation.

Centesimal The adjective centesimal means based on the number one hundred (100).

Centi 1. The numeric prefix centi- means one hundred (100). This prefix is derived from the Latin cardinal number centum (C).

2. The traditional metric prefix centi- (prefix symbol: c-) denotes a multiple of one one-hundredth (1/100 = 0.010).

Centigrade The adjective centigrade means having one hundred (100) steps or intervals. The Celsius temperature scale was previously known as the centigrade temperature scale.

Centillion The noun and adjective centillion means multiplied by 10^{303} in the American number naming system, but means multiplied by 10^{600} in the Chuquet number naming system.

Centioctogesimal The adjective centioctogesimal means based on the number one hundred eighty (180).

Centipede The noun centipede literally means having one hundred (100) feet. A centipede is an insect of the class Chilopoda.

Centivigesimal The adjective centivigesimal means based on the number one hundred twenty (120).

Central angle An angle whose vertex is located at the center of a circle.

Central limit theorem The means of a relatively large (>30) number of random samples from any population (not necessarily a normal distribution) will be approximately normally distributed with the population mean being their mean and variance being the (population variance / n). This approximation will improve as the sample size (the number of samples) increases.

Central tendency of a set of numbers A number considered to be the most typical or representative of that set.

Centroid The point of intersection of the medians of a triangle.

Centurion The noun centurion literally means one who commands one hundred (100). A centurion was a Roman officer commanding a century, an infantry unit of about one hundred (100) soldiers.

Century The noun century means one hundred (100). Most commonly, century means one hundred (100) years.

The Christian Era began with the year 1 A.D., thus the Twenty-First (21st) Century includes the years 2001 through 2100. In contrast, the Twenty-One Hundreds (2100's) include the years 2100 through 2199.

Chain rule The mathematical version states (f(g(x))'=f'(g(x))g'(x)

$$\frac{df}{dx} = \frac{df}{du}\frac{du}{dx}$$

Chaos Chaos is the breakdown of predictability, or a state of disorder .

Chi square (χ^2) The statistic Chi Square (χ^2) is what statisticians call an enumeration statistic. Rather than measuring the value of each of a set of items, a calculated value of Chi Square compares the frequencies of various kinds (or categories) of items in a random sample to the frequencies that are expected if the population frequencies are as hypothesized by the investigator.

Chi square is often used to assess the goodness of fit between an obtained set of frequencies in a random sample and what is expected under a given statistical hypothesis. For example, Chi Square can be used to determine if there is reason to reject the statistical hypothesis that the frequencies in a random sample are as expected when the items are from a normal distribution.

In the equation shown below, O and E are Observed and Expected frequencies in each category.

$$X^2 = \sum \frac{(O-E)^2}{E}$$

Chi square test A statistical test often used for analyzing categorical data.

Chiliad The noun chiliad means a group of one thousand (1,000) or one thousand (1,000) years.

Chiliagon The noun chiliagon means a plane figure with one thousand (1,000) sides.

Chiliahedron The noun chiliahedron means a solid figure with one thousand (1,000) faces.

Chi-squared distribution A distribution derived from the normal distribution. Chi-squared (χ^2) is distributed with v degrees of freedom with mean = v and variance = 2v. (Chi-squared Distribution, a Lecture on Chi-Squared Significance Tests).

Chi-squared test The most commonly used test for frequency data and goodness-of-fit. In theory, it is nonparametric but because it has no parametric equivalent, it is not classified as such. It is not an exact test and with the current level of computing facilities, there is not much excuse not to use Fisher's exact test for 2x2 contingency table analysis instead of Chi-squared test. Also for larger contingency tables, the G-test (log-likelihood ratio test) may be a better choice. The Chi-square value is obtained by summing up the values (residual^2/fit) for each cell in a contingency. In this formula, residual is the difference between the observed value and its expected counterpart and fit is the expected value.

Cholesky factorization The Cholesky factorization of a positive semidefinite symmetric matrix A has the form:

A = L * L'

where L is a lower triangular matrix, or equivalently:

A = R' * R

where R is an upper triangular matrix.

If a matrix is symmetric, then it is possible to determine whether or not the matrix is positive definite simply by trying to compute its Cholesky factorization: if the matrix has a zero eigenvalue, then it is positive semidefinite, and the algorithm should theoretically spot this by computing a zero diagonal element; if the matrix actually has a negative eigenvalue, then at a particular point in the algorithm, the square root of a negative number will be computed.

Software to compute the Cholesky factorization often saves space by using symmetric matrix storage, and overwriting the original matrix A by its Cholesky factor L.

As long as the matrix A is positive definite, the Cholesky factorization can be computed from an LDL factorization, or, if no pivoting was done, from an LU factorization.

The Cholesky factorization can be used to compute the square root of the matrix.

The LINPACK routines SCHDC, SCHUD, SCHDD, SCHEX, SPOCO, SPOFA, SPODI, and SPOSL compute and use the Cholesky factorization.

Chord A straight line from one point on a circle to another point on the circle. The longest chord of a circle is the diameter.

Chuquet system The Chuquet number naming system or Chuquet System was devised by French physician and mathematician Nicolas Chuquet (1445-1488), and it is still widely used today. The Chuquet System uses a Latin-based numeric prefix to denote an integer power of one million (1,000,000 = 10^6 = 10 raised to power +06), thus:

(prefix)illion = $1{,}000{,}000^{(\text{prefix})} = 10^{6\times(\text{prefix})}$

Example: In the Chuquet System, one billion = $10^{6\times2} = 10^{12}$ = 10 raised to power +12 = 1,000,000,000,000.

Cipher Ciphers are codes for writing secret messages. Two simple types are shift ciphers and affine ciphers.

Circle A circle is the collection of points in a plane that are all the same distance from a fixed point. The fixed point is called the center. A line segment joining the center to any point on the circle is called a radius.

Example:

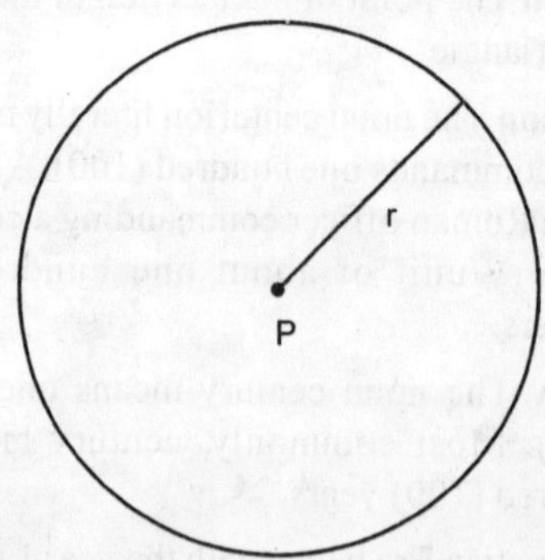

The blue line is the radius r, and the collection of red points is the circle.

Circle graph A graphical display that shows data as parts of a whole circle.

Circular functions Functions whose domains are angles measured in radians and whose ranges are values that correspond to analogous trigonometric functions.

Circumcenter The point of intersection of the perpendicular bisectors of a triangle.

Circumference Circumference is a number, usually the length of a circle, which is a geometric figure. The circumference of a circle is a meaningful expression, while cut the circumference of a circle in two points is not. Regretfully, the usage is wide spread: the word circumference is often used as a substitute for the word circle.

Circumference of a circle The distance around a circle. It is equal to Pi (π) imes the diameter of the circle. Pi or π is a number that is approximately 3.14159.

Example: What is the circumference of a circle having a diameter of 7.9 cm, to the nearest tenth of a cm? Using an approximation of 3.14159 for π, and the fact that the circumference of a circle is π times the diameter of the circle, the circumference of the circle is Pi × 7.9 ≅ 3.14159 × 7.9 = 24.81...cm, which equals 24.8 cm when rounded to the nearest tenth of a cm.

Class diagram Shows the static structure of an object-oriented model the object classes, their internal structure, and the relationships in which they participate.

Class interval In plotting a histogram, one starts by dividing the range of all values into non-overlapping intervals, called class intervals, in such a way that every piece of data is contained in some class interval.

Classification The ability to group according to features.

Closed ended questions Questions in interviews and on questionnaires that ask those responding to choose from among a set of specified responses.

Closed figure The boundary of a simple 2-dimensional region, including shapes with straight and curved sides.

Closed interval Closed interval is a piece of a straight line that includes its endpoints. On the number line, closed intervals [a,b] are defined as $\{x: a \le x \le b\}$. Closed intervals are closed sets.

Closure A set of numbers is closed under an operation if combining any two elements in the set by that operation yields a number in the set.

Co occurrence constraint A co-occurrence constraint specifies the minimum and maximum number of times an object or combination of objects can co-occur in the relationships of a relationship set with another object or combination of objects.

Code generators CASE tools that enable the automatic generation of program and database definition code directly from the design documents, diagrams, forms, and reports stored in the repository.

Coefficient of variation (CV) The standard deviation divided by the mean.

Coefficients The numbers in front of the letters in a mathematical expression, for example, in: 4d + $5t^2$ + 3s, the 4, 5, and 3 are coefficients for the d, t^2, and s.

Cofactor matrix The cofactor matrix of a square matrix A is generally used to define the adjoint matrix, or to represent the determinant.

For a given matrix A, the cofactor matrix is the transpose of the adjoint matrix:

cofactor (A) = (adjoint (A))'

The determinant det(A) can be represented as the product of each of the entries of any given row or column times their corresponding cofactor entries. In particular, consider the first row:

det(A) = A(1,1) * cofactor(A)(1,1) + A(1,2) * cofactor(A)(1,2) + ... + A(1,N) * cofactor(A)(1,N)

The formula for the (I,J) entry of the cofactor matrix of A is:

cofactor(A)(I,J) = $(-1)^{(I+J)}$ * det (M(A,I,J))

where M(A,I,J) is the minor matrix of A, constructed by deleting row I and column J.

Cofunction identities Fundamental identities that involve the basic trig functions of complementary angles.

Cofunctions Pairs of trigonometric functions of complimentary angles whose trigonometric ratios are equal.

Cohesion The extent to which a system or subsystem performs a single function.

Cohort effect The tendency for persons born in certain years to carry a relatively higher or lower risk of a given disease. This may have to be taken into account in case-control studies.

Collinear points A set of points lying on the same line.

Collinearity The condition in which the independent variables are (usually highly) correlated with each other.

Column totals The total number of observations occurring in a column of a contingency table.

Columns This section defines the names of the variable, the coefficients of the objective, and all the nonzero matrix elements $A_{i,j}$.

Combinations A set of items selected, without regard to order, from a given set of items.

Combinatorics The science that studies the numbers of different combinations, which are groupings of numbers. Combinatorics is often part of the study of probability and statistics.

Common ratio The common ratio is a constant.

Common side of two triangles A line that is shared by two triangles. It can be the hypotenuse of one triangle and the leg of the other.

Commutative Property a property of addition or multiplication in which the sum or product stays the.

Commutative property This property of both multiplication and addition states that you can rearrange the order of the numbers being added or reorder numbers being multiplied without changing the value of the expression. In mathematical terms, for all real numbers a and b, a+b=b+a and ab=ba.

Compass A tool used to draw circles and curved lines.

Compatible norms A matrix norm and a vector norm are compatible if it is true, for all vectors x and matrices A that

||A*x|| <= ||A|| * ||x||

In some texts, the word consistent is used in this sense, instead of compatible.

In particular, if you have not verified that a pair of norms are compatible, then the above inequality is not guaranteed to hold. For any vector norm, it is possible to define at least one compatible matrix norm, namely, the matrix norm defined by:

||A|| = supremum ||A*x|| / ||x||

where the supremum (roughly, the maximum) is taken over all nonzero vectors x. If a matrix norm can be derived from a vector norm in this way, it is termed a vector-bound matrix norm. Such a relationship is stronger than is required by compatibility.

If a matrix norm is compatible with some vector norm, then it is also true that

||A*B|| <= ||A|| * ||B||

where both A and B are matrices.

Complement of an event Event A is a complement of event B with respect to sample spaceS if and only if A and 8 include all outcomes of S, and A and B have no outcomes in common.

Complementary angles Two angles are called complementary angles if the sum of their degree measurements equals 90 degrees. One of the complementary angles is said to be the complement of the other.

Example: These two angles are complementary.

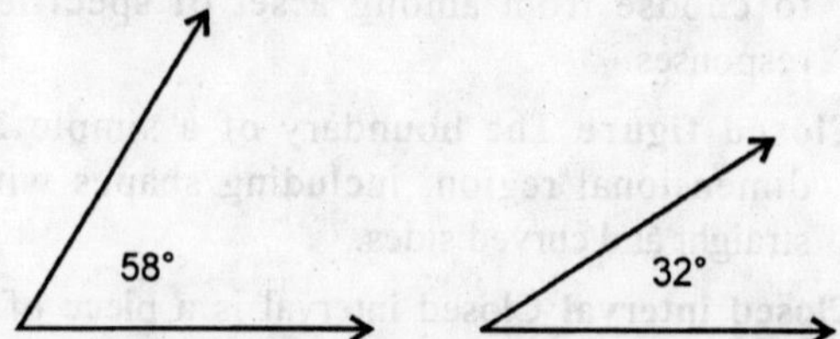

Note that these two angles can be pasted together to form a right angle!

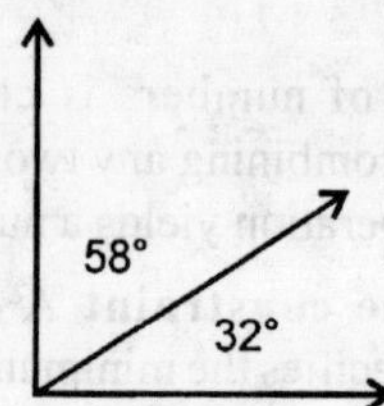

Completeness Correlation, also called correlation coefficient, is a numeric measure of the strength of linear relationship between two random variables (one can use it to quantify, for example, how shoe size and height are correlated in the population). An example is the Pearson product-moment correlation coefficient, which is found by dividing the covariance of the two variables by the product of their standard deviations. Independent variables have a correlation of 0.

The Covariance between two random variables X and Y, with expected values E(X) = μ and E(Y) = ν is defined as the expected value of random variable (X " μ)(Y " ν), and is written cov(X, Y). It is used for measuring correlation.

A data set is a sample and the associated data points.

A data point is a typed measurement - it can be a boolean value, a real number, a vector (in which case it's also called a data vector), etc.

A Distribution function is the function that gives the probability distribution of a random variable. It cannot be negative, and its integral on the probability space is equal to 1.

Completing the square A method used to solve a quadratic equation in which a number is added to both sides of the equation so that one side is a perfect square.

Complex conjugate For all real numbers a and b, a + bi and a - bi are complex conjugates.

Complex fraction A fraction that has frac- tions in either the numerator or the denominator or both.

Complex Number A complex number is a number that can be represented as an ordered pair consisting of a real number and an imaginary number. A complex number may be expressed as a+b·i, where a and b are real numbers and i is the imaginary unit. The complex number can also be represented as a plane vector $r \cdot e^{i \cdot \theta}$, where $r = (a^2+b^2)^{½}$, $\theta = \tan^{-1}(b/a)$, and i is the imaginary unit.

Complex plane A coordinate system for complex numbers.

Component An irreducible part or aggregation of parts that makes up a system; also called a subsystem.

Component diagram Shows the software components or modules and their dependencies.

Component vectors The horizontal and vertical component vectors of a given vector.

Components of an algebraic vector The ordered pair of numbers representing the vector.

Components or an ordered pair The individual numbers in an ordered pair.

Composition of functions Applying one function to another. For instance $\sin\left(\sqrt{x}\right)$ is the composition of sin *x* with $\sqrt{x}$.

Compound inequality A sentence linking two conditions with and or or. At least one of the conditions is an inequality.

Compound quantity A compound quantity is a quantity expressed in two or more different but related units of measure, such as dollars and cents or hours, minutes, and seconds.

Currency with a decimal subunit is usually represented as a decimal number, e.g., $139.64 CAD, $1,060.00 MXN, £62.39 GBP, €89.53 EUR. Dates in year, month, and day number may following one of several date formats: 2000–12–31 (ISO 8601), 31/12/2000 (d/m/yyyy), 31/12/00 (d/m/yy), 31/12 (d/m), 12/31/2000 (m/d/yyyy), 12/31/00 (m/d/y), and 12/31 (m/d). Time in hours, minutes, and seconds may following one of several time formats: 19:30:00 (ISO 8601), 19:30 (hh:mm), 19.30 (hh.mm), 19h30 (hhhmm), 1930 (hhmm), 7:30 p.m. (h:mm a.m./p.m.), 12:39.36 (m:ss.ss). Angles in degrees, minutes, and seconds are usually followed by their suffix symbols, e.g., 39° 502 083 N.

Compound symmetry The condition with constant variances on the main diagonal of a matrix, and constant covariances off the main diagonal.

Computer aided software engineering (CASE) Software tools that provide automated support for some portion of the systems development process.

Concave up A curve is concave up when it is a concave shape, meaning curved like the inside of a bowl, with the two ends of the curve pointing up.

Conceptual data model A detailed model that shows the overall structure of organizational data while being independent of any database management system or other implementation considerations.

Concordant pairs A pair of observations that are ordered in the same direction on two variables.

Concrete Physical objects used to represent mathematical situations.

Concrete class A class that can have direct instances.

Concurrent lines Lines that have a common point.

Condition A condition is a logical statement about the state of the system, the state or existence of an object, or the existence of relationships between objects.

Condition number The condition number of the coefficient matrix A of a linear system is a (nonnegative) number used to estimate the amount by which small errors in the right hand side b, or in A itself, can change the solution x.

This analysis ignores arithmetic roundoff, which is hard to analyze, and focusses on easily measurable quantities known beforehand, and how they will amplify or diminish the roundoff errors.

Small values of the condition number suggest that the algorithm will not be sensitive to errors, but large values indicate that small data or arithmetic errors may explode into enormous errors in the answer.

The condition number is defined in terms of a particular matrix norm. Many different matrix norms may be chosen, and the actual value of the condition number will vary depending on the norm chosen. However, the general rule that large condition numbers indicate sensitivity will hold true no matter what norm is chosen.

The condition number for a matrix A is usually defined as

$$\text{condition}(A) = \|A\| * \|A^{-1}\|.$$

If A is not invertible, the condition number is infinite.

Simple facts about the condition number:

The condition number is always at least 1;

The condition number of the identity matrix is 1;

The condition number of any orthogonal or unitary matrix is 1;

LINPACK routines such as SGECO return RCOND, an estimate of the reciprocal of the condition number in the L1 matrix norm.

Turing's M condition number, M(A), for a matrix of order N, is defined as

$$M(A) = N * \max |A_{i,j}| * \max |A^{-1}_{i,j}|.$$

Turing's N condition number, N(A) is

$$N(A) = \text{Frob}(A) * \text{Frob}(A^{-1}) / N$$

where Frob(A) is the Frobenius matrix norm.

The Von Neumann and Goldstine P condition number is

P(A) = | lambda_Max / lambda_Min |

where lambda_Max and lambda_Min are the eigenvalues of largest and smallest magnitude, which is equivalent to using the spectral radius of A and A^{-1}.

There is also a condition number defined for the eigenvalue problem, which attempts to estimate the amount of error to be expected when finding the eigenvalues of a matrix A.

Condition stub That part of a decision table that lists the conditions relevant to the decision.

Conditional (fixed-effects) logistic regression The conditional logistic regression (CLR) model is used in studies where cases and controls can be matched (as pairs) with regard to variables believed to be associated with the outcome of interest. The model creates a likelihood that conditions on the matching variable. It is the preferred method for the analysis of nested case-control studies when matching is done at the individual level (there may be more than one control per case). In economic analysis, it is called fixed-effects logit for panel data.

Conditional distribution The distribution of one variable for a fixed level of another variable.

Conditional equation An equation that is valid for a limited number of values of the variable.

Conditional means The means for one variable at individual levels of a second variable.

Conditional odds The odds of success given some level of another variable.

Conditional probability Conditional probability is the probability of an event occurring given that another event also occurs. It is expressed as P(A/B). It reads Probability of Event A on condition of Event B. P(A/B) = P(A and B)/P(B), where P(B) is the probability of event B and P(A and B) is the joint probability of A and B .

Conditional trigonometric equations True for only a limited number of replacement values.

Cone The term cone appears in at least two contexts.

In geometry, a cone is formed by a family of straight lines through the same point tracing a (commonly planar) curve. The point is called the apex or vertex of the cone.

In functional analysis, a cone is defined for an ordered set and consists of all elements that exceed a given one.

Confidence interval An interval, with limits at either end, with a specified probability of including the parameter being estimated.

Confidence limits The upper and lower bounds of an interval with a specified probability of including the parameter being estimated.

Configuration management The process of ensuring that only authorized changes are made to a system.

Confounded Two variables are said to be confounded when they are varied simultaneously and their effects cannot be separated.

Confounding factor In a well designed psychology experiment an investigator will randomly assign subjects to two or more groups and except for differences in the experimental procedure applied to each group, the groups will be treated exactly alike. Under these circumstances any differences between the groups that are statistically significant are attributed to differences in the treatment conditions. This, of course assumes that except for the various treatment conditions the groups were, in fact, treated exactly alike.

Unfortunately, however, it is always possible that despite an experimenter's best intentions there was some unsuspected systematic differences in the way the groups were treated in addition to the intended treatment conditions. Statisticians describe systematic differences of this sort as confounding factors or confounding variables.

If, for example, subjects in one group are simultaniously tested in a room with the heat set at 70 degrees whereas subjects in another group are simultaniously tested in a nearby identically appointed room with the heat set at 60 degrees, the obtained differences in performance could be attributed to any of three factors. It could be due to the random assignment of subjects (i.e. to chance). It could be due to the different temperatures in the two rooms. It could, however, be due to some confounding factor such as differences in ambient illumination that result from unnoticed differences in the orientation of each room with respect to the sun. In any experiment an appropriate statistical test can help in the decision as to whether or not to attribute the results to chance, but only the most careful analysis of the actual conditions of the experiment can suggest whether or not the results might be due to a confounding factor.

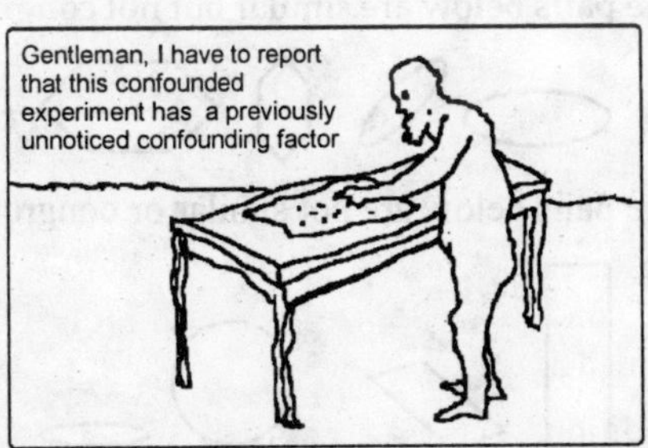

Confounding variable A variable that is associated with both the outcome and the exposure variable.

A classic example is the relationship between heavy drinking and lung cancer. Here, the data should be controlled for smoking as it is related to both drinking and lung cancer. A positive confounder is related to exposure and response variables in the same direction (as in smoking); a negative confounder shows an opposite relationship to these two variables (age in a study of association between oral contraceptive use and myocardial infarction is a negative confounder). The data should be stratified before analysing it if there is a confounding effect. Mantel-Haenszel test is designed to analyse stratified data to control for a confounding variable. Alternatively, a multivariable regression model can be used to adjust for the effects of confounders.

Congruency Geometric figures having the same size and shape; all corresponding parts of congruent figures have the same measure.

Congruent figures Two figures are congruent if they have the same shape and size.

Example: The following pairs of figures below are congruent. Note that if two figures are congruent, they must be similar.

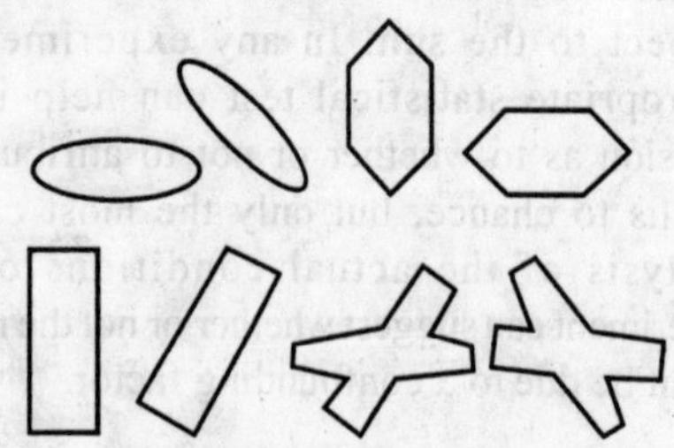

The pairs below are similar but not congruent.

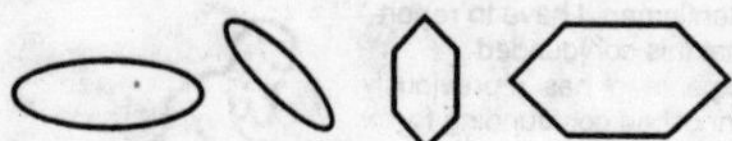

The pairs below are not similar or congruent.

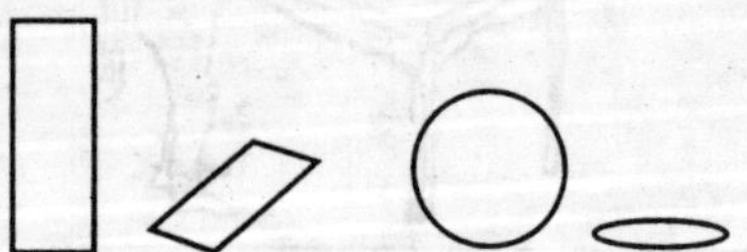

Congruent matrix Congruent matrices A and B are related by a nonsingular matrix P such that

A = P' * B * P.

Congruent matrices have the same inertia.

Congruence is of little interest by itself, but the case where P is also an orthogonal matrix is much more important.

Congruent Two figures are congruent to one another if they have the same size and shape.

Conic section A curve that has a sec- ond-degree equation and is defined in terms of the distance of its points from fixed points and/or lines. This includes circles, parabolas. ellipses and hyperbolas.

Conjugate Two binomials of the form (a + Nt/b) and (a - N/b). where a and b are rational numbers.

Conjugate of a complex number The conjugate of a complex number z = a + b * i is the complex number

conjugate (z) = a – b * i.

The conjugate is frequently represented by placing a bar over the quantity, or occasionally a star after it, as in z*.

The complex conjugate can be used in a formula for the norm or magnitude of a complex number, which must always be a real nonnegative value: norm (z) = sqrt (z * conjugate (z)) = sqrt (a^2 + b^2).

For complex vectors, an inner product with the correct properties may be defined as:

V dot W = (V, W) = sum (I = 1 to N) conjugate (V(I)) * W(I).

This inner product is computed in the BLAS function CDOTC, for example, and yields another relationship with the Euclidean vector norm:

|| V || = sqrt (V .V)

Conjunctive matrix Two (complex) matrices A and B are said to be conjunctive if there is some nonsingular matrix P so that

A = (conjugate (P))' * B * P,

This is the extension to complex matrices of the concept of a congruent real matrix.

Connected set A set that can't be split into a union of two sets each of which is both open and closed.

Conservation The concept that physical properties remain constant even as appearance and form changes.

Conservative test A test where the chance of type I error (false positive) is reduced and type II error risk is increased. Thus, these tests tend to give larger P values compared to non-conservative (liberal) tests for the same comparison.

Consistency An axiomatic theory is consistent if it's impossible (in the confines of the theory) to prove simultaneously a statement and its negation. The Godel's Theorem states that any (sufficiently powerful) consistent axiomatic theory is incomplete.

Consistent system of equations A system that has at least one solution.

Constant A fixed number, like 3 or $\sqrt{2}$. To be distinguished from a variable, which has no single value.

Constant difference The difference be- tween successive terms in a sequence.

Constant functions Functions that stay the same no matter what the variable does are called constant functions.

Constant linear function A function whose equation can be written in the form y = b and whose graph is a horizontal line.

Constant of inverse variation The number k in the inverse variation Y = K/X.

Constant of variation The number k in the direct variation y = kx or in the inverse variation Y = K/X.

Constant term of a polynomial The term a which can be considered the coefficient of x.

Constant term of a quadratic equation The term c in the form V = ax– + bx + r.

Constraint A constraint restricts possible occurences or allowable combinations in relationship sets, interactions, object classes, etc. There are Co-occurrence Constraints, Object-class Cardinality Constraints, Participation Constraints, and General Constraints.

Context diagram A data flow diagram of the scope of an organizational system that shows the system boundaries, external entities that interact with the system, and the major information flows between the entities and the system.

Contingency coefficient A coefficient, based on chi-square, reflecting the degree of relationship exhibited in a contingency table.

Contingency table A two-dimensional table in which each observation is classified on the basis of two variables simultaneously.

Continuity A mapping f:A–>B is continuous if images $f(a_1)$ and $f(a_2)$ of two close points a_1 and a_2 are also close to each other. This is only meaningful if for both spaces A and B the notion of closeness has been defined. The latter is defined in terms of neighborhoods.

Continuous A polynomial function whose graph is smooth without jumps.

Continuous graph In a graph, a continuous line with no breaks in it forms a continuous graph.

Continuous interaction A continuous interaction is one where the origin object is continuously sending information to the destination object.

Continuous variables Variables that take on any value between two extremes.

Continuum Any set that may be brought into 1–1 correspondence with the set of real numbers. Examples: a finite line segment, a square, a circle, a disk.

Contrast A contrast is combinations of treatment means, which is also called the main effect in ANOVA. It measures the change in the mean response when there is a change between the levels of one factor.

Converge If there is a real number that is the value of the infinite series, the series converges.

Conversion For fractions : The process of changing the numerator and the denominator of a fraction to make a new fraction of equal value. Example: 1/2 to 2/4. For units of measurement : The Process of changing a unit of measurement to a different unit of measurement while keeping the value the same. Example: 2 feet to inches, 2 feet = 24 inches.

Convex A figure is convex if every line segment drawn between any two points inside the figure lies entirely inside the figure. A figure that is not convex is called a concave figure.

Example: The following figures are convex.

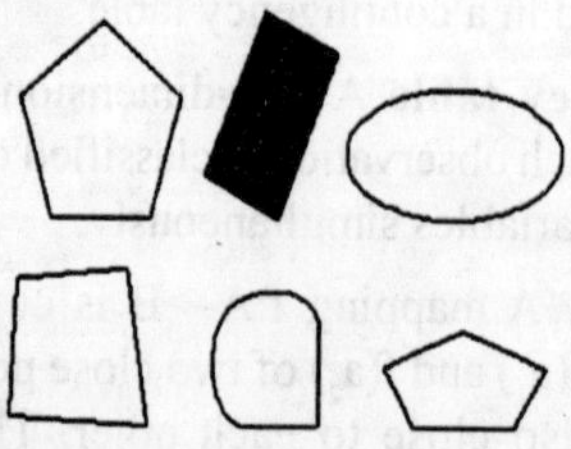

The following figures are concave. Note the red line segment drawn between two points inside the figure that also passes outside of the figure.

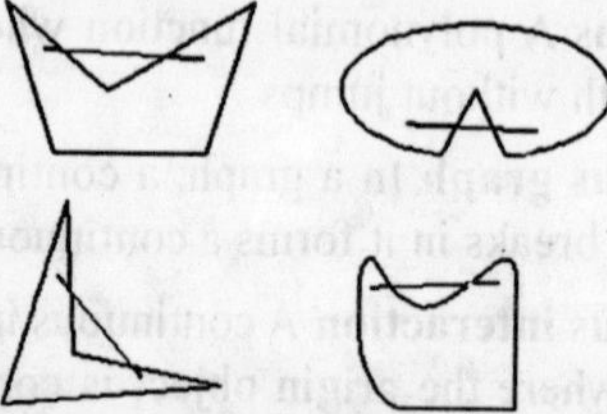

Convex polygonal region A region that has a polygon with no indentations as a boundary.

Cook statistics A diagnostic influence statistics in regression analysis designed to show the influential observations. Cook's distance considers the influence of the i th value on all n fitted values and not on the fitted value of the i th observation. It yields the shift in the estimated parameter from fitting a regression model when a particular observation is omitted. All distances should be roughly equal; if not, then there is reason to believe that the respective case(s) biased the estimation of the regression coefficients. Relatively large Cook statistics (or Cook's distance) indicates influential observations. This may be due to a high leverage, a large residual or their combination. An index plot of residuals may reveal the reason for it. The leverages depend only on the values of the explanatory variables (and the model parameters). Cook statistics depends on the residuals as well. Cook statistics may not be very satisfactory in binary regression models. Its formula uses the standardised residuals but the modified Cook statistics uses the deletion residuals.

Cookie crumbs A technique for showing a user where they are in a Web site by placing a series of tabs on a Web page that show a user where they are and where they have been.

Cook's D A measure of the influence of an observation in multiple regression.

Coordinate Coordinates are pairs of numbers that are used to determine points in a plane, relative to a special point called the origin. The origin has coordinates (0,0). We can think of the origin as the center of the plane or the starting point for finding all other points. Any other point in the plane has a pair of coordinates (x,y). The x value or x-coordinate tells how far left or right the point is from the point (0,0), just like on the number line (negative is left of the origin, positive is right of the origin). The y value or y-coordinate tells how far up or down the point is from the point (0,0), (negative is down from the origin, positive is up from the origin). Using coordinates, we may give the location of any point we like by simply using a pair of numbers.

Example: The origin below is where the x-axis and the y-axis meet. Point A has coordinates (2,3), since it is 2 units to the right and 3 units up from the origin. Point B has coordinates (3,0), since it is 3 units to the right, and lies on the x-axis. Point C has coordinates (6.3,9), since it is 6.3 units to the right, and 9 units up from the origin. Point D has coordinates (9,–2.5); it is 9 units to the right, and 2.5 units down from the origin. Point E has coordinates (–4,–3); it is 4 units to the left, and 3 units down from the origin. Point F has coordinates (–7,5.5); it is 7 units to the left, and 6 units up from the origin. Point G has coordinates (0,–7) since it lies on the y-axis 7 units below the origin.

Coordinate axes Two perpendicular lines selected in a plane.

Coordinate plane (Cartesian) A plane with a point selected as an origin, some length selected as a unit of distance, and two perpendicular lines that intersect at the origin, with positive and negative direction selected on each line. Traditionally, the lines are called x (drawn from left to right, with positive direction to the right of the origin) and y (drawn from bottom to top, with positive direction upward of the origin). Coordinates of a point are determined by the distance of this point from the lines, and the signs of the coordinates are determined by whether the point is in the positive or in the negative direction from the origin.

Coordinates of a point The numbers of an ordered pair.

Coprime Two integers n and m with no common factors are said to be mutually prime or coprime. By definition, gcd(n, m) = 1.

Corrective maintenance Changes made to a system to repair flaws in its design, coding, or implementation.

Correlation Given a pair of related measures (X and Y) on each of a set of items, the correlation coefficient (r) provides an index of the degree to which the paired measures co-vary in a linear fashion. In general r will be positive when items with large values of X also tend to have large values of Y whereas items with small values of X tend to have small values of Y.

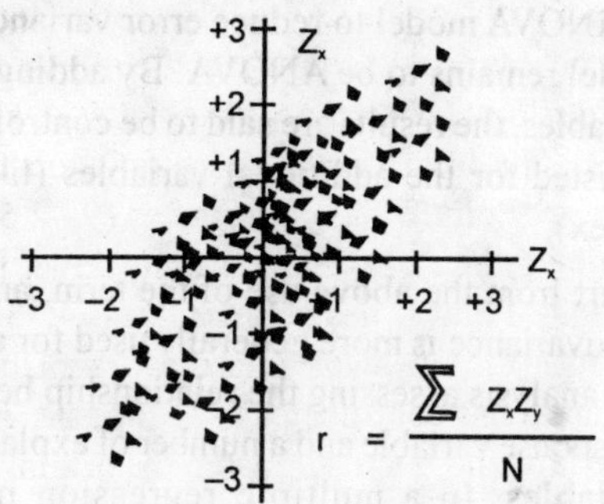

Correspondingly, r will be negative when Items with large values of X tend to have small values of Y whereas items with small values of X tend to have large values of Y. The value of r is calculated by first converting the Xs and Ys into their respective Z Scores and, keeping track of which Z Score goes with which item, determining the value of the mean Z Score product. Numerically, r can assume any value between −1 and +1 depending upon the degree of the relationship. Plus and minus one indicate perfect positive and negative relationships whereas zero indicates that the X and Y values do not co-vary in any linear fashion.

Correlation coefficient See Pearson's correlation coefficient (r), Spearman's rank correlation (rho) and Multiple regression correlation coefficient (R^2).

Correlational measures A measure of the degree of relationship between two variables that are each at least ordinal. As contrasted with measures of association.

Corresponding angles For any pair of parallel lines 1 and 2, that are both intersected by a third line, such as line 3 in the diagram below, angle A and angle C are called corresponding angles. Corresponding angles have the same degree measurement. Angle B and angle D are also corresponding angles.

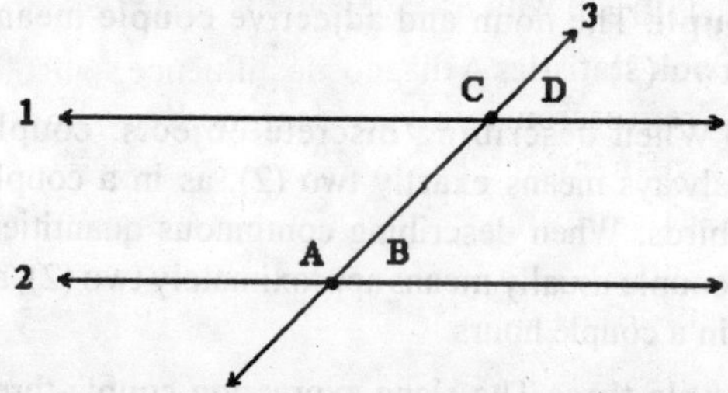

Corresponding elements of matrices For two matrices with the same dimensions. the elements in the same row and the same column.

Cosecant One of the functions of an angle. It is found by dividing the length of the hypotenuse of a right triangle by the length of the opposite side of a reference angle.

Cosine One of the functions of an angle. It is found by dividing the length of the adjacent side of a reference angle by the length of the hypotenuse.

Cotangent One of the functions of an angle. It is found by dividing the length of the adjacent side of a reference angle by the length of the opposite side of a reference angle.

Coterminal Two angles in standard position that share a terminal side.

Count 1. The verb count means to enumerate by assigning sequential integers.

2. The verb count means to tally by counting.

3. The noun and adjective count means the act of counting.

4. The noun and adjective count means the total number of objects in a set.

Count data Data representing counts or the number of observations in each category.

Counter-balancing An arrangement of treatment conditions designed to balance out practice effects.

Counting technique Methods to determine the number of possible outcomes of an event. Some of the methods are tree diagram, list, rules for multiplication, combinations, and permutations.

Countless The adjective countless means a very large number of individuals or discrete objects.

Couple The noun and adjective couple means two (2).

When describing discrete objects, couple always means exactly two (2), as in a couple birds. When describing continuous quantities, couple usually means approximately two (2), as in a couple hours.

Couple-three The slang expression couple-three means at least two (2) and at most three (3).

Coupling The extent to which subsystems depend on each other.

Covariance (covariation) It is a measure of the association between a pair of variables: the expected value of the product of the deviation of two random variables from their respective means. It is also called a measure of 'linear dependence' between the two random variables. If the two variables are independent (no linear correlation), then their covariance is zero but for non-zero values covariance is unstandardised (unlike correlation coefficient); there is no limit to possible values. Because of this, it is difficult to compare covariances. A negative value means that for small values of X, there are large values of Y (inverse association). It is calculated as the mean sum-of-products using each x_i and y_i values, their means m_x and m_y, and (n–1). (The covariance standardised to lie between –1 and +1 is Pearson's correlation coefficient.) .

Covariance matrix A matrix of variances and covariances among variables.

Covariance models Models containing some quantitative and some qualitative explanatory variables, where the chief explanatory variables of interest are qualitative and the quantitative variables are introduced primarily to reduce the variance of the error terms. [Models in which all explanatory variables are qualitative are called analysis of variance -ANOVA- models.] Analysis of covariance -ANCOVA- combines features of ANOVA and regression. It augments the ANOVA model containing the factor effects with one or more additional quantitative variables that are related to the response variable. The intention is to make the analysis more precise by reducing the variance of the error terms. Each continuous quantitative variable added to the ANOVA model is called a concomitant variable (and sometimes also covariates). If qualitative variables are added to an ANOVA model to reduce error variance, the model remains to be ANOVA. By adding extra variables, the results are said to be controlled or adjusted for the additional variables (like age or sex).

Apart from the above use of the term, analysis of covariance is more generally used for almost any analysis assessing the relationship between a response variable and a number of explanatory variables. In a multiple regression model, additional variables, which are known not to have any effect on the response variable, such as age and sex, are sometimes added to the model to adjust the response for these variables (age and sex in this case). Such variables are called confounders (or covariates). When the response is normally distributed, this is the preferred method over a simple t-test when the two groups compared differ, say, in their age and sex distribution (or any other confounding variable). The result is then controlled (or adjusted) for age and sex. When such adjustments are made, the regression coefficient for the significant effect variable will be (most probably) different from the one obtained from a univariable model involving only that variable (say the effect of a disease on pulse rate as compared to healthy controls).

Covariate Generally used to mean explanatory variable, less generally an additional explanatory variable is of no interest but included in the model. More specifically, it denotes a 'continuous' explanatory variable which is unaffected by treatments and has a linear relationship to the response. The intention is to produce more precise estimates of the effect of the explanatory variable of main interest. In the analysis, a model is first fitted using the covariate (and the blocking factor, if any). Then the main explanatory variable is added and its additional effect is assessed statistically. Whether the use of a covariate is wise can be judged by checking its effect on the residual (error) mean square (variance). If the addition of covariate reduces it remarkably, it will improve the analysis.

Cramer's coefficient of association (C) Also known as contingency coefficient. While Chi-squared is used to determine significance of an association (and varies by sample size for the same association), Cramer's C is a measure of association varying from 0 (no association) to 1 (perfect association) that is independent of the sample size. However, it seldom reaches its upper limit. It allows direct comparison of the degree of association between different contingency tables. It is calculated directly from the Chi-squared value and the total sample size as (?²/?²+N) ½.

Cramer's V A measure of the strength association for any size of contingency tables. It can be seen as a correction of the Chi-squared value for sample size. For a 2x2 table, Cramer's V is equal to the Phi coefficient. Cramer's V is most useful for large contingency tables.

Create interaction A create interaction creates a new obect in the model.

Critical path scheduling A scheduling technique whose order and duration of a sequence of task activities directly affect the completion date of a project.

Critical path The shortest time in which a project can be completed.

Critical periods Times of increased and favored sensitivity to particular aspects of development.

Critical point A value of x that makes the derivative f'(x) of a function either equal to 0 or nonexistent. It comes up either in graphing functions, telling you where the critical changes in the graph occur, or in applied max/min problems, where it tells where the potential maxima or minima are occurring.

Critical region The critical region of the sampling distribution of a statistics is also known as the alpha region.

It is the area, or areas, of the sampling distribution of a statistic that will lead to the rejection of the hypothesis tested when the hypothesis is true.

Critical value The critical value or values of a statistic is the value(s) of that statistic that cut off the proportion of the sampling distribution designated as the alpha region(s).

Crore The adjective and noun crore means one hundred lakh or ten million (10,000,000 usually punctuated 1,00,00,000 in India.)

Cross correlation The correlation between one predictor and all other predictors.

Cross cultural research Research designed to reveal variations existing across different groups of people.

Cross gender behaviors Behaviors stereotypical of the opposite sex.

Cross life cycle CASE CASE tools designed to support activities that occur across multiple phases of the systems development life cycle.

Cross section A cross-section of a space figure is the shape of a particular two-dimensional slice of a space figure.

Example: The circle on the right is a cross-section of the cylinder on the left.

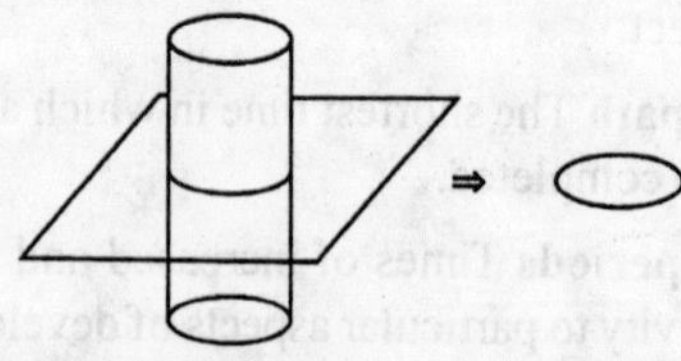

The triangle on the right is a cross-section of the cube on the left.

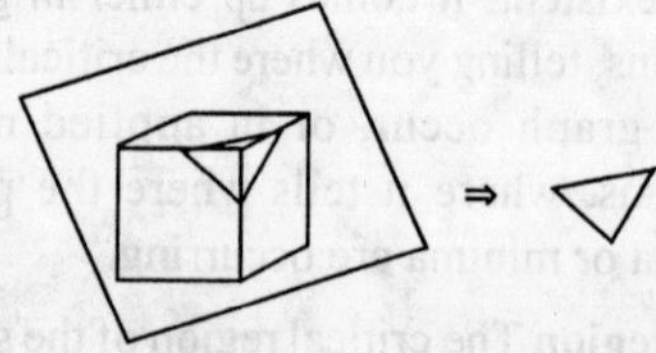

Cross-sectional data Data collected at one point in time (as opposed to longitudinal/cohort data for example).

Cross sectional study A study in which a number of different-age individuals with the same trait or characteristic of interest are studied at a single time.

Cross sequential study A study in which individuals in a cross-sectional sample are tested more than once over a specified period of time.

Cross validation The result of taking a regression equation from one set of data, applying it to a new set of data, and examining the correlation between the predicted and obtained values on the new set of data.

Crossover design A clinical trial design during which each subject crosses over from receiving one treatment to another one.

Crumb In computer terminology, the noun crumb means a pair (2) of binary units having two to the second power ($2^2 = 4$) possible states. A crumb is equal to two (2) bits or half (1/2) a nibble.

Crystallized intelligence The ability to use learned information collected throughout a life span.

Cube A cube is a three-dimensional figure having six matching square sides. If L is the length of one of its sides, the volume of the cube is $L^3 = L \times L \times L$. A cube has six square-shaped sides. The surface area of a cube is six times the area of one of these sides.

Example: The space figure pictured below is a cube. The grayed lines are edges hidden from view.

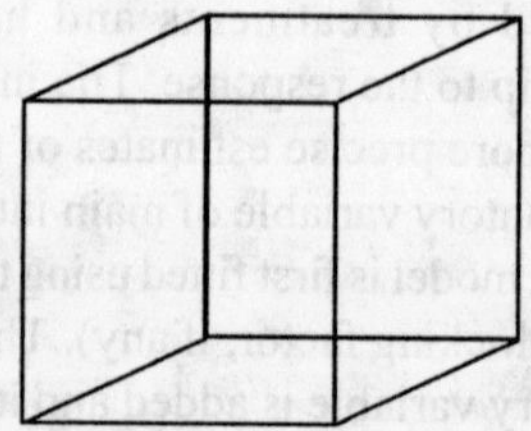

Example: What is the volume and surface are of a cube having a side-length of 2.1 cm?

Its volume would be $2.1 \times 2.1 \times 2.1 = 9.261$ cubic centimeters.

Its surface area would be $6 \times 2.1 \times 2.1 = 26.46$ square centimeters.

Cubic polynomial A polynomial of degree 3.

Culminating phase A phase of early adulthood that ranges from ages 33 to 45.

Culture free IQ tests Tests without cultural content.

Curve A curve is a continuous mapping of the segment [0, 1] into another space - container of the curve. A curve may not look as a line. For example, there are space filling curves.

Curve fitting The sketching of a line or curve to best describe a relationship between two variables on a scatter plot.

Curvilinear relationship A situation that is best represented by something other than a straight line.

Cycle The graph of a periodic function through one fundamental period.

Cyclic reduction Cyclic reduction is a method for solving a linear system A*x=b in the special case where A is a tridiagonal matrix.

On a parallel computer, this method solves the system in LOG(N) steps where N is the order of A. A standard Gauss elimination method for a tridiagonal system would require roughly N steps instead.

A tridiagonal system has some very special properties that will allow us to carry this operation out. Consider this system of 7 equations:

A11 x1 + A12 x2 = y1

A21 x1 + A22 x2 + A23 x3 = y2

A32 x2 + A33 x3 + A34 x4 = y3

A43 x3 + A44 x4 + A43 x5 = y4

A54 x4 + A55 x5 + A56 x6 = y5

A65 x5 + A66 x6 + A67 x7 = y6

A76 x6 + A77 x7 = y7

The first equation can be used to eliminate the coefficient A21 in the second equation, and the third equation to eliminate the coefficient A23 in the second equation. This knocks out variables x1 and x3 in the second equation, but adds x4 into that equation.

By the same method, x3 and x5 can be eliminated from the equation for x4, and so on. By eliminating the odd variables from the even equations, a smaller tridiagonal system system is derived, with half the equations and variables.

If elimination is applied to this set, the number of equations is again reduced by half; this reduction may be repeated until a single equation in one variable is reached. Backsubstitution then produces the values of all the variables.

The reason this method might have an advantage over Gauss elimination is that, at each step of the elimination phase, the parts of the step are independent. If many computer processors are available, then each can be working on a separate portion of the elimination. If the number of processors is large enough, the system can really be solved in LOG(N) time.

Cyclic tridiagonal matrix A cyclic tridiagonal matrix is a generalization of a tridiagonal matrix which includes an extra last entry in the first row, and an extra first entry in the last row.

An example of a cyclic tridiagonal matrix:

–2 1 0 0 1

1 –2 1 0 0

0 1 –2 1 0

0 0 1 –2 1

1 0 0 1 –2

A cyclic tridiagonal matrix is not a tridiagonal matrix. If the matrix is constant along the three generalized diagonals, a cyclic tridiagonal matrix is a circulant matrix. A cyclic tridiagonal matrix can arise in situations where a periodic boundary condition is applied.

It is very disappointing that a cyclic tridiagonal matrix is not a tridiagonal matrix, since there are so many good methods for solving tridiagonal linear systems. One way to solve a cyclic tridiagonal system is to use the Sherman Morrison Formula and view the matrix as a rank one perturbation of a tridiagonal matrix. Another approach is to view it as a border banded matrix.

A cyclic tridiagonal matrix may also be called a periodic tridiagonal matrix.

Cylinder A cylinder is a space figure having two congruent circular bases that are parallel. If L is the length of a cylinder, and r is the radius of one of the bases of a cylinder, then the volume of the cylinder is $L \times pi \times r^2$, and the surface area is $2 \times r \times pi \times L + 2 \times pi \times r^2$.

Example: The figure pictured below is a cylinder. The grayed lines are edges hidden from view.

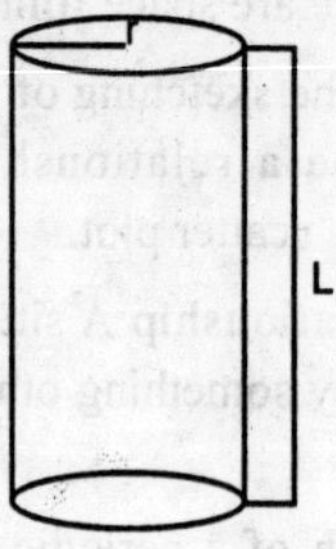

D

D The traditional metric prefix symbol d- (metric prefix: deci-) denotes a multiple of one tenth (1/10 = 0.100).

Da The traditional metric compound prefix symbol da- (metric prefix: deka-) denotes a multiple of ten (10).

Data Raw facts about people, objects, and events in an organization.

Database A shared collection of logically related data designed to meet the information needs of multiple users in an organization.

Data flow Data in motion, moving from one place in a system to another.

Data flow diagram (DFD) A graphic that illustrates the movement of data between external entities and the processes and data stores within a system.

Data oriented approach A strategy of information systems development that focuses on the ideal organization of data rather than on where and how they are used.

Data store Data at rest, which may take the form of many different physical representations.

Data structure Data structure is both a programming language construct that unifies several related attributes and a basic data type. Realization of the utility of data structures in programming led eventually to the modern object oriented trend. Among the basic data types are the stack, list, queue, tree.

Data type A coding scheme recognized by system software for representing organizational data.

Db The symbol dB represents a logarithmic power measurement in decibels.

De Moivre's theorem A theorem involving powers of complex numbers.

Dea The nonstandard prefix dea- denotes a multiple of ten to the thirtieth power (10^{30} = 10 raised to power +30 = 1,000,000,000,000,000,000,000, 000,000,000).

Debrief To give information concerning research that has just been completed.

Deca The prefix deca- means ten (10).

Decad The noun and adjective decad means a group of ten (10) members.

Decade The noun decade means ten (10) or ten (10) years.

The Christian Era began with the year 1 A.D., thus the Two Hundred First (201st) Decade includes the years 2001 through 2010. In contrast, the Two Thousand Tens (2010's) include the years 2010 through 2019.

Decagon A ten-sided polygon. The sum of the angles of a decagon is 1440 degrees.

Examples: A regular decagon:

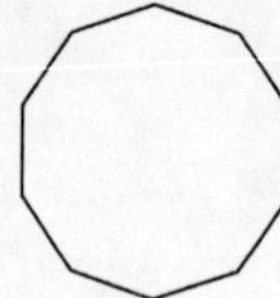

An irregular decagon:

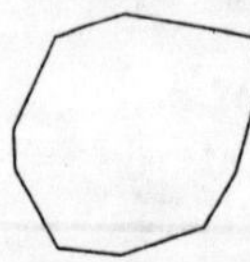

Decagram 1. The noun and adjective decagram means an interconnected star figure with ten (10) points formed by stellating a decagon.

2. The noun decagram means a word or character sequence consisting of ten (10) characters.

Decahedron The noun and adjective decahedron means a solid figure with ten (10) planar faces.

Decalogue The noun decalogue means ten (10) rules. Capitalized, Decalogue means the Ten Commandments. More generally, decalogue means a basic set of binding rules.

Decameter The noun decameter means a verse consisting of ten (10) metrical feet.

Decathlon The noun decathlon means an athletic competition consisting of ten (10) separate events.

December The month name December literally means the tenth (10th) month. The Roman year began on the first (1st) day of March.

Decemmillenary 1. The noun and adjective decemmillenary means a group of ten thousand (10000).

2. The noun and adjective decemmillenary means a period of ten thousand (10000) years.

3. The noun and adjective decemmillenary means a ten thousand (10000) year anniversary.

4. The adjective and noun decemmillenary means ten thousandth (10,000th) in rank or order.

Decemmillennium The noun and adjective decemmillennium means a period of ten thousand (10000) years.

Decemmillesimal The adjective decemmillesimal means based on the number ten thousand (10,000).

Decennary The noun and adjective decennary means a group of ten (10) or a period of ten (10) years.

Decennial 1. The adjective and noun decennial means recurring every ten (10) years.

2. The adjective and noun decennial means lasting ten (10) years.

3. The noun and adjective decennial means a ten (10) year anniversary.

Decennium The noun and adjective decennium means a period of ten (10) years.

Decennoval The adjective decennoval means based on the number nineteen (19).

Deci 1. The numeric prefix deci- means ten (10). This prefix is derived from the Latin cardinal number decem (X).

2. The traditional metric prefix deci- (prefix symbol: d-) denotes a multiple of one tenth (1/10 = 0.100).

Decibel 1. The noun and adjective decibel (symbol: dB) is a logarithmic comparison of power levels equal to a ratio of ten to the one-tenth power ($10^{0.1}$) or approximately 1.258 925 411 794. Ten (10) decibels equal one (1) bel or a power ratio of ten (10), and twenty (20) decibels equal two (2) bels or a power ratio of one hundred (100).

2. The noun and adjective decibel (symbol: dB) is a logarithmic measure of sound power density relative to a power density of ten to the negative twelfth power (10^{-12} = 10 raised to power –12 = 0.000 000 000 000) watts per square meter (W/m^2) = 1 picowatt per square meter (pW/m^2). Ten (10) decibels are a sound level of 10 pW/m^2, and twenty (20) decibels are a sound level of 100 pW/m^2.

Decile The noun decile means a frequency distribution divided into ten (10) equal parts.

Decillion The noun and adjective decillion means multiplied by 10^{33} in the American number naming system, but means multiplied by 10^{60} in the Chuquet number naming system.

Decimal The adjective decimal means based on the number ten (10).

Decimal format The compound noun and adjective decimal format means the base ten (base 10) number notation system.

Decimal fraction A fraction in which the denominator is 10 or a power of 10 (such as 100, 1000, 10,000, etc.). The denominator is seldom used. Instead, a dot called a decimal point is placed in front of the numerator. The denominator can be inferred by the number of decimal places behind the decimal point. If there is one number to the right of the decimal point, the denominator is 10. If there are two numbers to the right of the decimal point, the denominator is 100. There will be the same number of zeroes in the denominator as there are decimal places.

Decimal number A fraction where the denominator is a power of ten and is therefore expressed using a decimal point. For example: 0.37 is the decimal equivalent of 37/100.

Decimal point The compound noun and adjective decimal point means the punctuation mark (either a period (.) or a comma(,)) placed to the right of the units (1s) digit of a real number in standard decimal format.

Decimate The verb decimate literally means to kill one out of ten (1/10 = 0.100). More generally, decimate means to destroy a substantial portion. Decimate is often erroneously used to mean to exterminate or to destroy completely. The common expression completely decimate is both illogical and confusing. Origin: Roman military tacticians reasoned that a military unit could continue to fight effectively if less than 10% of its members were killed. A unit that suffered more than one in ten killed usually had to reorganize before continuing a campaign. Generals of the Roman Imperial Army later used this rule of thumb to punish cowardly or disobedient legions. The members of the unit to be punished by decimation were lined up unarmed in a single file between two fully armed legions of loyal soldiers. The soldiers were counted in order from one to ten, and each unlucky tenth soldier was executed on the spot. The survivors of this ruthless punishment were returned to duty to face death another day.

Decimation The noun decimation literally means the killing of one out of ten (1/10 = 0.100). More generally, decimation means the destruction of a substantial portion.

Decision tree Graphical representation of decisions involved in the choice of statistical procedures.

Decomposition The process of breaking the description of a system down into small components; also called functional decomposition.

Decreasing function A function such that for any numbers a and b in the domain. if a > b. then f(a) < f(b).

Decuple The verb, adjective, adverb, and noun decuple means multiply by ten (10).

Decurion The noun decurion literally means one who commands ten (10). A decurion was a Roman officer commanding a cavalry unit of about ten (10) soldiers.

Deductive reasoning A series of logical steps in which a conclusion is drawn directly from a set of statements that are known or assumed to be true. (e.g., if 5 + 4 = 9 and 6 + 3 = 9 , then 5 + 4 = 6 + 3).

Deep six The noun and verb deep six means disposal, especially by throwing overboard. A minimum water depth of six (6) fathoms (approximately 11 meters) was once considered adequate for disposal of rubbish at sea.

Default value The value a field will assume unless an explicit value is entered for that field.

Defective matrix A defective matrix is a (square) matrix that does not have a full set of N linearly independent eigenvectors.

For every eigenvalue, there is always at least one eigenvector, and eigenvectors corresponding to distinct eigenvalues are linearly independent. If the N eigenvalues of a matrix are distinct, then it surely has N linearly independent eigenvectors, and so cannot be defective.

Conversely, if a matrix is defective, then it must have at least one repeated eigenvalue, that is, an eigenvalue of algebraic multiplicity greater than 1. A matrix is defective if and only if its Jordan Canonical Form has at least one nonzero entry on the superdiagonal.

Thus, a simple example of a defective matrix is:

1 1

0 1

which has the single eigenvalue of 1, with algebraic multiplicity 2, but geometric multiplicity 1. The only eigenvector is (0, 1).

If a matrix is not defective, then its eigenvectors form a basis for the entire linear space. In other words, any vector y can be written as

y = X * c

where X is the array of eigenvectors of A.

If a matrix A is not defective, then it is similar to its diagonal eigenvalue matrix:

$A = X * LAMBDA * X^{-1}$

and the similarity transformation matrix X is actually the eigenvector matrix.

This in turn allows us to make interesting statements about the inverse, tranpose, and powers of A. For instance, we see that

$A^2 = (X * LAMBDA * X^{-1}) * (X * LAMBDA * X^{-1})$

$= X * LAMBDA^2 * X^{-1}$

leading us to the statement that for a nondefective matrix, the square has the same eigenvectors, and the square of the eigenvalues.

Deficient A number is deficient if sigma(n) < 2n, or, equivalently, if index(n) < 2. Another way of saying this is that sum of the proper divisors of n is less than n.

Every divisor of a deficient number is deficient.

Definite integral The definite integral of a function f(x) over an interval $a \le x \le b$ $a \le x \le b$ is a number, sometimes thought of as the area under a graph. Not to be confused with the indefinite integral, which gives a function.

Deflation Deflation is a technique for removing a known eigenvalue from a matrix, in order to facilitate the determination of other eigenvalues.

For example, the power method is able to estimate the eigenvalue of largest modulus of a matrix A. Once this is computed, it might be desired to find the next largest eigenvalue. Deflation can be used to essentially create a new matrix, A', which has the same eigenvalues as A, except that the largest eigenvalue has been dropped (and the order of A' reduced by 1) or the largest eigenvalue is replaced by 0. In either case, the power method applied to A' will produce the next largest eigenvalue of A.

To eliminate a known eigenvalue, lambda, it is necessary to know its eigenvector x, which we will assume has been scaled to have unit Euclidean norm. By the properties of eigenvectors, we know that

A * x = lambda * x.

Now define the matrix A^ so that:

A^ = A - lambda * x * x'.

Now x is an eigenvector of A^ with eigenvalue 0, because: A^ * x = (A – lambda * x * x') * x = A * x – lambda * x * x' * x = lambda * x – lambda * x = 0

If the power method is being employed, then the new iteration should try to factor out any component of the eigenvector x; otherwise, small errors in the computation of the first eigenvalue and eigenvector will interfere with the next results.

Theoretically, this process may be repeated as often as desired, eliminating each eigenvalue as it is discovered. Practically, however, accumulated errors in the eigenvalues and eigenvectors make the computation more and more unreliable with each step of deflation. Thus, if more than a few eigenvalues are desired, it is more appropriate to use a standard technique.

Degree In angular measurement, the noun and adjective degree (suffix symbol: -° {degree symbol}) means one three-hundred-sixtieth (1/360) of a circle.

The suffix symbol -° {degree symbol} denotes an angular measurement in degrees. One degree is equal to one three-hundred-sixtieth (1/360) of a circle. The degree symbol is also used as part of the compound symbols that represent units of relative temperature, e.g., -°C and -°F.

Degree of a monomial The sum of the exponents of the variables.

Degree of a polynomial The degree of the term (or terms) of highest degree.

Degrees of freedom (DF) Statisticians use the terms degrees of freedom to describe the number of values in the final calculation of a statistic that are free to vary. Consider, for example the statistic s-square.

To calculate the s-square of a random sample, we must first calculate the mean of that sample and then compute the sum of the several squared deviations from that mean. While there will be n such squared deviations only (n - 1) of them are, in fact, free to assume any value whatsoever. This is because the final squared deviation from the mean must include the one value of X such that the sum of all the Xs divided by n will equal the obtained mean of the sample. All of the other (n - 1) squared deviations from the mean can, theoretically, have any values whatsoever. For these reasons, the statistic s-square is said to have only (n - 1) degrees of freedom.

$$S_x^2 = \frac{\sum (x - \bar{x})^2}{n-1}$$

Degrees: measuring angles We measure the size of an angle using degrees.

Example: Here are some examples of angles and their degree measurements.

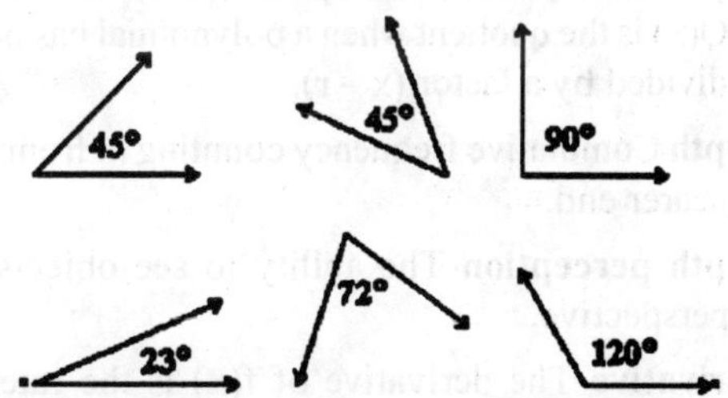

Deka The traditional metric prefix deka- (compound prefix symbol: da-) denotes a multiple of ten (10).

Deletion (or deleted) residual A modified version of the standardised residual, which uses an estimate of μ^2 from a regression in which point i has been deleted. It is used in calculation of the modified Cook's distance instead of standardised residuals. Deletion residuals are also known as likelihood residuals.

Deliverable An end product in a phase of the SDLC.

Delivery Expelling the baby and placenta from the vagina.

Delta A value used in power tables that combines gamma (?) and the sample size.

Denominator In a rational number, the number below the fraction bar that indicates how many parts the whole is divided into.

Denormalization The process of splitting or combining normalized relations into physical tables based on affinity of use of rows and fields.

Density Height of the curve for a given value of X- closely related to the probability of an observation in an interval around X.

Dependent events Two events A and B are dependent if and only if (A and then B) * P(A) – P(B).

Dependent system A system whose gr-aphs coincide and that has infinitely many solutions.

Dependent variable If the value of a de- pends on the value of b, or if a is usually defined in terms of b, a is called the dependent variable. and it is graphed on the vertical axis.

Depressed equation The equation Q(x) = 0. where Q(x) is the quotient when a polynomial has been divided by a factor (x – r).

Depth Cumulative frequency counting in from the nearer end.

Depth perception The ability to see objects in perspective.

Derivative The derivative of f(x) is the rate of change of f(x). Geometrically, it also represents the slope of the tangent line to the graph of the function y=f(x) at the point (x,f(x)), but that's a mouthful.

Descending order A polynomial is written in descending order when the term with highest degree is written first, followed by the term with next highest degree second, and so on.

Descriptive statistics Summary of available data. Examples are male-to-female ratio in a group; numbers of patients in each subgroup; the mean weight of male and female students in a class, etc. Only when the distribution is symmetric, mean and standard deviation can be used. Otherwise (as in survival data), mean and percentiles/range should be used to describe the data.

Design matrix A matrix of coded or dummy variables representing group membership.

Design strategy A particular approach to developing an information system. It includes statements on the system's functionality, hardware and system software platform, and method for acquisition.

Desk checking A testing technique in which the program code is sequentially executed manually by the reviewer.

Determinant A real number represented by a square array of numbers.

Determinant of a matrix The determinant of a matrix is the determinant with the same corresponding elements.

Deuce The noun deuce means two (2). In parlor games, deuce means a playing card, die, or domino that represents the number two or a throw of dice that totals two. In tennis, deuce means a tie that requires two successive points or games to win.

Developmentalists Researchers who study human development.

Developmentally disabled Persons who are slowed or delayed in development or progress, especially because of subnormal intellectual functioning and social skills.

Deviance A measure for judging the degree of matching of the model to the data when the parameter estimation is carried out by maximising the likelihood (as in GLMs). The deviance has asymptotically a Chi-squared distribution with df equal to the difference in the number of parameters in the two models being compared. The total deviance compares the fit of the saturated model to the null model, thus, expresses the total variability around a fitted line which can be decomposed to explained and unexplained (error) variability. The residual deviance in a GLM analysis of deviance table corresponds to RSS in an ANOVA table, and regression deviance corresponds to ESS. The residual deviance measures how much fit to the data is lost (in likelihood terms) by modelling compared to the saturated model. This will be a small value if the model is good (and it will be zero for the saturated model containing all main effects and all interactions). The regression deviance measures how much better is the model taking into account the explanatory variables compared to the simplest model ignoring all of them and only containing a constant (the mean of the responses). This measurement is again made in terms of log-likelihood and the bigger the regression deviance the better fits the model (i.e., the regression effect is strong). In likelihood terms, the residual deviance can be expressed as follows:

$$D = -2 [\ln L_c - \ln L_n] \text{ or } D = -2 \ln [L_c / L_n]$$

where L_n is the likelihood of the null model. The bigger the regression deviance the better the model including this particular variable.

For purposes of assessing the significance of an independent variable, the value of D with an without the independent variable is compared (note the nested character of the two sets). This is called the deviance difference. If a variable is dropped, the residual deviance difference, if a variable is added, the regression deviance difference is compared with the $?^2$ distribution.

Deviance difference In generalised linear modelling, models that have many explanatory variables may be simplified, provided information is not lost in this process. This is tested by the difference in deviance between any two nested models being compared:

G = D (for the model without the variable) - D (for the model with the variable)

The smaller the difference in (residual) deviance, the smaller the impact of the variables removed. This can be tested by the $?^2$ test.

In the simplest example (in simple linear regression), if the log-likelihood of a model containing only a constant term is L_0 and the model containing a single independent variable along with the constant term is L_1, multiplying the difference in these log-likelihoods by -2 gives the deviance difference, i.e., $G = -2(L_1 - L_0)$. G statistics (likelihood ratio test) can be compared with the $?^2$ distribution on df = 1, and it can be decided whether the model with the independent variable has a regression effect (if $P < 0.05$, it has). The same method can be used to detect the effect of the interaction by adding any interaction to the model and obtaining the regression deviance. If this deviance is significantly higher than the one without the interaction in the model, there is interaction [the coefficient for the interaction, however, does not give the odds ratio in logistic regression].

Deviation scores Data in which the mean has been subtracted from each observation.

DFD completeness The extent to which all necessary components of a data flow diagram have been included and fully described.

DFD consistency The extent to which information contained on one level of a set of nested data flow diagrams is also included on other levels.

Df_{total} Degrees of freedom associated with $SS_{total} = N - 1$.

Di The prefix di- means two (2).

Diagonal dominance A matrix is (column) diagonally dominant if, for every column, the sum of the absolute values of the offdiagonal elements is never greater than the absolute value of the diagonal element.

The matrix is strictly diagonally dominant if the offdiagonal sum is always strictly less than the absolute value of the diagonal element.

The same definitions can be used to consider rows instead of columns. The terms column diagonally dominant and row diagonally dominant may be used, if necessary, to specify which case is being considered.

A strictly diagonally dominant matrix cannot be singular, by Gershgorin's Theorem.

A diagonally dominant matrix which is also irreducible cannot be singular.

Here is a diagonally dominant matrix which is not strictly diagonally dominant:

–2 1 0 0 0

1 –2 1 0 0

0 1 –2 1 0

0 0 1 –2 1

0 0 0 1 –2

For a linear system A * x = b, if the matrix A is strictly diagonally dominant, then both Jacobi iteration and Gauss Seidel iteration are guaranteed to converge to the solution.

Diagonal For a polygon in the plane, any line segment joining non-adjacent vertices. For a polyhedron in space, a line segment joining two vertices not in the same face.

Diagonal matrix A diagonal matrix is one whose only nonzero entries are along the main diagonal. For example:

3 0 0

0 4 0

0 0 7

Simple facts about a diagonal matrix A:

A is singular if and only if any diagonal entry is zero;

The eigenvalues of A are the diagonal entries, and the eigenvectors are the columns of the identity matrix;

The determinant of A is the product of the diagonal entries;

Diagonalizable matrix A diagonalizable matrix is any (square) matrix A which is similar to a diagonal matrix:

$A = P * D * P^{-1}$.

This concept is important in the study of eigenvectors To see the relationship, post-multiply the equation by P:

$A * P = P * D$.

Looking at the columns of P as eigenvectors, and the diagonal entries of D as eigenvalues, this shows that a matrix is diagonalizable exactly when it has N linearly independent eigenvectors.

In certain cases, not only is a matrix diagonalizable, but the matrix P has a special form. The most interesting case is that of any (real) symmetric matrix A; not only can such a matrix be diagonalized, but the similarity matrix is orthogonal:

$A = Q * D * Q^{-1} = Q * D * Q'$,

This fact can be interpreted to show that not only does every symmetric matrix have a complete set of eigenvectors, but the eigenvectors and eigenvalues are real, and the eigenvectors are pairwise orthogonal.

Similarly, a complex matrix that is a Hermitian has a complete set of eigenvectors, and the eigenvalues are real. (The eigenvectors in this case will not generally be real).

Simple facts: If the eigenvalues of A are distinct, then it can be diagonalized;

A matrix is diagonalizable if, and only if, it has a set of N linearly independent eigenvectors; in other words, if it is not defective.

If a matrix is normal, then it is not only diagonalizable, but the transformation matrix is unitary.

Diameter The word Diameter has several meanings in mathematics. For a circle, diameter is either the line inside the circle through its center, or the length of such a line. As a number, the diameter is twice the radius of the circle. As a line segment, a diameter joins two points on the circle lying opposite each other. All diameters of a circle meet in its center. The definition of diameter as a line segment in a circle extends to arbitrary convex shapes and triangulations.

Dichotomous variables Variables that can take on only two different values.

Dicker The noun dicker means ten (10).

Dicta The prefix dicta- means two hundred (200).

Difference identities for tangent Identities involving the tangents of differences of angles.

Difference identity for cosine one of the trigonometric addition identities.

Difference identity for sine one of the trigonometric addition identities.

Difference scores The set of scores representing the difference between the subjects' performance on two occasions. Also known as gain scores.

Differentiable function A function is differentiable at a point if its derivative exists at that point. For instance, $f(x)= x^2$ is differentiable everywhere, whereas $g(x) = |x|$ is differentiable everywhere except $x= 0$.

Differential equation An equation that involves derivatives, as in

$$6\frac{dy}{dx} + y - x = 25$$

These equations govern most of the physical world.

Differential A differential is also a small change in a variable. For instance, dy is a small change in the value of y. Although dy often occurs as part of the symbol for the derivative and as part of the symbol for the integral, and although in those other guises, dy plays a role very similar to the one intended when we call it a differential, it is best to just think of the differential dy as a very small change in y.

Digit 1. The noun and adjective digit means a single symbol used to represent a number.

2. The noun and adjective digit means the position occupied by a digit in a positional number notation system.

3. The adjective digit means the total number of digits in a number.

Dilation A transformation which produces a figure similar to the original by proportionally shrinking or stretching the figure.

Dilia The prefix dilia- means two thousand (2000).

Dimensional analysis A method of converting units within a measurement system.

Dimensions of a matrix The number of rows and the number of columns.

Dimerous The adjective dimerous means consisting of two (2) parts or segments.

Dimeter The noun dimeter means a verse consisting of either two (2) metrical feet or two (2) dipodies.

Diophantine equation An equation which is required to be solved by integers.Thus the diophantine equation $2x^2 - 3y^2 = 5$ has the integer solution x=2,y=1.

Direct variation A relationship in which the ratio of two variables is constant.

Directed line segment A line segment of a given length and a given direction.

Directed network A network that indicates the direction of flow on each arc.

Directional test A test that rejects extreme outcomes in only one specified tail of the distribution. Also known as a one-tailed test.

Directrix A line used in the definition of a conic section.

Discipline Treatment that corrects or punishes and is intended to control or to establish habits of self-control.

Discontinuous graph A line in a graph that is interrupted, or has breaks in it forms a discontinuous graph.

Discordant pairs A pair of observations that are ordered in opposite directions on two variables.

Discount rate The interest rate used to compute the present value of future cash flows.

Discrete mathematics The study of mathematical properties of sets and systems that have only a finite number of elements.

Discrete variable A variable of countable number of integer outcomes. Examples include -ordinal multinomial- several prognostic outcomes (such as poor, median and good) as a function of treatment modalities, stage of the disease, age etc., or -multinomial- people's choices of hospitals (hospital A, B or C) as a function of their income level, age, education etc. A discrete variable may be binomial: diseased or non-diseased in a cohort or case-control study. The nature of the outcome variable as discrete or continuous is crucial in the choice of a regression model .

Discriminant analysis A procedure for developing a procedure for optimally discriminating between two groups. This technique often being replaced with logistic regression.

Discriminant The expression $b^2 - 4ac$ of the quadratic formula that determines several characteristics of the roots of a quadratic equation.

Disjoint events Two events are disjoint if they can't both happen at the same time (in other words, if they have no outcomes in common). Equivalently, two events are disjoint if their intersection is the empty set.

Disordinal interaction An interaction in which group differences reverse their sign at some level of the other variable.

Dispersion The degree to which individual data points are distributed around the mean.

Disruptive technologies Technologies that enable the breaking of long-held business rules that inhibit organizations from making radical business changes.

Distance The vertical distance between a point and the regression line. Usually known as the residual.

Distance formula For points P1(X1, Y1) and P2(Y1, Y2), $P1P2 = (X2–X1)^2 + (Y2–Y1)^2$.

Distortion When a subject does not respond honestly to questions.

Distressor A negative event, such as a death or loss of a job, that is stressful.

Distribution-free tests Statistical tests that do not rely on parameter estimation or precise distributional assumptions.

Distributive property A property which establishes a relationship between multiplication and addition such that multiplication distributes across the addition [i.e., a(b+c) = ab + ac].

Diverge If there is no real number that is the value of an infinite series, the series diverges.

Divisibility (rules of) Special tests to determine if a particular integer is a factor of a given number, (e.g., a number is divisible by 10 if it ends in a 0).

Division bar of a fraction The line between the numerator and the denominator of a fraction.

Division The inverse operation of multiplication .

Divorce The legal and formal dissolution of a marriage.

Dodeca The prefix dodeca- means twelve (12).

Dodecad The noun and adjective dodecad means a group of twelve (12) members.

Dodecagon The noun and adjective dodecagon means a plain figure with twelve (12) straight sides.

Dodecagram 1. The noun and adjective dodecagram means an interconnected star figure with twelve (12) points formed by stellating a dodecagon.

2. The noun dodecagram means a word or character sequence consisting of twelve (12) characters.

Dodecahedron The noun and adjective dodecahedron means a solid figure with twelve (12) planar faces.

Dodecillion The noun and adjective dodecillion is an alternate form of duodecillion.

Dodrancentennial 1. The adjective and noun dodrancentennial means recurring every seventy-five (75) years.

2. The adjective and noun dodrancentennial means lasting seventy-five (75) years.

3. The noun and adjective dodrancentennial means a seventy-five (75) year anniversary.

Dodranmillenary 1. The noun and adjective dodranmillenary means a group of seven hundred fifty (750).

2. The noun and adjective dodranmillenary means a period of seven hundred fifty (750) years.

3. The noun and adjective dodranmillenary means a seven hundred fifty (750) year anniversary.

4. The adjective and noun dodranmillenary means seven hundred fiftieth (750th) in rank or order.

Dodranmillennial 1. The adjective and noun dodranmillennial means recurring every seven hundred fifty (750) years.

2. The adjective and noun dodranmillennial means lasting seven hundred fifty (750) years.

3. The noun and adjective dodranmillennial means a seven hundred fifty (750) year anniversary.

Dodranmillennium The noun and adjective dodranmillennium means a period of seven hundred fifty (750) years.

Doheptaconta The prefix doheptaconta- means seventy-two (72).

Doicosa The prefix doicosa- means twenty-two (22).

Domain The set of input values for a function.

Domain of a function A function's domain is just the set of all values for x that it makes sense to plug into f(x). For instance, the domain of $f(x) = \sqrt{r}$ is all $x \geq 0$.

Domain of a relation The set of first components of the ordered pairs in the relation.

Dooctaconta The prefix dooctaconta- means eighty-two (82).

Dot product A process of combining two vectors yielding a single number.

Dotplot A distribution that represents the frequencies of individual points by stacking dots about the axis—similar to a histogram.

Double The verb, adjective, adverb, and noun double means multiply by two (2).

Double agent The compound noun double agent means a person who serves as an agent of one government while pretending to be the agent of another, often adversarial, government.

Double angle identities Useful in writing trig functions involving double angles as trig functions of single angles.

Double angle identities for tangent Useful in writing trig functions involving double angles as functions of single numbers.

Double cross 1. The suffix symbol –3 {double prime} denotes a linear measurement in inches. One inch is equal to 0.025 400 meter.

2. The suffix symbol –3 {double prime} denotes an angular measurement in seconds. One second is equal to one three-thousand-six-hundredth (1/3,600) of a degree (suffix symbol: –° {degree symbol}) or one one-million-two-hundred-ninety-six-thousandth (1/1,296,000) of a circle.

Double root Two roots that are alike.

Double-entendre The compound noun double-entendre means a word or phrase having two (2) or more meanings, especially when a meaning is risqué.

Dovigintillion The noun and adjective dovigintillion is an alternate spelling of duovigintillion.

Downshift matrix The downshift matrix A circularly shifts all vector entries or matrix rows down 1 position.

Example:

0 0 0 1

1 0 0 0

0 1 0 0

0 0 1 0

Facts about the downshift matrix A:

A is a permutation matrix;

A is an N-th root of the identity matrix;

A is persymmetric;

A is a circulant matrix;

any circulant matrix, generated by a column vector v, can be regarded as the Krylov matrix $(v, A*v, A^2*v, ..., A^{n-1}*v$.

the inverse of the downshift matrix is the upshift matrix.

Dual The adjective dual means having two (2) aspects or components.

Dual axle The compound noun and adjective dual axle means a truck axle with two (2) wheels mounted on both (2) sides for a total of four (4) wheels.

Dual tandem The compound noun and adjective dual tandem means a pair (2) of dual truck axles mounted in tandem for a total of eight (8) wheels.

Dummy variables A binary variable that is used to represent a given level of a categorical variable. In genetic data analysis, for example, it is created for each allele at a multiallelic locus. The most common choices as dummy variables are 0 and 1. For each level of the variable, 1 denotes having the feature and 0 denotes all other levels. Also called indicator variables. If an indicator variable is used, the regression coefficient gives the change per unit compared to the reference value. Creating dummy variables on Stata: Stata & UCLA Statistics: STATA Dummy Variables.

Dunn Sidák test A test similar to the Bonferroni test which is based on a more precise inequality and has slightly more power.

Dunnett's test When ANOVA shows a significant difference between groups, if one of the groups is a control (reference) group, Dunnett's Test is used as a post hoc test. This multiple comparison test can be used to determine the significant differences between a single control group mean and the remaining treatment group means in an analysis of variance setting. It is one of the least conservative post hoc tests.

Duodenary 1. The adjective and noun duodenary means twelfth (12th) in order or importance.

2. The adjective duodenary means having twelve (12) parts.

Duononaginta The numeric prefix duononaginta- means ninety-two (92). This prefix is derived from the Latin cardinal number duo et nonaginta (LXXXXII).

Duononagintennary The noun and adjective duononagintennary means a group of ninety-two (92) or a period of ninety-two (92) years.

Duononagintennial 1. The adjective and noun duononagintennial means recurring every ninety-two (92) years.

2. The adjective and noun duononagintennial means lasting ninety-two (92) years.

3. The noun and adjective duononagintennial means a ninety-two (92) year anniversary.

Duononagintennium The noun and adjective duononagintennium means a period of ninety-two (92) years.

Duononagintillion The noun and adjective duononagintillion means multiplied by 10^{279} in the American number naming system, but means multiplied by 10^{552} in the Chuquet number naming system.

Duooctoginta The numeric prefix duooctoginta- means eighty-two (82). This prefix is derived from the Latin cardinal number duo et octoginta (LXXXII).

Duooctogintennary The noun and adjective duooctogintennary means a group of eighty-two (82) or a period of eighty-two (82) years.

Duooctogintennial 1. The adjective and noun duooctogintennial means recurring every eighty-two (82) years.

2. The adjective and noun duooctogintennial means lasting eighty-two (82) years.

3. The noun and adjective duooctogintennial means an eighty-two (82) year anniversary.

Duooctogintennium The noun and adjective duooctogintennium means a period of eighty-two (82) years.

Duooctogintillion The noun and adjective duooctogintillion means multiplied by 10^{249} in the American number naming system, but means multiplied by 10^{492} in the Chuquet number naming system.

Duoquadraginta The numeric prefix duoquadraginta- means forty-two (42). This prefix is derived from the Latin cardinal number duo et quadraginta (XXXXII).

Duoquadragintennary The noun and adjective duoquadragintennary means a group of forty-two (42) or a period of forty-two (42) years.

Duoquadragintennial 1. The adjective and noun duoquadragintennial means recurring every forty-two (42) years.

2. The adjective and noun duoquadragintennial means lasting forty-two (42) years.

3. The noun and adjective duoquadragintennial means a forty-two (42) year anniversary.

Duoquadragintennium The noun and adjective duoquadragintennium means a period of forty-two (42) years.

Duoquadragintillion The noun and adjective duoquadragintillion means multiplied by 10^{129} in the American number naming system, but means multiplied by 10^{252} in the Chuquet number naming system.

Duoquinquaginta The numeric prefix duoquinquaginta- means fifty-two (52). This prefix is derived from the Latin cardinal number duo et quinquaginta (LII).

Duoquinquagintennary The noun and adjective duoquinquagintennary means a group of fifty-two (52) or a period of fifty-two (52) years.

Duoquinquagintennial 1. The adjective and noun duoquinquagintennial means recurring every fifty-two (52) years.

2. The adjective and noun duoquinquagintennial means lasting fifty-two (52) years.

3. The noun and adjective duoquinquagintennial means a fifty-two (52) year anniversary.

Duoquinquagintennium The noun and adjective duoquinquagintennium means a period of fifty-two (52) years.

Duoquinquagintillion The noun and adjective duoquinquagintillion means multiplied by 10^{159} in the American number naming system, but means multiplied by 10^{312} in the Chuquet number naming system.

Duoseptuaginta The numeric prefix duoseptuaginta- means seventy-two (72). This prefix is derived from the Latin cardinal number duo et septuaginta (LXXII).

Duoseptuagintennary The noun and adjective duoseptuagintennary means a group of seventy-two (72) or a period of seventy-two (72) years.

Duoseptuagintennial 1. The adjective and noun duoseptuagintennial means recurring every seventy-two (72) years.

2. The adjective and noun duoseptuagintennial means lasting seventy-two (72) years.

3. The noun and adjective duoseptuagintennial means a seventy-two (72) year anniversary.

Duoseptuagintennium The noun and adjective duoseptuagintennium means a period of seventy-two (72) years.

Duoseptuagintillion The noun and adjective duoseptuagintillion means multiplied by 10^{219} in the American number naming system, but means multiplied by 10^{432} in the Chuquet number naming system.

Duosexaginta The numeric prefix duosexaginta- means sixty-two (62). This prefix is derived from the Latin cardinal number duo et sexaginta (LXII).

Duosexagintennary The noun and adjective duosexagintennary means a group of sixty-two (62) or a period of sixty-two (62) years.

Duosexagintennial 1. The adjective and noun duosexagintennial means recurring every sixty-two (62) years.

2. The adjective and noun duosexagintennial means lasting sixty-two (62) years.

3. The noun and adjective duosexagintennial means a sixty-two (62) year anniversary.

Duosexagintennium The noun and adjective duosexagintennium means a period of sixty-two (62) years.

Duosexagintillion The noun and adjective duosexagintillion means multiplied by 10^{189} in the American number naming system, but means multiplied by 10^{372} in the Chuquet number naming system.

Duotrigesimal The adjective duotrigesimal means based on the number thirty-two (32).

Duotriginta The numeric prefix duotriginta- means thirty-two (32). This prefix is derived from the Latin cardinal number duo et triginta (XXXII).

Duotrigintacentenary 1. The noun and adjective duotrigintacentenary means a group of three thousand two hundred (3200).

2. The noun and adjective duotrigintacentenary means a period of three thousand two hundred (3200) years.

3. The noun and adjective duotrigintacentenary means a three thousand two hundred (3200) year anniversary.

4. The adjective and noun duotrigintacentenary means three thousand two hundredth (3200th) in rank or order.

Duotrigintacentennial 1. The adjective and noun duotrigintacentennial means recurring every three thousand two hundred (3200) years.

2. The adjective and noun duotrigintacentennial means lasting three thousand two hundred (3200) years.

3. The noun and adjective duotrigintacentennial means a three thousand two hundred (3200) year anniversary.

Duotrigintacentennium The noun and adjective duotrigintacentennium means a period of three thousand two hundred (3200) years.

Duotrigintennary The noun and adjective duotrigintennary means a group of thirty-two (32) or a period of thirty-two (32) years.

Duotrigintennial 1. The adjective and noun duotrigintennial means recurring every thirty-two (32) years.

2. The adjective and noun duotrigintennial means lasting thirty-two (32) years.

3. The noun and adjective duotrigintennial means a thirty-two (32) year anniversary.

Duotrigintennium The noun and adjective duotrigintennium means a period of thirty-two (32) years.

Duotrigintillion The noun and adjective duotrigintillion means multiplied by 10^{99} in the American number naming system, but means multiplied by 10^{192} in the Chuquet number naming system.

Duple The adjective and noun duple means multiplied by two (2).

Duplex The noun and adjective duplex means having two (2) parts.

Duplicate 1. The noun and adjective duplicate means either of two (2) things exactly alike.

2. The adjective duplicate means consisting of two (2) identical or corresponding parts or examples.

3. The noun and adjective duplicate means a second (2nd or 2^{d}) instance or copy of something.

4. The verb duplicate means to make double (2).

5. The verb duplicate means to create a second (2nd or 2^{d}) instance or copy of an object.

Duplicate is often used generically to mean reproduce, replicate, or copy, but properly duplicate means the production of only a single (1) copy.

6. The verb duplicate means to create something equal to the original (1st) instance.

7. The verb duplicate means to do something a second (2nd or 2^{d}) time.

Duplicitous The adjective and adverb duplicitous means having a contradictory doubleness, especially belying true intentions by deception.

Duplicity The noun duplicity means the state of being two (2) at once. Most commonly, duplicity means a contradictory doubleness, especially belying true intentions by deception.

Duumvir The noun duumvir means one (1) of two (2) ruling men. (In Latin, duumvir specifically refers to an adult male.) The plural of duumvir is duumviri.

Duumvirate The noun and adjective duumvirate means a ruling body of two (2) men.

Dyad The noun and adjective dyad means a group of two (2) members.

Dyagram The noun and adjective dyagram means a word or character pair consisting of two (2) characters.

E

E The Latin lower case letter e represents Napier's Constant, the base of the natural, hyperbolic, or Naperian logarithms. Napier's Constant is a transcendental number approximately 2.718 281 828 459.

Eccentricity A ratio used in some definitions of conic sections.

Ecological fallacy The aggregation bias, which is the unfortunate consequence of making inferences for individuals from aggregate data. It results from thinking that relationships observed for groups necessarily hold for individuals. The problem is that it is not valid to apply group statistics to an individual member of the same group.

Ecological study Analyses based on data grouped to the municipal, provincial or national level.

Economic feasibility A process of identifying the financial benefits and costs associated with a development project.

Edwards' test A statistical test for seasonality that looks for a one-cycle sinusoidal deviation from the null distribution.

Effect modification The situation in which a measure of effect changes over values of another variable (the association estimates are different in different subpopulations of the sample). The relative risk or odds ratio associated with exposure will be different depending on the value of the effect modifier. For example if in a disease association study, the odds ratios are different in different age groups or in different sexes, age or sex are effect modifiers. Effect modification is highly related to statistical interaction in regression models. If where an exposure decreases risk for one value of the effect modifier and increases risk for another value of effect modifier, this is called crossover.

Effect size (d) The difference between two population means divided by the standard deviation of either population.

Effective exponent The effective exponent of 2^n is (n + 1) minus the sum of all primes p dividing (n + 1) for which 2^p – 1 is a Mersenne prime.

Effective sample size The sample size needed in equal-sized groups to achieve the available power when we have groups of unequal sizes. It will generally be less than the total number of subjects in the unequal groups.

Efficiency The degree to which repeated values for a statistic cluster around the parameter.

Ei The compound binary prefix symbol Ei- (binary prefix: exbi-) denotes a multiple of two to the sixtieth power (2^{60} = 1,152,921,504,606,846,976) as defined by the International Electrotechnical Commission (IEC).

Eigen (value or state) Possible values for a parameter of an equation for which the solutions will be compatible with the boundary conditions.In quantum mechanics the energy eigenvalues for the Schrödinger equation are the possible energy levels for the system.

Eigenvalues (latent values) In multivariate statistics, eigenvalues give the variance of a linear function of the variables. Eigenvalues measure the amount of the variation explained by each principal component (PC) and will be largest for the first PC and smaller for the subsequent PCs. An eigenvalue greater than 1 indicates that PCs account for more variance than accounted by one of the original variables in standardized data. This is commonly used as a cut-off point for which PCs are retained.

Eigenvectors A nonzero vector x is an eigenvector of the square matrix A if

A * x = lambda * x

for some scalar value lambda, called the associated eigenvalue.

Sometimes this eigenvector is more particularly described as a right eigenvector, so that we may also consider left eigenvectors, that is, vectors y for which it is true that

y * A = A' * y = mu * y

for some scalar mu.

For every eigenvalue of a matrix, there is at least one eigenvector. Every nonzero multiple of this eigenvector is also an eigenvector, but in an uninteresting way. If, and only if, an eigenvalue is a repeated root, then there may be more than one linearly independent eigenvector associated with that eigenvalue. In particular, if an eigenvalue is repeated 3 times, then there will be 1, 2 or 3 linearly independent eigenvectors corresponding to that eigenvalue.

Facts about eigenvectors:

If x is an eigenvector, so is s*x, for any nonzero scalar s.

If A is singular, then it has an eigenvector associated with the eigenvalue 0, so that A * x = 0.

If x is a right eigenvector for eigenvalue lambda, and y is a left eigenvector for eigenvalue mu, and lambda and mu are distinct, then x and y are orthogonal. This property is sometimes call biorthogonality.

Eight The noun, pronoun, and adjective eight (8) is the standard name for the cardinal number that follows seven (7). Eight equals two to the third power (2^3). Eight corresponds with the ordinal number eighth (8th).

Eighteen The <>noun, pronoun, and adjective </ >eighteen (18) is the standard name for the cardinal number that follows seventeen (17). Eighteen corresponds with the ordinal number eighteenth (18th).

Eighteen-wheeler The compound noun eighteen-wheeler (18-wheeler) means a tractor and trailer combination freight truck with eighteen (18) wheels. Most commonly, a tractor and trailer rig consists of a tractor with two (2) wheels for steering and tandem dual drive axles and a trailer with tandem dual trailing axles mounted near the rear. Eighties 1. The plural noun and adjective eighties (80s) means an integer multiple of eighty (80).

2. The plural noun and adjective eighties (80's) means the range of numbers at least eighty (80) but less than ninety (90). When referring to dates, eighties means a span of ten years in which the tens (10s) digit of the year number is eight (8).

Eighty eight The noun and adjective eighty-eight (88) is the standard name for the cardinal number that follows eighty-seven (87). Eighty-eight corresponds with the ordinal number eighty-eighth (88th).

Eighty-five The noun and adjective eighty-five (85) is the standard name for the cardinal number that follows eighty-four (84). Eighty-five corresponds with the ordinal number eighty-fifth (85th).

Eighty-four The noun and adjective eighty-four (84) is the standard name for the cardinal number that follows eighty-three (83). Eighty-four corresponds with the ordinal number eighty-fourth (84th).

Eighty-nine The noun and adjective eighty-nine (89) is the standard name for the cardinal number that follows eighty-eight (88). Eighty-nine corresponds with the ordinal number eighty-ninth (89th).

Eighty-one The noun and adjective eighty-one (81) is the standard name for the cardinal number that follows eighty (80). Eighty-one corresponds with the ordinal number eighty-first (81st).

EISPACK A package of routines for handling the standard and generalized eigenvalue problems. The beginning user who is not interested in trying to learn the details of EISPACK and simply wants the answer to an eigenvalue problem quickly should call one of the main driver routines. Each of these is tailored to handle a given problem completely with a single subroutine call. For more advanced work, it may be worth investigating some of the underlying routines.

Driver routines to solve A*x=lambda*x include:

CG, complex general matrix.

CH, complex Hermitian matrix.

RG, real general matrix.

RS, real symmetric matrix.

RSB, real symmetric band matrix.

RSP, real symmetric, packed storage matrix.

RSPP, real symmetric packed storage matrix, some eigenvectors.

RST, real symmetric tridiagonal matrix.

RT, real sign-symmetric tridiagonal.

For the generalized eigenvalue problem:

RGG solves A*x=lambda*B*x for A and B real, general matrices.

RSG solves A*x=lambda*B*x for A real symmetric, B real positive definite.

RSGAB solves A*B*x=lambda*x for A real symmetric, B real positive definite.

RSGBA solves B*A*x=lambda*x for A real symmetric, B real positive definite.

Elapsed time The amount of time between a beginning time and an ending time.

Electronic commerce (EC) Internet-based communication to support day-to-day business activities.

Electronic data interchange (EDI) The use of telecommunications technologies to transfer business documents directly between organizations.

Electronic performance support system (EPSS) Component of a software package or application in which training and educational information is embedded. An EPSS may include a tutorial, expert system, and hypertext jumps to reference material.

Elementary column operations Elementary column operations are a simple set of matrix operations that can be used to carry out Gauss elimination, Gauss Jordan elimination, or the reduction of a matrix to column echelon form.

Restricting the operations to a simple set makes it easy to:

Guarantee that each step is legitimate;

Record each step using a simple notation;

compute the inverse of the total set of operations.

The three elementary column operations include:

interchange any two columns;

multiply any column by a nonzero value;

add a multiple of any column to another column.

Each of these operations may be represented by an elementary matrix, and the transformation of the original matrix A to the reduced matrix B can be expressed as postmultiplication by a concatenation of elementary matrices:

$B = A * E(1) * E(2) * \ldots * E(k)$

which may be abbreviated as:

$B = A * C$

Since C will be guaranteed to be invertible, we also know that,

$B * C^{-1} = A$

which yields a factorization of A.

Elementary row operations Elementary row operations are a simple set of matrix operations that can be used to carry out Gauss elimination, Gauss Jordan elimination, or the reduction of a matrix to row echelon form.

Restricting the operations to a simple set makes it easy to:

Guarantee that each step is legitimate;

Record each step using a simple notation;

Compute the inverse of the total set of operations.

The three elementary row operations include:

Interchange any two rows;

Multiply any row by a nonzero value;

add a multiple of any row to another row.

Each of these operations may be represented by an elementary matrix, and the transformation of the original matrix A to the reduced matrix B can be expressed as premultiplication by a concatenation of elementary matrices:

B = E(k) * E(k–1) * ... * E(2) * E(1) * A

which may be abbreviated as:

B = C * A.

Since C will be guaranteed to be invertible, we also know that,

C^{-1} * B = A

which yields a factorization of A.

Eleven The noun, pronoun, and adjective eleven (11) is the standard name for the cardinal number that follows ten (10). Eleven corresponds with the ordinal number eleventh (11th).

Ellipse Step on a circle until it squawks, and you've got an ellipse. It's a bit longer than it is wide. The general formula is like the formula for a circle only with a few extra a's and b's thrown in, as in $x^2/a^2 + y^2/b^2 = 1$.

Ellipsis A symbol indicating that the pattern-continues in a sequence.

Ellipsoids An ellipsoid is an N dimensional generalization of an ellipse. The formula for an ellipsoid may be written as:

sum (I = 1 to N, J = 1 to N) $A_{i,j} * X_i * X_j = 1$.

where A is a positive definite symmetric matrix.

A principal axis of an ellipsoid is any N dimensional point X on the ellipsoid such that the vector from the origin to X is normal to the ellipsoid.

In the general case, there are exactly N principal axes (plus their negatives). In degenerate cases, there may be an entire plane of vectors that satisfy the requirement, but it is always possible to choose a set of N principal axes which are linearly independent.

Moreover, in the general case, the principal axes are pairwise orthogonal vectors, and in the degenerate case, may be chosen pairwise orthogonal.

Moreover, it is always true that the principal axes are eigenvectors of the matrix A of ellipsoid coefficients. The length of the principal axis vector associated with an eigenvalue lambda(I) is 1 / Sqrt (lambda(I)).

These facts have a strong relationship to the formulation of the conjugate gradient method.

EM Algorithm A method for calculating maximum likelihood estimates with incomplete data. E (expectation)-step computes the expected values for missing data and M (maximisation)-step computes the maximum likelihood estimates assuming complete data. It was first used in genetics (Ceppellini R et al, 1955) to estimate allele frequency for phenotype data when genotypes are not fully observable (this requires the assumption of HWE and calculation of expected genotypes from phenotype frequencies).

Empirical P value A P value obtained by a simulation program such as Monte Carlo statistics. This is less likely to be affected by multiple comparisons.

Empirical rule In variables normally distributed, 68% of the data values are within 1SD of the mean; 95% are within 2SD of the mean; and 99.7% (nearly all) are within 3SD of the mean.

Encapsulation The technique of hiding the internal implementation details of an object from its external view.

End point convention In histograms, one needs to decide where to count values that are on the exact boundary between two intervals: either in the left or in the right interval. Let readers of the histogram know which side is chosen .

End users Non-information-system professionals in an organization. End users often request new or modified software applications, test and approve applications, and may serve on project teams as business experts.

Endpoints An endpoint is a point used to define a line segment or ray. A line segment has two endpoints; a ray has one.
Example: The endpoints of line segment DC below are points D and C, and the endpoint of ray MN is point M below:

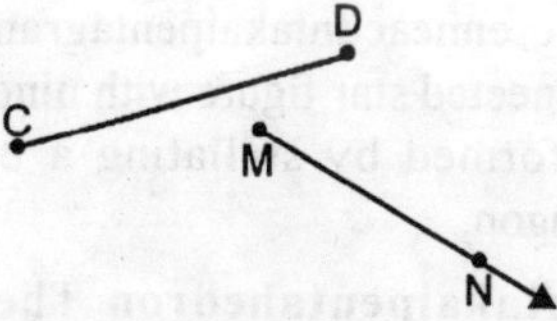

Ennea The prefix ennea- means nine (9).

Enneaconta The prefix enneaconta- means ninety (90).

Enneacontad The noun and adjective enneacontad means a group of ninety (90) members.

Enneacontadi The prefix enneacontadi- means ninety-two (92). Enneacontadi- is an alternate form of the prefix enneacontakaidi-.

Enneacontaennea The prefix enneacontaennea- means ninety-nine (99). Enneacontaennea- is an alternate form of the prefix enneacontakaiennea-.

Enneacontagon The noun and adjective enneacontagon means a plain figure with ninety (90) straight sides.

Enneacontagram The noun and adjective enneacontagram means an interconnected star figure with ninety (90) points formed by stellating a enneacontagon.

Enneacontahedron The noun and adjective enneacontahedron means a solid figure with ninety (90) planar faces.

Enneacontahena The prefix enneacontahena- means ninety-one (91). Enneacontahena- is an alternate form of the prefix enneacontakaihena-.

Enneacontahexa The prefix enneacontahexa- means ninety-six (96). Enneacontahexa- is an alternate form of the prefix enneacontakaihexa-.

Enneacontakaidi The prefix enneacontakaidi- means ninety-two (92).

Enneacontakaidigon The noun and adjective enneacontakaidigon means a plain figure with ninety-two (92) straight sides.

Enneacontakaidigram The noun and adjective enneacontakaidigram means an interconnected star figure with ninety-two (92) points formed by stellating a enneacontakaidigon.

Enneacontakaidihedron The noun and adjective enneacontakaidihedron means a solid figure with ninety-two (92) planar faces.

Enneacontakaidyad The noun and adjective enneacontakaidyad means a group of ninety-two (92) members.

Enneacontakaiennea The prefix enneacontakaiennea- means ninety-nine (99).

Enneacontakaiennead The noun and adjective enneacontakaiennead means a group of ninety-nine (99) members.

Enneacontakaienneagon The noun and adjective enneacontakaienneagon means a plain figure with ninety-nine (99) straight sides.

Enneacontakaienneagram The noun and adjective enneacontakaienneagram means an interconnected star figure with ninety-nine (99) points formed by stellating a enneacontakaienneagon.

Enneacontakaienneahedron The noun and adjective enneacontakaienneahedron means a solid figure with ninety-nine (99) planar faces.

Enneacontakaihena The prefix enneacontakaihena- means ninety-one (91).

Enneacontakaihenad The noun and adjective enneacontakaihenad means a group of ninety-one (91) members.

Enneacontakaihenagon The noun and adjective enneacontakaihenagon means a plain figure with ninety-one (91) straight sides.

Enneacontakaihenagram The noun and adjective enneacontakaihenagram means an interconnected star figure with ninety-one (91) points formed by stellating a enneaconta-kaihenagon.

Enneacontakaihepta The prefix enneacontakaihepta- means ninety-seven (97).

Enneacontakaiheptad The noun and adjective enneacontakaiheptad means a group of ninety-seven (97) members.

Enneacontakaiheptagon The noun and adjective enneacontakaiheptagon means a plain figure with ninety-seven (97) straight sides.

Enneacontakaiheptagram The noun and adjective enneacontakaiheptagram means an interconnected star figure with ninety-seven (97) points formed by stellating a enneaconta-kaiheptagon.

Enneacontakaiheptahedron The noun and adjective enneacontakaiheptahedron means a solid figure with ninety-seven (97) planar faces.

Enneacontakaihexa The prefix enneacontakaihexa- means ninety-six (96).

Enneacontakaihexad The noun and adjective enneacontakaihexad means a group of ninety-six (96) members.

Enneacontakaihexagram The noun and adjective enneacontakaihexagram means an interconnected star figure with ninety-six (96) points formed by stellating a enneacontakaihexagon.

Enneacontakaihexahedron The noun and adjective enneacontakaihexahedron means a solid figure with ninety-six (96) planar faces.

Enneacontakaiocta The prefix enneacontakaiocta- means ninety-eight (98).

Enneacontakaioctad The noun and adjective enneacontakaioctad means a group of ninety-eight (98) members.

Enneacontakaioctagon The noun and adjective enneacontakaioctagon means a plain figure with ninety-eight (98) straight sides.

Enneacontakaioctagram The noun and adjective enneacontakaioctagram means an interconnected star figure with ninety-eight (98) points formed by stellating a enneaconta-kaioctagon.

Enneacontakaioctahedron The noun and adjective enneacontakaioctahedron means a solid figure with ninety-eight (98) planar faces.

Enneacontakaipenta The prefix enneacontakaipenta- means ninety-five (95).

Enneacontakaipentad The noun and adjective enneacontakaipentad means a group of ninety-five (95) members.

Enneacontakaipentagon The noun and adjective enneacontakaipentagon means a plain figure with ninety-five (95) straight sides.

Enneacontakaipentagram The noun and adjective enneacontakaipentagram means an interconnected star figure with ninety-five (95) points formed by stellating a enneaconta-kaipentagon.

Enneacontakaipentahedron The noun and adjective enneacontakaipentahedron means a solid figure with ninety-five (95) planar faces.

Enneacontakaitetra The prefix enneacontakaitetra- means ninety-four (94).

Enneacontakaitetrad The noun and adjective enneacontakaitetrad means a group of ninety-four (94) members.

Enneacontakaitetragon The noun and adjective enneacontakaitetragon means a plain figure with ninety-four (94) straight sides.

Enneacontakaitetragram The noun and adjective enneacontakaitetragram means an interconnected star figure with ninety-four (94) points formed by stellating a enneacontakaitetragon.

Enneacontakaitetrahedron The noun and adjective enneacontakaitetrahedron means a solid figure with ninety-four (94) planar faces.

Enneacontakaitri The prefix enneacontakaitri- means ninety-three (93).

Enneacontakaitriad The noun and adjective enneacontakaitriad means a group of ninety-three (93) members.

Enneacontakaitrigon The noun and adjective enneacontakaitrigon means a plain figure with ninety-three (93) straight sides.

Enneacontakaitrigram The noun and adjective enneacontakaitrigram means an interconnected star figure with ninety-three (93) points formed by stellating a enneacontakaitrigon.

Enneacontakaitrihedron The noun and adjective enneacontakaitrihedron means a solid figure with ninety-three (93) planar faces.

Enneacontaocta The prefix enneacontaocta- means ninety-eight (98). Enneacontaocta- is an alternate form of the prefix enneacontakaiocta-.

Enneacontapenta The prefix enneacontapenta- means ninety-five (95). Enneacontapenta- is an alternate form of the prefix enneacontakaipenta-.

Enneacontatetra The prefix enneacontatetra- means ninety-four (94). Enneacontatetra- is an alternate form of the prefix enneacontakaitetra-.

Enneacontatri The prefix enneacontatri- means ninety-three (93). Enneacontatri- is an alternate form of the prefix enneacontakaitri-.

Ennead The noun and adjective ennead means a group of nine (9) members.

Enneadeca The prefix enneadeca- means nineteen (19). Enneadeca- is an alternate form of the prefix enneakaideca-.

Enneagon The noun and adjective enneagon means a plain figure with nine (9) straight sides.

Enneagram 1. The noun and adjective enneagram means an interconnected star figure with nine (9) points formed by stellating a enneagon.

2. The noun enneagram means a word or character sequence consisting of nine (9) characters.

Enneahedron The noun and adjective enneahedron means a solid figure with nine (9) planar faces.

Enneakaideca The prefix enneakaideca- means nineteen (19).

Enneakaidecad The noun and adjective enneakaidecad means a group of nineteen (19) members.

Enneakaidecagon The noun and adjective enneakaidecagon means a plain figure with nineteen (19) straight sides.

Enneakaidecagram The noun and adjective enneakaidecagram means an interconnected star figure with nineteen (19) points formed by stellating a enneakaidecagon.

Enneakaidecahedron The noun and adjective enneakaidecahedron means a solid figure with nineteen (19) planar faces.

Entering the adult world A stage in the novice phase of early adulthood, ranging from ages 22 to 28.

Enterprise solutions software or enterprise resource planning (ERP) system A system that integrates individual traditional business functions into a series of modules so that a single transaction occurs seamlessly within a single information system rather than in several separate systems.

Entity A person, place, object, event, or concept in the user environment about which the organization wishes to maintain data.

Entity instance (instance) A single occurrence of an entity type.

Entity relationship diagram (E R diagram) A detailed, logical, and graphical representation of the entities, associations, and data elements for an organization or business area.

Entity type A collection of entities that share common properties or characteristics.

Enumerate The verb enumerate means to assign a number to an object. Most commonly enumerate means to count.

Environment Everything external to a system that interacts with the system.

Ephemeral The adjective and noun ephemeral means lasting only one (1) day. More generally, ephemeral means lasting a comparatively short time.

Epidemiologic flaws and fallacies Beware of confounders, selection bias, response bias, variable observer, Hawthorne effect (changes caused by the observer in the observed values), diagnostic accuracy bias, regression to the mean, significance Turkey, nerd of nonsignificance, cohort effect, ecological fallacy, Berkson bias (selection bias in hospital-based studies) and others (discussed in M Michael III et al. Biomedical Bestiary. Little, Brown and Company, 1984; and in Bias & Confounding in Molecular Epidemiology).

Equality An equality is a statement of two mathematical objects being equal. Like equations, equalities are written as two mathematical objects connected by the equality sign =. The meaning of two objects being equal depends very much on the nature of the objects. E.g., two matrices are equal iff they have identical dimensions and all their corresponding elements are equal. Two triangles are equal if there exists a distance preserving transformation that maps one on the other (Nowadays equal triangles more often than not are called congruent.) Equalities are sometimes called unconditional equations.

Equally likely In probability, when there are the same chances for more than one event to happen, the events are equally likely to occur. For example, if someone flips a coin, the chances of getting heads or tails are the same. There are equally likely chances of getting heads or tails.

Equally likely outcomes Events in a sample space that have the same probability of occurring.

Equally weighted means An analysis of variance in which cell means all carry the same weight in determining row and column means, regardless of the number of subjects in each cell.

Equation An equation is a statement that two mathematical object, e.g. algebraic expressions, are equal under certain conditions. Equations are usually and sometimes tacitly accompanied by the requirement to establish these conditions. In the case where mathematical objects contain arbitrary variables, such a requirement may mean finding those values of the variables that turn the equation into equality. Equations are written as two mathematical objects connected by the equality sign =. Equations are sometimes called more explicitly as conditional equations.

Equilateral triangle or equiangular triangle A triangle having all three sides of equal length. The angles of an equilateral triangle all measure 60 degrees.

Examples:

Isosceles triangle A triangle having two sides of equal length.

Examples:

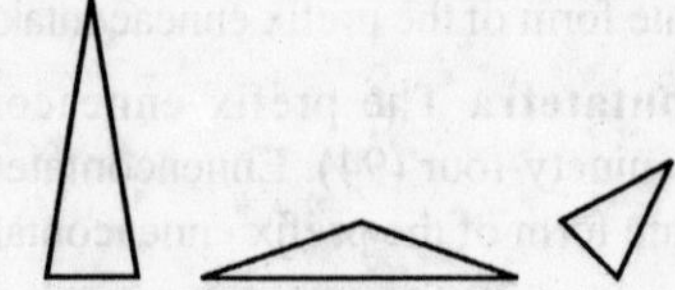

Scalene triangle A triangle having three sides of different lengths.Examples:

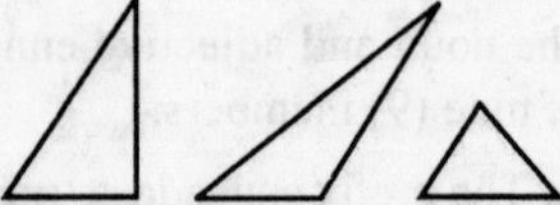

Equilibration The technique of balancing the rows or columns of a matrix by rescaling them.

Consider, for instance, the fact that the following two equations are equivalent:

0.0001 * x + 0.0001 * y = 0.0001

and

1000 * x + 1000 * y = 1000

However, the large coefficients in the second equation will bias a Gauss elimination routine to choose that equation as its pivot. Actually, it's more important in this case that the chosen row be as linearly independent as possible from the other rows, and this is more likely to occur if we ensure that all the rows start out with an equal norm. This can be done very simply, by finding the element of maximum absolute value in each row and dividing that row (and its right hand side) by that value. Such a technique is called row equilibration. It is not necessary that the rows have precisely the same norm; it is desirable that the norms of the rows be maintained within some controlled range.

Equilibration is useful in many areas of linear algebra, including eigenvalue calculations. In some cases, column equilibration is preferred, and in other cases, the norms of both the rows and columns are to be controlled.

Equilibrium Piaget's term for the basic process underlying the human ability to adapt; the search for balance between self and the world.

Equivalent matrix Matrices A and B are said to be equivalent if there are nonsingular matrices P and Q so that

A = P * B * Q.

Simple facts about equivalence:

Every matrix is equivalent to a diagonal matrix;

If A and B are equivalent, then they are both singular or both nonsingular.

Equivalence is a very loose concept of relatedness. A stronger and more useful concept is similarity.

Equivalent Numbers or expressions that have the same value.

Equivalent transformation A transformation that results in an equation equivalent to the original.

Equivalent vectors Two vectors that have the same magnitude and direction.

Error rate per comparison (PC) The probability of making a Type I error on any specific comparison when using multiple comparison procedures.

Error terms Residuals in regression models. Shown as W_i or $?_i$. Their expected value is zero, thus, they vary around zero with a variance equal to $?^2$: N(0, $?^2$). They are assumed to be normally distributed, have equal variance for all fitted values, and independent. Normality is a reasonable assumption in many cases. The assumption of equal variance implies that every observation on the dependent variable contains the same amount of information. The impact of heterogeneous variances is a loss of precision of estimates compared to the precision that would have been realised if the heterogeneous variances had been taken into account. Transformation of the dependent variable may help to homogenize the unequal variances. Correlated errors are most frequent in time sequence data and they also cause the loss in precision in the estimates.

Error variance The square of the standard error of estimate.

Errors of prediction The differences between Y and $\hat{Y}$.

Escapees Values for C in the Julia Set or Mandelbrot set where at each iteration the resulting value grows larger and larger, approaching infinity.

Estimate The best guess arrived at after considering all the information given in a problem.

Estimation The process of finding a number close to an exact amount.

Estimator A function of the known data that is used to estimate an unknown parameter; an estimate is the result from the actual application of the function to a particular set of data. The mean can be used as an estimator.

Eta squared A measure of the magnitude of effect. Also known as the correlation ratio.

Euclidean algorithm The method for finding remainders by multiplying the divisor by the quotient and subtracting that amount from the number being divided. For example, when finding the remainder for 25 divided by 4, the quotient is 6, so one multiplies 6 times 4 (giving 24) and then subtracts 25 from 24, leaving 1 as the remainder.

Euclidean geometry The geometry (plane and solid) based on Euclid's postulates.

Euler-Mascheroni constant The Euler-Mascheroni Constant, also known as Euler's Constant, (symbol: γ {Greek lower case letter gamma}) is approximately 0.577 215 664 902. Swiss mathematician Leonhard Euler (1707-1783) discovered γ in 1734. Italian priest and mathematician Lorenzo Mascheroni made the first precise calculation of γ in 1790.

Even function A function is even if f(–x) = f(x).

Event An event is a change within the system, including the creation or deletion of objects or relationships, the starting or stopping of an activity, and the reception of messages.

Event-Based Trigger

Every The adjective every means each member of a group without exception.

Exa The metric prefix exa- (prefix symbol: E-) denotes a multiple of ten to the eighteenth power (10^{18} = 10 raised to power +18 = 1,000,000,000, 000,000,000) as defined by Le Système International d'Unités (SI).

Exabinary The adjective exabinary means two to the sixtieth power (2^{60} = 1,152,921,504,606,846, 976).

Exbi The binary prefix exbi- (prefix symbol: Ei-) denotes a multiple of two to the sixtieth power (2^{60} = 1,152,921,504,606,846,976) as defined by the International Electrotechnical Commission (IEC).

Exbin The adjective and noun exbin means two to the sixtieth power (2^{60} = 1,152,921,504,606, 846,976).

Exception An exception is a system event or condition which is not part of normal system behavior. For example, a printer running out of paper during a print job is an exception.

Exchange matrix Constructed from an identity matrix by reversing the order of the columns.

For example, the matrix J of order 4:

0 0 0 1

0 0 1 0

0 1 0 0

1 0 0 0

Facts about the exchange matrix J:

J is a square root of the identity matrix;

J is a permutation.

J is symmetric.

J is persymmetric.

J is a Hankel matrix.

J is an anticirculant matrix.

The vector J*x comprises the entries of x in reverse order.

For any matrix A, the matrix J*A has the same rows but in reversed order.

For any matrix A, the matrix A*J has the same columns but in reversed order.

The exchange matrix is also called the anti-identity or counter-identity matrix.

Exhaustive A set of events that represents all possible outcomes.

Expanded notation The sum of terms representing a quantity.

Expected frequencies The expected value for the number of observations in a cell if H_0 is true.

Expected value The amount that is predicted to be gained, using the calculation for average expected payoff.

Experiment An event is a subset of the sample space, to which a probability can be assigned. For example, on rolling a die, getting a five or a six is an event (with a probability of one third if the die is fair).

Experimental hypothesis Another name for the research hypothesis.

Experimental probability The chances of something happening, based on repeated testing and observing results. It is the ratio of the number of times an event occurred to the number of times tested. For example, to find the experimental probability of winning a game, one must play the game many times, then divide the number of games won by the factor.

Any of the numbers or symbols in mathematics that when multiplied together form a product.

For example, 3 is a factor of 12, because 3 can be multiplied by 4 to give 12. Similarly, 5 is a factor of 20, because 5 times 4 is 20 .

Experimental research Research based on, tested by, or having the nature of an experiment.

Experimenter bias When researchers' expectations about what should or should not happen in a study sway the results.

Explained (regression) sum of squares (ESS) The measure of between treatments sum of squares (variability) in ANOVA. If the means of treatment groups are different, the ESS would be greater than RSS to yield a high variance ratio. The bigger the ESS, the better explained the data by the model.

Explicit definition A definition used to state the relationship between a term t. of a sequence and the number n of that term.

Explicit relationship A sequence rule using the number of the term to define the function [e.g., in the sequence 3, 6, 9,..., the explicit rule is f(n) = 3n where n is the number of the term and f(n) is the value of the term].

Exploratory data analysis An initial look at the data with minimal use of formal mathematics or statistical methods, but more with an informal graphical approach. Scatter plots, correlation matrices and contingency tables (for binary data) can be used to get an initial idea for relationships between explanatory variables (for collinearity) or between an explanatory variable(s) and a response variable(s) (correlation). In ANOVA, normality can be checked by box-plots. It gives some indication of which variables should be in the model and which one of them should be put into the model first, and whether linear relationship is adequate.

Exponent A number which is placed to the right of and above another number (base). the value of the exponent determines how many times the base is used as a factor (e.g., 34= 3 x 3 x 3 x3; {3 is the base and is used as a factor 4 times} the exponent is 4) exponential function: a function whose general equation is a y=ab x or y=ab kx, where a, b, and k stand for constants.

Exponential distribution The (continuous) distribution of time intervals between independent consecutive random events like those in a Poisson process.

Exponential equation An equation with the variable in an exponent.

Exponential family A family of probability distributions in the form

$$f(x) = \exp\{a(x)b(x) + c(x) + d(x)\}$$

where x is a parameter and a, b, c, d are known functions. This family includes the normal distribution, binomial distribution, Poisson distribution and gamma distribution as special cases.

Exponential function This is the function $f(x) = e^x$.It is its own derivative.

Exponential growth A function experiences exponential growth if it is at least as big as a function of the form CK^x, where $C > 0$ and $K > 1$. For example, the function 2^x experiences exponential growth. Notice that the function 2^x doubles in value each time x increases by one. So although 2^1 is only 2, 2^{10} is already 1024, and 2^{20} is 1,048,576.

Exposure In an epidemiologic study, exposure may represent an environmental exposure, an intervention or a the presence of a marker (biomarker/genetic marker).

Expression A mathematical phrase with no equal sign such as 6, 3n, or + 4.

External documentation System documentation that includes the outcome of structured diagramming techniques such as data flow and entity-relationship diagrams.

External storage algorithms A method of solving a problem that is too large to be loaded into computer memory as a whole.

Instead, the problem is solved incrementally, with most of the problem data residing, at any one time, in computer files, also called disk storage or external storage. It is a considerable difficulty just to rewrite an algorithm that can handle a situation where, say, part of a matrix is in one place, and part is in another, remote place.

However, such algorithms must also be aware that data transfers between memory and disk are very slow. Hence, if the algorithm is to be of any use, it must do as much processing as possible on the portion of the data that resides in memory, and read the external problem data into memory as rarely as possible, and in large contiguous chunks.

External validity The ability to generalize the results from this experiment to a larger population.

Extraneous variable A variable unrelated to an experiment (such as room temperature or noise level) that may interfere with the results of the experiment.

Extranet Internet-based communication to support business-to-business activities.

Extrema (or extremum) (extrema is the plural form.) A word for either a maximum or a minimum.

Extreme value, local or global A function f:A->B is numeric if B is a set of numbers. For a numeric function it's possible to compare its values at different points a ∈ A. Extreme is a value which, in some sense is either maximum or minimum. If f(a) exceeds all other values of f then we say it's a global extreme, maximum. If it's only larger than values of f for points near a, the maximum is local.

F The metric prefix symbol f- (metric prefix: femto-) denotes a multiple of ten to the negative fifteenth power (10^{-15} = 0.000 000 000 000 001) as defined by Le Système International d'Unités (SI).

F distribution A continuous probability distribution of the ratio of two independent random variables, each having a Chi-squared distribution, divided by their respective degrees of freedom. The commonest use is to assign P values to mean square ratios (variance ratios) in ANOVA. In regression analysis, the F-test can be used to test the joint significance of all variables of a model.

F test The F test employs the statistic (F) to test various statistical hypotheses about the mean (or means) of the distributions from which a sample or a set of samples have been drawn.

Face A plane surface of a three-dimensional figure.

Factor A categorical explanatory variable with small number of levels such that each item classified belongs to exactly one level for that category. If the factor is 'sex', the levels are 'male' and 'female'; if the factor is 'drug received', the levels are 'drug A', 'drug B', 'drug C', etc. A set of factor levels, uniquely defining a single treatment, is called a cell. A cell may have just one observation (no replication) or multiple observations (replications).

Factorial n! = n × (n–1) × ...3 × 2 × 1, that is to say, the factorial of an integer n is simply the product of all of the integers from 1 to n.

Factorial analysis of variance An analysis in which the treatments differ in terms of two or more factors (with several levels) as opposed to treatments being different levels of a single factor as in one-way ANOVA.

Factorial design An experimental design in which every level of every variable is paired with every level of every other variable.

Factorial experiments In some data, the explanatory (predictor) variables are all categorical (i.e., factors) and the response variable is quantitative. When there are two or more categorical predictor variables, the data are called factorial. The different possible values of the factors are often assigned numerical values known as levels.

Factorial notalion For each positive integer n, n! = n(n – 1)(n – 2) ... 3 x 2 x 1. Also,0! = 1.

Factors Another word for independent variables in the analysis of variance.

False The adjective, adverb, and noun FALSE is a logical value. FALSE and TRUE are often used as a binary logic pair.

Familywise error rate The probability that a family of comparisons contains at least one Type I error.

Feasibility study Determines if the information system makes sense for the organization from an economic and operational standpoint.

Femto The metric prefix femto- (prefix symbol: f-) denotes a multiple of ten to the negative fifteenth power (10^{-15} = 10 raised to power –15 = 0.000 000 000 000 001) as defined by Le Système International d'Unités (SI).

Fermat's theorem The statement that the equation $x^n + y^n = x^n$ has no positive integer solution x,y,z if n is an integer greater than 2.Fermat's last problem is currently one of the most famous of the unsolved problems.

Few The noun and positive adjective few means a relatively small number of individuals or discrete objects.

The adjectives few/fewer/fewest are used with count nouns and collection nouns (integers), while the adjectives little/less/least are used with mass nouns (non-discrete quantities).

Example: I know my automobile uses less fuel because I make fewer stops to refill.

Fewer The comparative adjective fewer means a smaller number of individuals or discrete objects. s usage note at few.

Fewest The superlative adjective fewest means the smallest number of individuals or discrete objects.

Fibonacci numbers A set of numbers formed by adding the last two numbers to get the next in the series: 0, 1, 1, 2, 3, 5, 8, 13. Named for Leonardo of Pisa, an Italian mathematician of the Middle Ages, who called himself Fibonacci, short for filius Bonacci which means son of Bonacci. The original problem he investigated in1202 A.D. was about how fast rabbits could breed under ideal circumstances. His research led to the construction of this unique set of numbers.

Fibonacci numbers A set of numbers formed by adding the last two numbers to get the next in the series: 0, 1, 1, 2, 3, 5, 8, 13. Named for Leonardo of Pisa, an Italian mathematician of the Middle Ages, who called himself Fibonacci, short for filius Bonacci which means son of Bonacci. The original problem he investigated in1202 A.D. was about how fast rabbits could breed under ideal circumstances. His research led to the construction of this unique set of numbers.

Field A field is a ring in which multiplication is a group operation. In France (and sometimes elsewhere in Europe), the multiplicative group need not be commutative. In the US and Russia it must be.

Fifteen The noun, pronoun, and adjective fifteen (15) is the standard name for the cardinal number that follows fourteen (14). Fifteen corresponds with the ordinal number fifteenth (15th).

Fifth cousin The compound noun fifth (5th) cousin means an individual sharing a common pair of great-great-great-great-grandparents with the referenced individual.

Fifth element The compound noun fifth (5th) element means the fifth (5th) and most important element of the universe.

Fifth essence The compound noun fifth (5th) essence means the fifth (5th) and highest essence of an object or being.

Fifty The noun, pronoun, and adjective fifty (50) is the standard name for the cardinal number that follows forty-nine (49). Fifty corresponds with the ordinal number fiftieth (50th).

Fifty-eight The noun and adjective fifty-eight (58) is the standard name for the cardinal number that follows fifty-seven (57). Fifty-eight corresponds with the ordinal number fifty-eighth (58th).

Fifty-first The compound adjective, adverb, and noun fifty-first (51st) is the standard name for the ordinal number that follows fiftieth (50th). Fifty-first corresponds with the cardinal number fifty-one (51).

Fifty-five The noun and adjective fifty-five (55) is the standard name for the cardinal number that follows fifty-four (54). Fifty-five corresponds with the ordinal number fifty-fifth (55th).

Fifty-four The noun and adjective fifty-four (54) is the standard name for the cardinal number that follows fifty-three (53). Fifty-four corresponds with the ordinal number fifty-fourth (54th).

Fifty-nine The noun and adjective fifty-nine (59) is the standard name for the cardinal number that follows fifty-eight (58). Fifty-nine corresponds with the ordinal number fifty-ninth (59th).

Fifty-one The noun and adjective fifty-one (51) is the standard name for the cardinal number that follows fifty (50). Fifty-one corresponds with the ordinal number fifty-first (51st).

Fifty-second The compound adjective, adverb, and noun fifty-second (52nd or 52^{d}) is the standard name for the ordinal number that follows fifty-first (51st). Fifty-second corresponds with the cardinal number fifty-two (52).

Fifty-seven The noun and adjective fifty-seven (57) is the standard name for the cardinal number that follows fifty-six (56). Fifty-seven corresponds with the ordinal number fifty-seventh (57th).

Fifty-six The noun and adjective fifty-six (56) is the standard name for the cardinal number that follows fifty-five (55). Fifty-six corresponds with the ordinal number fifty-sixth (56th).

Fifty-something The compound noun and adjective fifty-something means at least fifty (50) but less than sixty (60). Fifty-something most commonly means between fifty (50) and sixty (60) years of age.

Fifty-three The noun and adjective fifty-three (53) is the standard name for the cardinal number that follows fifty-two (52). Fifty-three corresponds with the ordinal number fifty-third (53rd or 53^{d}).

Fifty-two The noun and adjective fifty-two (52) is the standard name for the cardinal number that follows fifty-one (51). Fifty-two corresponds with the ordinal number fifty-second (52nd or 52^{d}).

File organization A technique for physically arranging the records of a file.

Final transition A final tansition is a transition with no subsequent state conjunction.

Finite sequence A sequence that has a last term.

Finite The adjective finite means a quantity that can be bound by limits.

First aid The compound noun and adjective first (1st) aid means emergency medical treatment administered to an injured or sick person before professional medical care is available.

First class The compound noun, adjective, and adverb first (1st) class means the best class.

First order interaction The interaction of two variables. Also known as a simple interaction.

First person The compound noun and adjective first (1st) person denotes the speaker or writer referring to himself or herself.

First string The noun and adjective first (1st) string means the group of players designated to start games for a sports team, usually the best player at each position.

First stringer The noun first-stringer means a player on the first string.

Fisher's exact test An exact significance test to analyse 2x2 tables for any sample size. It is a misconception that it is suitable only for small sample sizes. This arises from the demanding computational procedure for large samples, which is no longer an issue. It is the only test for a 2x2 table when an expected number in any cell is smaller than 5.

Fisher's Least Significant Difference Test (LSD) A multiple comparison technique that requires a significant overall F, and that involves standard t tests between pairs of means. Also known as the protected t test.

Fitted line The line that best approximates the graph of the relationship between two vari- ables, if that relationship is nearly linear.

Five The noun, pronoun, and adjective five (5) is the standard name for the cardinal number that follows four (4). Five corresponds with the ordinal number fifth (5th).

Five-star The compound adjective five-star (5-star) means of highest quality based upon the one (1) to five (5) star rating system often used to evaluate accomodations, restaurants, customer service, etc.

Fixed marginal totals The situation in which the marginal totals in a contingency table are known before the data are collected and are not subject to sampling error.

Fixed model Anova An analysis of variance model in which the levels of the independent variable are treated as fixed (i.e. have been selected by the experimenter).

Fixed variable A variable that takes on a specific set of values. An independent variable whose levels are assigned by the experimenter.

Floor For a real number r, its floor value [r] is defined as the largest integer no greater than r. Thus [5]=[5.1]=5 and [–5]=–5 while [–5.1]=–6.

Focus The fixed point or points (foci) used to define a conic section.

Fold The suffix -fold means multiplied by the prefix number. Thus, twofold (2–fold) means doubled and threefold (3–fold) means tripled.

Folding When we talk about folding a plane figure, we mean folding it as if it were a piece of paper in that shape. We might fold this into a solid figure such as a box, or fold the figure flat along itself.

Example: Folding the figure on the left into a box:

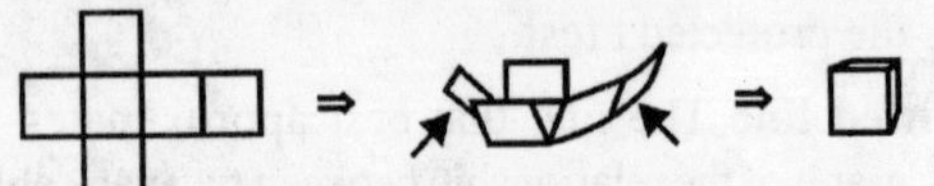

Folding the figure on the left flat along the dotted line:

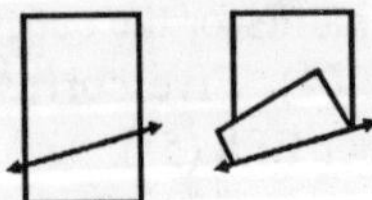

Folio The noun folio means a book with leaves printed on one half (1/2) of the printer sheet.

Foreign key An attribute that appears as a nonprimary key attribute in one relation and as a primary key attribute (or part of a primary key) in another relation.

Form A business document that contains some predefined data and may include some areas where additional data are to be filled in. An instance of a form is typically based on one database record.

Form and report generators CASE tools that support the creation of system forms and reports in order to prototype how systems will look and feel to users.

Formal system The official way a system works as described in organizational documentation.

Formula An equation that follows a rule. Formulas contain variables that when replaced by numbers, allow unknown quantities to be found. An example of a formula is the Pythagorean Theorem, a squared + b squared = c squared.

Forties 1. The plural noun and adjective forties (40s) means an integer multiple of forty (40).

2. The plural noun and adjective forties (40's) means the range of numbers at least forty (40) but less than fifty (50). When referring to dates, forties means a span of ten years in which the tens (10s) digit of the year number is four (4).

Fortieth 1. The adjective, adverb, and noun fortieth (40th) is the standard name for the ordinal number that follows thirty-ninth (39th). Fortieth corresponds with the cardinal number forty (40).

2. The adjective and noun fortieth (1/40) means divided by forty (40).

Fortnightly The adjective fortnightly means recurring every fourteen (14) days.

Forty The noun, pronoun, and adjective forty (40) is the standard name for the cardinal number that follows thirty-nine (39). Forty corresponds with the ordinal number fortieth (40th).

Forty-eight The noun and adjective forty-eight (48) is the standard name for the cardinal number that follows forty-seven (47). Forty-eight corresponds with the ordinal number forty-eighth (48th).

Forty-fifth The compound adjective, adverb, and noun forty-fifth (45th) is the standard name for the ordinal number that follows forty-fourth (44th). Forty-fifth corresponds with the cardinal number forty-five (45).

Forty-first The compound adjective, adverb, and noun forty-first (41st) is the standard name for the ordinal number that follows fortieth (40th). Forty-first corresponds with the cardinal number forty-one (41).

Forty-five The noun and adjective forty-five (45) is the standard name for the cardinal number that follows forty-four (44). Forty-five corresponds with the ordinal number forty-fifth (45th).

Forty-four The noun and adjective forty-four (44) is the standard name for the cardinal number that follows forty-three (43). Forty-four corresponds with the ordinal number forty-fourth (44th).

Forty-nine The noun and adjective forty-nine (49) is the standard name for the cardinal number that follows forty-eight (48). Forty-nine corresponds with the ordinal number forty-ninth (49th).

Forty-one The noun and adjective forty-one (41) is the standard name for the cardinal number that follows forty (40). Forty-one corresponds with the ordinal number forty-first (41st).

Forty-second The compound adjective, adverb, and noun forty-second (42nd or 42d) is the standard name for the ordinal number that follows forty-first (41st). Forty-second corresponds with the cardinal number forty-two (42).

Forty-seven The noun and adjective forty-seven (47) is the standard name for the cardinal number that follows forty-six (46). Forty-seven corresponds with the ordinal number forty-seventh (47th).

Forty-Six The noun and adjective forty-six (46) is the standard name for the cardinal number that follows forty-five (45). Forty-six corresponds with the ordinal number forty-sixth (46th).

Forty-something The compound noun and adjective forty-something means at least forty (40) but less than fifty (50). Forty-something most commonly means between forty (40) and fifty (50) years of age.

Forty-three The noun and adjective forty-three (43) is the standard name for the cardinal number that follows forty-two (42). Forty-three corresponds with the ordinal number forty-third (43rd or 43d).

Forty-two The noun and adjective forty-two (42) is the standard name for the cardinal number that follows forty-one (41). Forty-two corresponds with the ordinal number forty-second (42nd or 42d).

Four colour theorem It is desired to colour a political map on a plane so that countries sharing a common boundary are coloured differently. Experience of map-makers has shown that four colours suffice.This assertion in its precise formulation was an unsolved problem in mathematics for over a hundred years.It has now been proved by an assist from a computer.

Four The noun, pronoun, and adjective four (4) is the standard name for the cardinal number that follows three (3). Four equals two to the second power (2^2). Four corresponds with the ordinal number fourth (4th).

Fourscore and ten The compound noun and adjective fourscore and ten means ninety (90).

Fourscore The noun and adjective fourscore means eighty (80).

Four-star The compound adjective four-star (4-star) means of highest quality based upon the one (1) to four (4) star rating system often used to evaluate movies, etc.

Fourteen The noun, pronoun, and adjective fourteen (14) is the standard name for the cardinal number that follows thirteen (13). Fourteen corresponds with the ordinal number fourteenth (14th).

Fourth 1. The adjective, adverb, and noun fourth (4th) is the standard name for the ordinal number that follows third (3rd or 3^{d}). Fourth corresponds with the cardinal number four (4).

2. The adjective and noun fourth (1/4) means divided by four (4).

Fourth cousin The compound noun fourth (4th) cousin means an individual sharing a common pair of great-great-great-grandparents with the referenced individual.

Fourth world The compound noun and adjective Fourth (4th) World means the poorest nations of the world, more specifically, those nations whose per capita gross domestic product (GDP) is less than about one fifth (20%) the average per capita gross domestic product of all the Earth's population (about $1,600 USD per person per year in 2003.)

Fractal erm coined by Benoit Mandelbrot in 1975, referring to objects built using recursion, where some aspect of the limiting object is infinite and another is finite, and where at any iteration, some piece of the object is a scaled down version of the previous iteration.

Fractiles A generic name for statistics such as deciles, percentiles, and quartiles.

Fraction A number that expressed a portion of a whole. The denominator of a fraction represents the number of the portions the whole has been divided into, and the numerator expresses the number of the portions measured. The fraction 1/4 could be stated as 1 our of 4 parts of the whole.

Fractional equation An equation with a vari- able in the denominator of a fraction.

Fractional part For a real number r, its fractional part is defined as {r}=r–[r], where [r] is the floor value of r.

Fractionally perfect number (FPFN) A number whose index is a simple fraction. Here, simple usually means that the denominator is less than ten.

Fractions, ruler The portions into which the English units of measurement such as inches can be divided — for example, halves, fourths, eighths, sixteenths, and thirty-seconds.

Free group Groups have generators, such elements that all other elements of a group could be obtained from generators and their inverses using the group operation. A group is said to be free if no relation exists between its generators other than between an element and its inverse. The additive group of integers is free with a single generator 1. The multiplicative group of all positive rational numbers has prime numbers as its generators. From the Fundamental Theorem of Arithmetic representation of integers in the form $p^a q^b \ldots r^c$ where p, q, r are all different primes, is unique. Therefore the group is free. The group of sliding motions in the sliders puzzle is not free. Indeed, for the sequence $S_{j+}S_{+i}S_{j-}S_{-i}$ we obviously have the following identity $(S_{j+}S_{+i}S_{j-}S_{-i})^3$=Id, where Id is the unit element of the group, i.e. the element that leaves all the counters unchanged.

Frequency data Data representing counts or number of observations in each category.

Frequency distribution A distribution in which the values of the dependent variable are tabled or plotted against their frequency of occurrence.

Frequency table A display to show how often items, numbers, or a range of numbers occur.

Frequency The number of items occurring in a given category.

Frequency view An approach taken by mathematicians and scientists to determine the chances of an event happening by repeating the experiment many times and using the results to calculate the probability.

Friedman test This non-parametric test is an extension of Wilcoxon matched pairs signed ranks test which is concerned with more than two time periods of data collection or conditions, and groups of three or more matched subjects.

Friedman's rank test for k correlated samples A nonparametric test analogous to a standard one-way repeated-measures analysis of variance.

Frobenius matrix norm A matrix norm that has

the simple formula: ||A|| = the square root of the sum of the squares of all the entries of the matrix.

The Frobenius matrix norm is not a vector-bound matrix norm, although it is compatible with the L2 vector norm, and much easier to compute that the L2 matrix norm.

The Frobenius matrix norm is sometimes called the Schur matrix norm or Euclidean matrix norm

Function A mathematical function is a machine where you put in a real number (often denoted by a variable x, but sometimes by t or some other letter) and it spits out a new real number. For instance, $f(x) = x^2$. You put in the number 3 for x and it spits out the number 9. It's domain is the set of values that are legal to put in, and its range is the set of possible values it can spit out.

Function notation A notation that describes a function. for a function *f*, when x is a member of the domain, the symbol *f*(x) denotes the corresponding member of the range [e.g., an equation of a function might be $f(x) = x+3$].

Functional dependency A particular relationship between two attributes. For a given relation, attribute B is functionally dependent on attribute A if, for every valid value of A, that value of A uniquely determines the value of B. The functional dependence of B on A is represented by A'B.

Functions of an angle The name of a particular ratio of the sides of a right triangle. The name of the function will depend on which side is divided into which side. There are six possible ways to divide the sides of a right triangle into each other. There are six functions for each angle. The names are: sine, cosine, tangent, cosecant, secant, and cotangent.

Functions table A table of the functions of angles. This table allows one to find either the function of an angle or the degree of an angle.

Fundamental Theorem of Calculus This theorem is usually stated in two parts. One part states that finding areas under curves can be done by taking antiderivatives and plugging in the limits.

$\int_a^b f(x)dx = F(b) - F(a)$, where F(x) is any function whose derivative is f(x), F'(x) = f(x).This can also be stated as:

$$\int_a^b f'(x)dx = f(b) - f(a)$$

If you integrate the derivative of a function over an interval from a to b, you just get the original function evaluated at b minus the original function evaluated at a. The other part states that

$$\frac{d}{dx}\int_a^x f(t)dt = f(x)$$

Both parts show that derivatives and integrals are intimately related, and we don't just mean on a first name.

G The metric prefix symbol G- (metric prefix: giga-) denotes a multiple of ten to the ninth power (10^9 = 10 raised to power +09 = 1,000,000,000) as defined by Le Système International d'Unités (SI).

The Greek letter gamma represents Euler's Constant, also known as the Euler-Mascheroni Constant, γ is approximately 0.577 215 664 902.

G Statistics An application of the log-likelihood ratio statistics for the hypothesis of independence in an r x c contingency table. It can also be used to test goodness-of-fit.

Gambler's ruin A classical topic in probability theory. It is a game of chance related to a series of Bernoulli trials. There are variations of the game theory associated with problems of the random walk and sequential sampling.

Game theory The Game Theory studies winning strategies for parties involved in situations where their interest conflict with each other. Developed by John von Neumann, the theory has applications to real games (cards, chess, etc.), economics, commerce, politics and some say even military. J. Conway used his theory of surreal numbers to qualitatively evaluate game positions.

Gamma The symbol for the effect size.

Gamma function A statistical function closely related to factorials.

Gantt chart A graphical representation of a project that shows each task as a horizontal bar whose length is proportional to its time for completion.

Gap analysis The process of discovering discrepancies between two or more sets of data flow diagrams or discrepancies within a single DFD.

Gauss jordan elimination Gauss Jordan elimination is a method for solving a system of linear equations A * x = b for x, or for computing the inverse matrix of A.

Gauss elimination and Gauss Jordan elimination are very closely related. Gauss elimination reduces A to an upper triangular matrix, and saves the elimination factors in a lower triangular matrix. Gauss Jordan elimination proceeds relentlessly until A has been converted into the identity matrix.

Thus, unlike the Gauss elimination procedure, the Gauss Jordan elimination does not produce a factorization of the matrix, but only a solution to the linear system. This means that if a second linear system has to be solved, the matrix has to be set up and eliminated all over again.

The simplest way to describe Gauss Jordan is to note that to solve, say, the linear system A * x =

b, the right hand side is appended as an extra column of the matrix. Then, on step I of the elimination, we choose a pivot row, move it to row I, divide it through by the pivot value, and then eliminate the matrix entries in column I from all other rows, rather than simply from rows I+1 through N. When the process is completed, the solution x has overwritten the right hand side b that was stored in column N+1.

Several right hand sides can be handled at once, by appending all of them to the coefficient matrix; the inverse can be computed by appending a copy of the identity matrix to the coefficient matrix before beginning elimination.

Gauss Jordan elimination is primarily used as a teaching tool, and for small linear systems. In practical computation, standard Gauss elimination is universally preferred.

Gauss seidel iteration for linear equations The Gauss Seidel iteration for linear equations is an iterative method for solving linear systems of equations A*x=b. It is similar to the the Jacobi algorithm and the successive overrelaxation method (SOR).

The Gauss Seidel iteration should only be used for matrices which are symmetric and positive definite, or for a matrix which is strictly diagonally dominant.

Each step of the Gauss Seidel iteration begins with an approximate answer x, and produces a new approximation y. Each component of y is computed in order, using the formula:

```
y(i) = [ b(i)
       – a(i,1)*y(1)
       – a(i,2)*y(2)
       ...
       – a(i,i–1)*y(i–1)
       – a(i,i+1)*x(i+1)
       ...
       – a(i,n)*x(n) ] / a(i,i)
```

The calculation of each entry of y is dependent on the calculation of the entries of lower index. Thus, the value of y(1) is calculated first, and then y(2) is calculated based on the values of y(1) as well as x(3) through x(n), and so on.

The process is to be repeated until the residual error is small, or the change in the approximate solution is negligible.

The Gauss Seidel iteration can be considered in terms of its matrix splitting. That is, if we decompose the matrix A into its strictly lower triangular, diagonal, and strictly upper triangular parts:

$$A = L + D + U$$

then the method is equivalent to the iteration

$$(L + D) * xnew = b - U * x.$$

which means that the convergence of the algorithm can be understood in terms of the behavior of powers of the iteration matrix:

$$-(L + D)^{-1} * U,$$

which in turn may best be understood by looking at the eigenvalues.

If the original coefficient matrix A is symmetric, then it may be preferred to use the symmetric Gauss Seidel iteration or SGS. In this case, the iteration consists of pairs of Gauss Seidel steps. The odd steps are the same as the usual iteration. But in the even steps, the variables are solved for in reverse order. Each pair of such steps is a single step of the SGS iteration, which has the property that its iteration matrix is similar to a symmetric matrix (though not necessarily symmetric itself). Among other things, this means that SGS can be used as a preconditioner for certain other problems.

Gaussian distribution Another name for the normal distribution (GraphPad Gaussian Distribution Calculator).

General constraint A general constraint specifies any information constraining relationship set which cannot be specified as co-occurrence constraint, object-class cardinality constraint, or participation constraint.

General linear model A group of linear regression models in which the response variable is continuous and normally distributed, the response variable values are predicted from a linear combination of predictor variables, and the linear combination of values for the predictor variables is not transformed (i.e., there is no link function as in generalised linear models). Linear multiple regression is a typical example of general linear models whereas simple linear regression is a special case of generalised linear models with the identity link function.

General matrix storage A general matrix is one which has no special matrix structure or matrix symmetry.

In such a case, there are no space advantages to be gained by using a special matrix storage format, and so the matrix entries are stored using the standard two dimensional array format provided by the programming language.

For general matrices, the only remaining issue concerns the problem that occurs when the matrix storage must be set aside before the size of the matrix is known. In FORTRAN, for example, it is common to specify a maximum matrix size of, say, 100 by 100. If the actual problem to be solved is of size 25 by 25, then it may be necessary to describe the data with both the matrix order of 25, and the leading dimension of the storage array, which is 100. In LINPACK and LAPACK, variables containing leading dimension information have names like LDA, LDB and so on.

LAPACK and LINPACK provide routines, with the prefix SGE, which apply to matrices in general storage.

General solution Solutions defined over entire domain.

Generalised linear model (GLM) A model for linear and non-linear effects of continuous and categorical predictor variables on a discrete or continuous but not necessarily normally distributed dependent (outcome) variable. (Note that in the general linear model, the dependent (outcome) variable should be normally distributed). Normal, binary (or linear logistic; when the outcome variable is a proportion), binomial or Poisson (when the outcome variable is a count), exponential and gamma (when the outcome variable is continuous and non-negative) models are different versions of generalised linear models. Particular types of models arise by specifying an appropriate link function, variance and distribution. For example, normal linear regression corresponds to an identity link function, constant variance and a normal distribution. Logistic regression arises from a logit link function and a binomial distribution (the variance of the response (npq) is related to its mean (np): variance = mean (1 - (mean/n)). Loglinear models are used for binomial or Poisson counts. Standard techniques for analysing censored survival data, such as the Cox regression, can also be handled within the GLM framework.

Generalization A generalization is an object class which is a superset of another object class (or classes). Generalization models the is a relationship set since members of the specialization class (or classes) are always members of the generalization class. This means that members of the specialization class have all of the same properties of the generalization class including relationships with other objects as well as behaviour. Therefore, specialization plays an important role in the OR

Generalized permutation matrix A generalized permutation matrix is a square matrix A with at most one nonzero entry in each row, and in each column.

A standard permutation matrix, of course, has exactly one nonzero entry in each row and column, and that entry has value 1.

An interesting fact: if A is a nonsingular nonnegative matrix, then the inverse of A is also nonnegative if and only if A is a generalized permutation matrix. In other words, it's very hard for both A and A^{-1} to be nonnegative, and essentially can only happen if A is diagonal.

So suppose A >= 0 but A is not a generalized permutation matrix. For an arbitrary vector b >=

0 , can we say that the solution x of A * x = b is nonnegative? No, because we know that the inverse of A is not nonnegative. Therefore A^{-1} contains at least one negative entry, say entry (i,j). Choose b = E(J), where E(J) is the J-th Cartesian basis vector. Then x = A^{-1} * b and it is easy to see that x(i) is negative.

Generating function Independence or Statistical independence : Two events are independent if the outcome of one does not affect that of the other (for example, getting a 1 on one die roll does not affect the probability of getting a 1 on a second roll). Similarly, when we assert that two random variables are independent, we intuitively mean that knowing something about the value of one of them does not yield any information about the value of the other.

Generator The bent line-segment or figure that replaces the initiator at each iteration of a fractal.

Genetic distance A measurement of genetic relatedness of populations. The estimate is based on the number of allelic substitutions per locus that have occurred during the separate evolution of two populations.

Genetic distance estimation by phylip The most popular (and free) phylogenetics program PHYLIP can be used to estimate genetic distance between populations.

Geometric mean G = $(x_1.x_2...x_n)^{1/n}$ where n is the sample size. This can also be expressed as antilog ((1/n) Ó log x), which means the antilog of the mean of the logs of each value.

Geometric sequence A sequence with a constant ratio between two consecutive terms. Each successive term is found by multiplying the preceding term by the preceding term by the ratio. (e.g., 1, 2, 4, 8, 16, ...is a geometric sequence with a ratio of 2.).

Geometric series The indicated sum of the terms of a geometric sequence.

Geometric vector A quantity that can be represented by a directional line segment.

Gershgorin disks The method of Gershgorin disks provides an estimate of the size of the eigenvalues of a matrix. The accuracy of the estimate varies wildly, depending on the size of the elements of the matrix. It is most useful for matrices that are diagonally dominant or sparse. Gershgorin's theorem states that the eigenvalues of any matrix A lie in the space covered by the disks D_i:

D_i = (x: sqrt $(x - A(I,I))^2$ <= R_i)

where R_i is the sum of the absolute values of the off-diagonal elements of row I:

R_i = sum (J =/= I) | $A_{i,j}$ |.

The theorem may also be applied using columns instead of rows.

Gi The compound binary prefix symbol Gi- (binary prefix: gibi-) denotes a multiple of two to the thirtieth power (2^{30} = 1,073,741,824) as defined by the International Electrotechnical Commission (IEC).

Gibi The binary prefix gibi- (prefix symbol: Gi-) denotes a multiple of two to the thirtieth power (2^{30} = 1,073,741,824) as defined by the International Electrotechnical Commission (IEC).

Gibin The adjective and noun gibin means two to the thirtieth power (2^{30} = 1,073,741,824).

Giga The metric prefix giga- (prefix symbol: G-) denotes a multiple of ten to the ninth power (10^9 = 10 raised to power +09 = 1,000,000,000) as defined by Le Système International d'Unités (SI).

Gigabinary The adjective gigabinary means two to the thirtieth power (2^{30} = 1,073,741,824).

Global (extremum, maximum, minimum, warming) Another expression for the absolute extremum, maximum or minimum. It comes from the fact that this extremum is the most extreme extremum on the globe.

Golden ratio The Golden Ratio (symbol: φ {Greek lower case letter phi}) is the irrational number that is one greater than its multiplicative inverse, i.e., φ = 1 + 1/φ = $(1+5^{1/2})/2$ = approximately 1.618 033 988 750. The ancient Egyptians discovered φ, which they honored as their Sacred Ratio. The ancient Greeks used φ extensively in art and architecture and treasured as their Golden Ratio. The ancient Egyptians discovered φ, which they honored as their Sacred Ratio. The ancient Greeks treasured φ as their Golden Ratio. The ratio φ is used extensively in art and architecture.

Goodness-of-fit test A test for comparing observed frequencies with theoretically predicted frequencies.

Googol The noun googol means ten to the one hundredth power (10^{100} = 10 raised to power +100). The word googol was invented by Milton Sirotta, the 9-year-old nephew of mathematician Edward Kasner, who had asked his nephew what he thought such a large number should be called.

Googolplex The noun googolplex means ten to the googol power (10^{googol} = 10 raised to power +(10 raised to power +100)).

Goose egg The compound noun goose egg is a euphemism for the digit zero (0).

Grade In angular measurement, the noun and adjective grade means one four-hundredth (1/400 = 0.002 500) of a circle.

Gradient Gradient is a vector pointing in the direction of the maximal change of a function of two or more variables. Components of the gradient of a function in Cartesian coordinates are just partial derivatives of that function. In 2-dimensional space, the gradient is often confused with the slope of a function of one variable. Given a function y = f(x), it is always possible to look at its graph as 0-level curve of function F(x, y) = f(x) – y. In this case, the gradient of F, grad(F), is the vector (f'(x), –1), with the first component being the derivative of f(x) - read the slope of the graph of y = f(x). Thus the notions of the slope and that of the gradient are related. However, one of them is a scalar (slope), the other a vector (gradient).

Gram matrix Given a set of M vectors V_i, each of dimension N, the M by M Gram matrix is formed by all the pairwise dot products of the vectors:

$G_{i,j} = V_i \text{ dot } V_j$

The Gram matrix is a positive semidefinite symmetric. If any vector is linearly dependent on the others, the matrix is singular. If the vectors are linearly independent, the Gram matrix has full rank M. If the vectors form a basis for the space, the Gram matrix has rank N.

The Gram matrix can be used to construct a projection matrix into the space spanned by the vectors.

Gram Schmidt Orthogonalization Gram Schmidt orthogonalization is a process which starts with a set of N vectors X(I), each with M components, and produces a set of N2 orthonormal vectors Y which span the linear space of the original vectors.

N2 is less than N if the vectors X(I) are linearly dependent, and equal to N if they are linearly independent. Thus, one use for the Gram Schmidt process is simply to determine if a set of vectors are linearly dependent; a second is to determine the dimension of the space spanned by a linearly dependent set.

The Gram Schmidt process may be defined iteratively:

```
for I = 1 to N
N2    = I
Y(I)  = X(I)
for J = 1 to I–1
C(J)  = dot_product (X(I), Y(J))
Y(I)  = Y(I) – C(J) * Y(J)
end for
Norm= sqrt (dot_product (Y(I), Y(I)))
if (Norm= 0) exit
Y(I) = Y(I)/Norm

end for
```

Another way of looking at the process is to use the vectors to form the columns of a matrix A. The Gram Schmidt process can then be used to construct one version of the QR factorization of the matrix A:

A = Q * R

where Q is orthogonal and R is upper triangular.

Grand gross The compoun noun grand gross means one dozen dozen dozen (1728).

Grand total (ÓX) The sum of all of the observations.

Graph A pictorial representation of a function formed by plotting f(x) in the y direction on the x-y plane. Very useful, because while pictures say a thousand words, a graph gives an infinite number of function values.

Given a function f(x), the graph of the function is simply the set of points (x,y) in the Cartesian plane that satisfy the equation y=f(x). Most important property? Any vertical line can intersect the graph at most once, since a vertical line is defined by a particular value of x. But for a particular value of x, there is only one value of y such that y = f(x).

Graph of the function f The set of all the points on the coordinate plane of the form (x, f(x)) with x in the domain of f.

Graph's completeness A graph is complete if any two of its vertices are connected by exactly one edge. A complete graph with N vertices is often denoted as K_N.

Greatest common factor The common factor that is a multiple of each common factor.

Greatest integer function A step function that pairs a real number with the greatest integer that is less than or equal to the real number.

Gross The noun gross means one dozen dozen (144).

Group Group is the most fundamental and pervasive notion of the Higher or Abstract Algebra. It's a set along with a single operation defined on its elements. The group is called additive if the symbol for the operation is +. It's multiplicative if the symbol · of multiplication is used instead. But any other symbol can be used as well. There is always a unique element (1, for multiplicative, and 0, for additive, groups) that leaves elements invariant (unchanged) under the defined operation, like a+0=a. Also, for every element a there exists a unique inverse b such that, for example, in the case of the additive symbol, a+b=0 and b+a=0. Most often, however, the inverse is denoted as a^{-1}. Lastly, the group operation must be associative like in a·(b·c)=(a·b)·c. A group is commutative or Abelian if its operation is symmetric, like in a+b=b+a.

Growth Many things may grow in Mathematics. Functions may grow monotonously or in jumps. Complexity of a system may grow exponentially with the system size. Lately, due to the development of the Fractal Geometry, growth of crystals and other natural phenomena became a subject of scientific scrutiny.

H spread The range between the two hinges.

H The traditional metric prefix symbol h- (metric prefix: hecto-) denotes a multiple of one hundred (100).

Hadamard product The Hadamard product of matrices A and B is a matrix C created by elementwise multiplication:

$C_{i,j} = A_{i,j} * B_{i,j}$

The Hadamard product is defined for any pair of rectangular matrices, as long as they have the same shape, that is, the same number of rows, and the same number of columns.

Example of a Hadamard product:

1 2 3 * 7 8 9 = 7 16 27

4 5 6 10 11 12 40 55 72

Hadamard's inequality Hadamard's inequality provides an upper bound on the size of the determinant of a matrix. It is related to the fact that the determinant represents the volume of an N-dimensional parallelepiped.

Let ||C(I)|| designate the Euclidean norm of column I of the matrix A. Hadamard's inequality states that

det (A) <= ||C(1)|| * ||C(2)|| * ... * ||C(N)||,

with equality holding only if one of the C(I)'s is zero, (yielding the minimum possible value of 0), or if all the C(I)'s are pairwise orthogonal vectors, (yielding the largest possible value). The theorem may also be applied using rows instead of columns.

Half a hundred The compound noun and adjective half a hundred means fifty (50).

Half angle identities for tangent Useful in writing trig functions involving half angles as functions of single angles.

Half angle identities useful in writing trig functions involving half angles as trig functions of single angles.

Half century The compound noun half century means fifty (50) years.

Half decade The compound noun half decade means five (5) years.

Half dozen The compound noun and adjective half dozen means six (6).

Half life The length of time it takes half of an amount of radioactive material to decay.

Half The singular noun and adjective half (1/2) means divided by two (2). The plural of half is halves. A half coin has a value of one half of a dollar ($0.50).

Half-normal plot A diagnostic test for model inadequacy or revealing the presence of outliers. It compares the ordered residuals from the data to the expected values of ordered observations from a normal distribution. While the full-normal plots use the signed residuals, half-normal plots use the absolute values of the residuals. Outliers appear at the top right of the plot as distinct points, and departures from a straight line mean that the model is not satisfactory. It is appropriate to use a half-normal plot only when the distribution is symmetrical about zero because any information on symmetry will be lost.

Halves The plural noun and adjective halves (1/2) means divided by two (2). The singular of halves is half.

Hankel matrix A Hankel matrix is a matrix which is constant along each of its anti-diagonals.

Here is an example of a square Hankel matrix:

7 6 5 4

6 5 4 3

5 4 3 2

4 3 2 1

and a rectangular Hankel matrix:

7 6 5 4 3 2

6 5 4 3 2 1

5 4 3 2 1 0

Simple facts about a Hankel matrix A:

A is symmetric;

the inverse of A is symmetric, but need not be a Hankel matrix;

Compare the concepts of Toeplitz Matrix, an Anticirculant Matrix, and a Persymmetric Matrix.

Haplotype relative risk method This method uses non-inherited parental haplotypes of affected persons as the control group and thus eliminates the risks and bias associated with using unrelated individuals as controls in case-control association studies, as well as the higher cost .

Hardy-weinberg equilibrium (HWE) In an infinitely large population, gene and genotype frequencies remain stable as long as there is no selection, mutation, or migration. For a bi-allelic locus where the gene frequencies are p and q: $p^2+2pq+q^2 = 1$. HWE should be assessed in controls in a case-control study and any deviation from HWE should alert for genotyping errors (Lewis, 2002).

Harmonic mean In a set of n numbers. n divided by the sum of the reciprocals of the number.

Harmonic mean Of a set of numbers (y_1 to y_n), the harmonic mean is the reciprocal of the arithmetic mean of the reciprocal of the numbers [$H = N / (1/(y_1 + y_2 + ... y_n))$]. The harmonic mean is either smaller than or equal to the arithmetic mean. It is a measure of position.

Harmonic mean The number of elements to be averaged divided by the sum of the reciprocals of the elements.

Hashed file organization The address for each row is determined using an algorithm.

Hazard function (instantaneous failure rate, conditional failure, intensity, or force of mortality function): The function that describes the probability of failure during a very small time increment (assuming that no failures have occurred prior to that time). Hazard is the slope of the survival curve – a measure of how rapidly subjects are having the event (dying, developing an outcome etc).

Hazard rate It is a time-to-failure function used in survival analysis. It is defined as the probability per time unit that a case that has survived to the beginning of the respective interval will fail in that interval. Specifically, it is computed as the number of failures per time units in the respective interval, divided by the average number of surviving cases at the mid-point of the interval.

Hazard ratio (Relative Hazard) Hazard ratio compares two groups differing in treatments or prognostic variables etc. If the hazard ratio is 2.0, then the rate of failure in one group is twice the rate in the other group. The computation of the hazard ratio assumes that the ratio is consistent over time, and that any differences are due to random sampling. Before performing any tests of hypotheses to compare survival curves, the proportionality of hazards assumption should be checked (and should hold for the validity of Cox's proportional hazard models).

HBSMC Finite Element Matrix Storage The HBSMC finite element storage format is a special matrix storage format for those matrices in the collection which derive from a finite element problem.

Matrices arising in finite element applications are usually assembled from numerous small elemental matrices. The collection includes a few sparse matrices in original unassembled form. The storage of the individual unassembled matrices is based on the general sparse format, which stores a matrix as a list of matrix columns. The elemental representation stores the matrix as a list of elemental matrices. Each elemental matrix is represented by a list of the row/column indices (variables) associated with the element and by a small dense matrix giving the numerical values by columns, or in the symmetric case, only the lower triangular part. The lists of indices are held contiguously, just as for the lists of row indices in the standard format. The dense matrices are held contiguously in a separate array, with each matrix held by columns. Although there is not a one to one correspondence between the arrays of integer and numerical values, the representation does not hold the pointers to the beginning of the real values for each element. These pointers can be created from the index start pointers (ELTPTR) after noting that an element with NU variables has NU*NU real values, or (NU*(NU+1))/2 in the symmetric case.

We illustrate the elemental storage scheme with a small, 5 by 5 example:

5.0 0.0 0.0 1.0 2.0

0.0 4.0 3.0 0.0 6.0

0.0 3.0 7.0 8.0 1.0

1.0 0.0 8.0 9.0 0.0

2.0 6.0 1.0 0.0 10.0

generated from four elemental matrices:

1 4 1 5 2 3 5 3 4

1 (2.0 1.0) 1 (3.0 2.0) 2 (4.0 3.0 6.0) 3 (2.0 8.0)

4 (1.0 7.0) 5 (2.0 8.0) 3 (3.0 5.0 1.0) 4 (8.0 2.0)

5 (6.0 1.0 2.0)

where the variable indices are indicated by the integers marking the rows and columns. This matrix would be stored in the ELTPTR (location of first entry), VARIND (variable indices) and VALUES (numerical values) arrays as follows:

Subscripts: 1 2 3 4 5 6 7 8 9 10 11 12 13 14 15

ELTPTR: 1 3 5 8 10

VARIND: 1 4 1 5 2 3 5 3 4

VALUES: 2. 1. 7. 3. 2. 8. 4. 3. 6. 5. 1. 2. 2. 8. 2.

Heavy tailed distribution A distribution with a higher percentage of scores in the tails than we would expect in a normal distribution.

Hecatohedron The noun hecatohedron means a solid figure with one hundred (100) faces.

Hecta The prefix hecta- means one hundred (100).

Hectad The noun and adjective hectad means a group of one hundred (100) members.

Hecto The traditional metric prefix hecto- (prefix symbol: h-) denotes a multiple of one hundred (100).

Hectogon The noun and adjective hectogon means a plain figure with one hundred (100) straight sides.

Hectogram The noun and adjective hectogram means an interconnected star figure with one hundred (100) points formed by stellating a hectogon.

Hectohedron The noun and adjective hectohedron means a solid figure with one hundred (100) planar faces.

Help desk A single point of contact for all user inquiries and problems about a particular information system or for all users in a particular department.

Hemi The Greek-based numeric prefix hemi- means one half (1/2).

Hendeca The prefix hendeca- means eleven (11).

Hendecad The noun and adjective hendecad means a group of eleven (11) members.

Hendecagon The noun and adjective hendecagon means a plain figure with eleven (11) straight sides.

Hendecagram 1. The noun and adjective hendecagram means an interconnected star figure with eleven (11) points formed by stellating a hendecagon.

2. The noun hendecagram means a word or character sequence consisting of eleven (11) characters.

Hendecahedron The noun and adjective hendecahedron means a solid figure with eleven (11) planar faces.

Hepta The prefix hepta- means seven (7).

Heptaconta The prefix heptaconta- means seventy (70).

Heptacontad The noun and adjective heptacontad means a group of seventy (70) members.

Heptacontadi The prefix heptacontadi- means seventy-two (72). Heptacontadi- is an alternate form of the prefix heptacontakaidi-.

Heptacontaennea The prefix heptacontaennea- means seventy-nine (79). Heptacontaennea- is an alternate form of the prefix heptacontakaiennea-.

Heptacontagon The noun and adjective heptacontagon means a plain figure with seventy (70) straight sides.

Heptacontagram The noun and adjective heptacontagram means an interconnected star figure with seventy (70) points formed by stellating a heptacontagon.

Heptacontahedron The noun and adjective heptacontahedron means a solid figure with seventy (70) planar faces.

Heptacontahena The prefix heptacontahena- means seventy-one (71). Heptacontahena- is an alternate form of the prefix heptacontakaihena-.

Heptacontahepta The prefix heptacontahepta- means seventy-seven (77). Heptacontahepta- is an alternate form of the prefix heptacontakaihepta-.

Heptacontahexa The prefix heptacontahexa- means seventy-six (76). Heptacontahexa- is an alternate form of the prefix heptacontakaihexa-.

Heptacontakaidi The prefix heptacontakaidi- means seventy-two (72).

Heptacontakaidigon The noun and adjective heptacontakaidigon means a plain figure with seventy-two (72) straight sides.

Heptacontakaidigram The noun and adjective heptacontakaidigram means an interconnected star figure with seventy-two (72) points formed by stellating a heptacontakaidigon.

Heptacontakaidihedron The noun and adjective heptacontakaidihedron means a solid figure with seventy-two (72) planar faces.

Heptacontakaidyad The noun and adjective heptacontakaidyad means a group of seventy-two (72) members.

Heptacontakaiennea The prefix heptacontakaiennea- means seventy-nine (79).

Heptacontakaiennead The noun and adjective heptacontakaiennead means a group of seventy-nine (79) members.

Heptacontakaienneagon The noun and adjective heptacontakaienneagon means a plain figure with seventy-nine (79) straight sides.

Heptacontakaienneagram The noun and adjective heptacontakaienneagram means an interconnected star figure with seventy-nine (79) points formed by stellating a heptacontakaienneagon.

Heptacontakaienneahedron The noun and adjective heptacontakaienneahedron means a solid figure with seventy-nine (79) planar faces.

Heptacontakaihena The prefix heptacontakaihena- means seventy-one (71).

Heptacontakaihenad The noun and adjective heptacontakaihenad means a group of seventy-one (71) members.

Heptacontakaihenagon The noun and adjective heptacontakaihenagon means a plain figure with seventy-one (71) straight sides.

Heptacontakaihenagram The noun and adjective heptacontakaihenagram means an interconnected star figure with seventy-one (71) points formed by stellating a heptaconta-kaihenagon.

Heptacontakaihenahedron The noun and adjective heptacontakaihenahedron means a solid figure with seventy-one (71) planar faces.

Heptacontakaihepta The prefix heptacontakaihepta- means seventy-seven (77).

Heptacontakaiheptad The noun and adjective heptacontakaiheptad means a group of seventy-seven (77) members.

Heptacontakaiheptagon The noun and adjective heptacontakaiheptagon means a plain figure with seventy-seven (77) straight sides.

Heptacontakaiheptagram The noun and adjective heptacontakaiheptagram means an interconnected star figure with seventy-seven (77) points formed by stellating a heptacontakaiheptagon.

Heptacontakaiheptahedron The noun and adjective heptacontakaiheptahedron means a solid figure with seventy-seven (77) planar faces.

Heptacontakaihexa The prefix heptacontakaihexa- means seventy-six (76).

Heptacontakaihexad The noun and adjective heptacontakaihexad means a group of seventy-six (76) members.

Heptacontakaihexagon The noun and adjective heptacontakaihexagon means a plain figure with seventy-six (76) straight sides.

Heptacontakaihexagram The noun and adjective heptacontakaihexagram means an interconnected star figure with seventy-six (76) points formed by stellating a heptaconta-kaihexagon.

Heptacontakaihexahedron The noun and adjective heptacontakaihexahedron means a solid figure with seventy-six (76) planar faces.

Heptacontakaiocta The prefix heptacontakaiocta- means seventy-eight (78).

Heptacontakaioctad The noun and adjective heptacontakaioctad means a group of seventy-eight (78) members.

Heptacontakaioctagon The noun and adjective heptacontakaioctagon means a plain figure with seventy-eight (78) straight sides.

Heptacontakaioctagram The noun and adjective heptacontakaioctagram means an interconnected star figure with seventy-eight (78) points formed by stellating a heptacontakaioctagon.

Heptacontakaioctahedron The noun and adjective he-ptacontakaioctahedron means a solid figure with seventy-eight (78) planar faces.

Heptacontakaipenta The prefix heptacontakaipenta- means seventy-five (75).

Heptacontakaipentad The noun and adjective heptacontakaipentad means a group of seventy-five (75) members.

Heptacontakaipentagon The noun and adjective heptacontakaipentagon means a plain figure with seventy-five (75) straight sides.

Heptacontakaipentagram The noun and adjective heptacontakaipentagram means an interconnected star figure with seventy-five (75) points formed by stellating a heptacon-takaipentagon.

Heptacontakaipentahedron The noun and adjective heptacontakaipentahedron means a solid figure with seventy-five (75) planar faces.

Heptacontakaitetra The prefix heptacontakaitetra- means seventy-four (74).

Heptacontakaitetrad The noun and adjective heptacontakaitetrad means a group of seventy-four (74) members.

Heptacontakaitetragon The noun and adjective heptacontakaitetragon means a plain figure with seventy-four (74) straight sides.

Heptacontakaitetragram The noun and adjective heptacontakaitetragram means an interconnected star figure with seventy-four (74) points formed by stellating a heptacontakaitetragon.

Heptacontakaitetrahedron The noun and adjective heptacontakaitetrahedron means a solid figure with seventy-four (74) planar faces.

Heptacontakaitri The prefix heptacontakaitri- means seventy-three (73).

Heptacontakaitriad The noun and adjective heptacontakaitriad means a group of seventy-three (73) members.

Heptacontakaitrigon The noun and adjective heptacontakaitrigon means a plain figure with seventy-three (73) straight sides.

Heptacontakaitrigram The noun and adjective heptacontakaitrigram means an interconnected star figure with seventy-three (73) points formed by stellating a heptacontakaitrigon.

Heptacontakaitrihedron The noun and adjective heptacontakaitrihedron means a solid figure with seventy-three (73) planar faces.

Heptacontaocta The prefix heptacontaocta- means seventy-eight (78). Heptacontaocta- is an alternate form of the prefix heptacontakaiocta-.

Heptacontapenta The prefix heptacontapenta- means seventy-five (75). Heptacontapenta- is an alternate form of the prefix heptacontakaipenta-.

Heptacontatetra The prefix heptacontatetra- means seventy-four (74). Heptacontatetra- is an alternate form of the prefix heptacontakaitetra-.

Heptacontatri The prefix heptacontatri- means seventy-three (73). Heptacontatri- is an alternate form of the prefix heptacontakaitri-.

Heptacta The prefix heptacta- means seven hundred (700).

Heptad The noun and adjective heptad means a group of seven (7) members.

Heptadeca The prefix heptadeca- means seventeen (17).

Heptagon A seven-sided polygon. The sum of the angles of a heptagon is 900 degrees.

Examples: A regular heptagon:

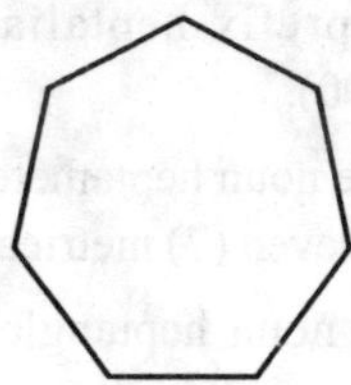

An irregular heptagon:

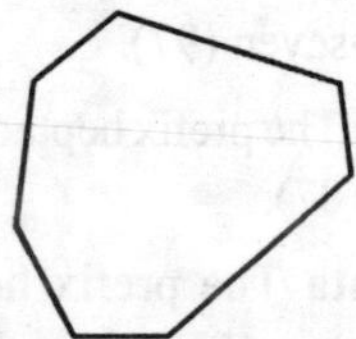

Heptagram 1. The noun and adjective heptagram means an interconnected star figure with seven (7) points formed by stellating a heptagon.

2. The noun heptagram means a word or character sequence consisting of seven (7) characters.

Heptahedron The noun and adjective heptahedron means a solid figure with seven (7) planar faces.

Heptaheptaconta The prefix heptaheptaconta- means seventy-seven (77).

Heptahexaconta The prefix heptahexaconta- means sixty-seven (67).

Heptaicosa The prefix heptaicosa- means twenty-seven (27).

Heptakaideca The prefix heptakaideca- means seventeen (17).

Heptakaidecad The noun and adjective heptakaidecad means a group of seventeen (17) members.

Heptakaidecagon The noun and adjective heptakaidecagon means a plain figure with seventeen (17) straight sides.

Heptakaidecagram The noun and adjective heptakaidecagram means an interconnected star figure with seventeen (17) points formed by stellating a heptakaidecagon.

Heptakaidecahedron The noun and adjective heptakaidecahedron means a solid figure with seventeen (17) planar faces.

Heptalia The prefix heptalia- means seven thousand (7000).

Heptameter The noun heptameter means a verse consisting of seven (7) metrical feet.

Heptangle The noun heptangle means a plane figure with seven (7) vertices.

Heptanonaconta The prefix heptanonaconta- means ninety-seven (97).

Heptaoctaconta The prefix heptaoctaconta- means eighty-seven (87).

Heptapentaconta The prefix heptapentaconta- means fifty-seven (57).

Heptatetraconta The prefix heptatetraconta- means forty-seven (47).

Heptathlon The noun heptathlon means an athletic competition consisting of seven (7) separate events.

Heptatriaconta The prefix heptatriaconta- means thirty-seven (37).

Heredity A property of a space is hereditary if each of its subspaces possesses this property. Being countable is a hereditary property. Having holes is not.

Hermite normal form A nonsingular integer matrix is in Hermite Normal Form if it is lower triangular, all entries are non-negative, and each row has a unique maximum element which is located on the main diagonal.

(In some definitions, the matrix is required to be upper triangular instead.)

Any nonsingular integer matrix can be transformed to Hermite Normal Form using a series of unimodular column operations:

add an integer multiple of one column to another;

exchange two columns;

multiple any column by –1.

For example, given the matrix A:

5 2 1

–4 2 4

0 –3 6

its Hermite normal form is:

1 0 0

4 6 0

6 15 30

Heron's formula A formula for finding the area of a triangle given the lengths of the three sides.

Hetereoscedastic data Data that have non-constant (heterogeneous) variance across the predicted values of y. In this case, residual graph will show varying variability across the fitted values. This is a regression diagnostic problem and should be fixed by transforming the data.

Heterogeneity of variance A situation in which samples are drawn from populations having different variances.

Heterogeneous subsamples Data in which the sample of observations could be subdivided into two or more distinct sets on the basis of some other variable.

Heuristics A term in computer science that refers to guesses made by a program to obtain approximately accurate results. Frequently used in phylogenetics and computational biology.

Hexa The prefix hexa- means six (6).

Hexaconta The prefix hexaconta- means sixty (60).

Hexacontad The noun and adjective hexacontad means a group of sixty (60) members.

Hexacontadi The prefix hexacontadi- means sixty-two (62). Hexacontadi- is an alternate form of the prefix hexacontakaidi-.

Hexacontaennea The prefix hexacontaennea- means sixty-nine (69). Hexacontaennea- is an alternate form of the prefix hexacontakaiennea-.

Hexacontagon The noun and adjective hexacontagon means a plain figure with sixty (60) straight sides.

Hexacontagram The noun and adjective hexacontagram means an interconnected star figure with sixty (60) points formed by stellating a hexacontagon.

Hexacontahedron The noun and adjective hexacontahedron means a solid figure with sixty (60) planar faces.

Hexacontahena The prefix hexacontahena- means sixty-one (61). Hexacontahena- is an alternate form of the prefix hexacontakaihena-.

Hexacontahepta The prefix hexacontahepta- means sixty-seven (67). Hexacontahepta- is an alternate form of the prefix hexacontakaihepta-.

Hexacontahexa The prefix hexacontahexa- means sixty-six (66). Hexacontahexa- is an alternate form of the prefix hexacontakaihexa-.

Hexacontakaidi The prefix hexacontakaidi- means sixty-two (62).

Hexacontakaidigon The noun and adjective hexacontakaidigon means a plain figure with sixty-two (62) straight sides.

Hexacontakaidigram The noun and adjective hexacontakaidigram means an interconnected star figure with sixty-two (62) points formed by stellating a hexacontakaidigon.

Hexacontakaidihedron The noun and adjective hexacontakaidihedron means a solid figure with sixty-two (62) planar faces.

Hexacontakaidyad The noun and adjective hexacontakaidyad means a group of sixty-two (62) members.

Hexacontakaiennea The prefix hexacontakaiennea- means sixty-nine (69).

Hexacontakaiennead The noun and adjective hexacontakaiennead means a group of sixty-nine (69) members.

Hexacontakaienneagon The noun and adjective hexacontakaienneagon means a plain figure with sixty-nine (69) straight sides.

Hexacontakaienneagram The noun and adjective hexacontakaienneagram means an interconnected star figure with sixty-nine (69) points formed by stellating a hexacontakaienneagon.

Hexacontakaienneahedron The noun and adjective hexacontakaienneahedron means a solid figure with sixty-nine (69) planar faces.

Hexacontakaihena The prefix hexacontakaihena- means sixty-one (61).

Hexacontakaihenad The noun and adjective hexacontakaihenad means a group of sixty-one (61) members.

Hexacontakaihenagon The noun and adjective hexacontakaihenagon means a plain figure with sixty-one (61) straight sides.

Hexacontakaihenagram The noun and adjective hexacontakaihenagram means an interconnected star figure with sixty-one (61) points formed by stellating a hexacontakaihenagon.

Hexacontakaihenahedron The noun and adjective hexacontakaihenahedron means a solid figure with sixty-one (61) planar faces.

Hexacontakaihepta The prefix hexacontakaihepta- means sixty-seven (67).

Hexacontakaiheptad The noun and adjective hexacontakaiheptad means a group of sixty-seven (67) members.

Hexacontakaiheptagon The noun and adjective hexacontakaiheptagon means a plain figure with sixty-seven (67) straight sides.

Hexacontakaiheptagram The noun and adjective hexacontakaiheptagram means an interconnected star figure with sixty-seven (67) points formed by stellating a hexacon-takaiheptagon.

Hexacontakaiheptahedron The noun and adjective hexacontakaiheptahedron means a solid figure with sixty-seven (67) planar faces.

Hexacontakaihexa The prefix hexacontakaihexa- means sixty-six (66).

Hexacontakaihexad The noun and adjective hexacontakaihexad means a group of sixty-six (66) members.

Hexacontakaihexagon The noun and adjective hexacontakaihexagon means a plain figure with sixty-six (66) straight sides.

Hexacontakaihexagram The noun and adjective hexacontakaihexagram means an interconnected star figure with sixty-six (66) points formed by stellating a hexacontakaihexagon.

Hexacontakaihexahedron The noun and adjective hexacontakaihexahedron means a solid figure with sixty-six (66) planar faces.

Hexacontakaiocta The prefix hexacontakaiocta- means sixty-eight (68).

Hexacontakaioctad The noun and adjective hexacontakaioctad means a group of sixty-eight (68) members.

Hexacontakaioctagon The noun and adjective hexacontakaioctagon means a plain figure with sixty-eight (68) straight sides.

Hexacontakaioctagram The noun and adjective hexacontakaioctagram means an interconnected star figure with sixty-eight (68) points formed by stellating a hexacontakaioctagon.

Hexacontakaioctahedron The noun and adjective hexacontakaioctahedron means a solid figure with sixty-eight (68) planar faces.

Hexacontakaipenta The prefix hexacontakaipenta- means sixty-five (65).

Hexacontakaipentad The noun and adjective hexacontakaipentad means a group of sixty-five (65) members.

Hexacontakaipentagon The noun and adjective hexacontakaipentagon means a plain figure with sixty-five (65) straight sides.

Hexacontakaipentagram The noun and adjective hexacontakaipentagram means an interconnected star figure with sixty-five (65) points formed by stellating a hexacontakaipentagon.

Hexacontakaipentahedron The noun and adjective hexacontakaipentahedron means a solid figure with sixty-five (65) planar faces.

Hexacontakaitetra The prefix hexacontakaitetra- means sixty-four (64).

Hexacontakaitetrad The noun and adjective hexacontakaitetrad means a group of sixty-four (64) members.

Hexacontakaitetragon The noun and adjective hexacontakaitetragon means a plain figure with sixty-four (64) straight sides.

Hexacontakaitetragram The noun and adjective hexacontakaitetragram means an interconnected star figure with sixty-four (64) points formed by stellating a hexacontakaitetragon.

Hexacontakaitetrahedron The noun and adjective hexacontakaitetrahedron means a solid figure with sixty-four (64) planar faces.

Hexacontakaitri The prefix hexacontakaitri- means sixty-three (63).

Hexacontakaitriad The noun and adjective hexacontakaitriad means a group of sixty-three (63) members.

Hexacontakaitrigon The noun and adjective hexacontakaitrigon means a plain figure with sixty-three (63) straight sides.

Hexacontakaitrigram The noun and adjective hexacontakaitrigram means an interconnected star figure with sixty-three (63) points formed by stellating a hexacontakaitrigon.

Hexacontakaitrihedron The noun and adjective hexacontakaitrihedron means a solid figure with sixty-three (63) planar faces.

Hexacontaocta The prefix hexacontaocta- means sixty-eight (68). Hexacontaocta- is an alternate form of the prefix hexacontakaiocta-.

Hexacontapenta The prefix hexacontapenta- means sixty-five (65). Hexacontapenta- is an alternate form of the prefix hexacontakaipenta-.

Hexacontatetra The prefix hexacontatetra- means sixty-four (64). Hexacontatetra- is an alternate form of the prefix hexacontakaitetra-.

Hexacontatri The prefix hexacontatri- means sixty-three (63). Hexacontatri- is an alternate form of the prefix hexacontakaitri-.

Hexacta The prefix hexacta- means six hundred (600).

Hexad The noun and adjective hexad means a group of six (6) members.

Hexadeca The prefix hexadeca- means sixteen (16).

Hexadecimal The adjective hexadecimal means based on the number sixteen (16).

A numeral system based on the number sixteen (16) is properly called sexdecimal. The more common term hexadecimal is a bizarre combination of the Greek-based prefix hexakaideca- (16) and the Latin-based term sexdecimal (16) in an apparent attempt to rid mathematics of the evil influence of sex-.

Hexagon A six-sided polygon. The sum of the angles of a hexagon is 720 degrees.

Examples: A regular hexagon:

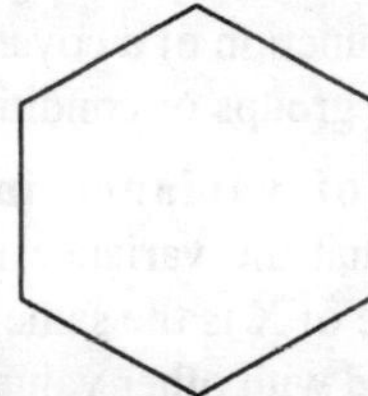

An irregular hexagon:

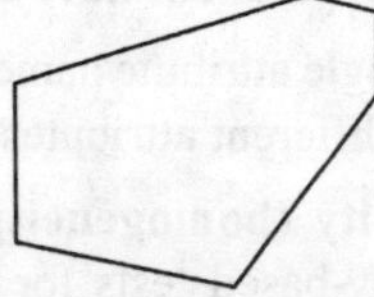

Hexagon The noun and adjective hexagon means a plain figure with six (6) straight sides.

Hexagram 1. The noun and adjective hexagram means an interconnected star figure with six (6) points formed by stellating a hexagon.

2. The noun hexagram means a word or character sequence consisting of six (6) characters.

Hexahedron The noun and adjective hexahedron means a solid figure with six (6) planar faces.

Hexaheptaconta The prefix hexaheptaconta- means seventy-six (76).

Hexahexaconta The prefix hexahexaconta- means sixty-six (66).

Hexaicosa The prefix hexaicosa- means twenty-six (26).

Hexakaideca The prefix hexakaideca- means sixteen (16).

Hexakaidecad The noun and adjective hexakaidecad means a group of sixteen (16) members.

Hexakaidecagon The noun and adjective hexakaidecagon means a plain figure with sixteen (16) straight sides.

Hexakaidecagram The noun and adjective hexakaidecagram means an interconnected star figure with sixteen (16) points formed by stellating a hexakaidecagon.

Hexakaidecahedron The noun and adjective hexakaidecahedron means a solid figure with sixteen (16) planar faces.

Hexalia The prefix hexalia- means six thousand (6000).

Hexameter The noun hexameter means a verse consisting of six (6) metrical feet.

Hexangle The noun hexangle means a plane figure with six (6) vertices.

Hexatriaconta The prefix hexatriaconta- means thirty-six (36).

Hierarchical (sequential) sums of squares Sums of squares in the analysis of variance where later terms in the model are adjusted only for terms that precede them.

Hierarchical log linear model A model in which the presence of an interaction requires the inclusion of any main effects that comprise that interaction.

High level interaction A high-level interaction is an interaction that is an abstraction of lower level interactions, objects, object classes, general interactions constraints, and notes.

High level object class A high-level object class is an object class that contains other object classes, relationship sets, constraints, and notes.

High level relationship set A high-level relationship set is a relationship set that is an abstraction containing other relationship sets, object classes, constraint, and notes.

High level state A high-level state is a state that is an abstraction of lower level states, transitions, constraints, and notes.

High level transition A high-level transition is a transition that is an abstraction of lower level transitions, states, constraints, and notes combined.

High level view A high-level view is the view of an interaction with the internal details omitted.

Higher order interaction The interaction of three or more variables.

Hinge location The location of the hinge in an ordered series.

Hinges (Quartiles) Those points that cut off the bottom and top quarter of a distribution.

Histogram A histogram is a way of graphically showing the characteristics of the distribution of items in a given population or sample. Here is a histogram of the measures of height in the illustrated population. In a histogram each measure is represented by a single block that is placed over the midpoint of the class interval into which the measure falls.

The measures of the heights in inches of the individuals shown below are: 56, 54, 64, 68, 38, 46, 62, 67 and 77.

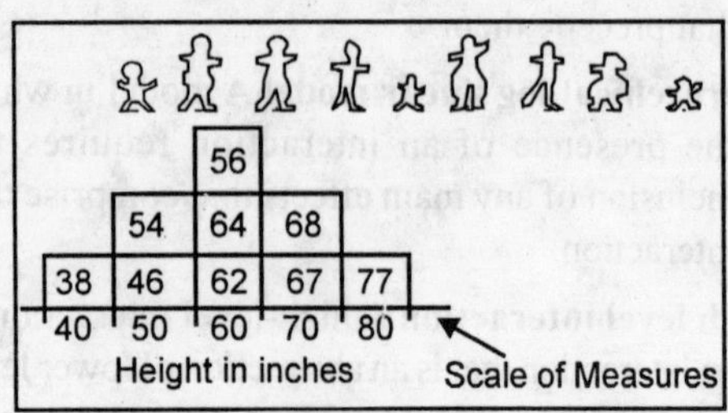

Historical fallacy The mistake of assuming that an association observed in cross-sectional data will be similar to that observed in longitudinal data or vice versa.

Holder norms (p-norms) The Holder norm of a vector, sometimes called the p-norm, is a vector norm of the form $||x||_p$ = (sum (1 <<= i <= n) $|x_i|^p)^{(1/p)}$ where p is any real value greater than or equal to 1.

Although the formula does not make sense, there is also a Holder norm for p equal to infinity; it's simply the maximum of the absolute values of the entries of the vector.

The L1 norm, the L2 norm, and the L Infinity norm are exactly the Hölder norms with p equal to 1, 2 and infinity.

Hole Torus has a hole; sphere does not. If every curve on a surface can be continuously shrunk into a point the surface has no holes. Attaching handles to the sphere in order to create surfaces with holes is an important and ubiquitous topological activity.

Homogeneity of regression The assumption that the regression line expressing the dependent variable as a function of a covariate is constant across several groups or conditions.

Homogeneity of variance in arrays The requirement that the variance in Y associated with one value of X is the same as the variance in Y associated with other values of X.

Homogeneity of variance The situation in which two or more populations have equal variances.

Homonym A single attribute name that is used for two or more different attributes.

Homoscedasticity (homogeneity of variance) Normal-theory-based tests for the equality of population means such as the t-test and analysis of variance, assume that the data come from populations that have the same variance, even if the test rejects the null hypothesis of equality of population means. If this assumption of homogeneity of variance is not met, the statistical test results may not be valid. Heteroscedasticity refers to lack of homogeneity of variances.

Hotelling's T^2 test This is a generalisation of Student's t-test for multivariate data. Designed to provide a global significance test for the difference between two groups with simultaneously measured multiple dependent/outcome variables and multiple explanatory/independent variables. It can also be used for one group with simultaneously measured multiple dependent outcome variables (another test similar to Hotelling's T^2 test is Mahalanobis's D^2 test).

Hour The noun hour means one twenty-fourth (1/24) of a day.

Origin: In antiquity, each night was divided into twelve (12) hours and each daytime was divided into twelve (12) hours. The duration of night hours and daylight hours changed with latitude and the seasons. Modern timekeeping has changed the definition of an hour to a fixed duration equal to thirty-six hundred (3,600) Universal Coordinated Time (UTC) seconds.

Hourly The adjective hourly means recurring once (1) each hour.

Hundred The noun and adjective hundred means multiplied by one hundred (100)).

Hundredth 1. As part of the name of an ordinal number, the noun and adjective hundredth means a multiple of one hundred (100).

2. As part of the name of a rational fraction, the noun and adjective hundredth means divided by one hundred (1/100 = 0.010).

Hundredweight The noun and adjective hundredweight (abbreviated cwt.) means one hundred (100) pounds.

Hyperbola The set of all points in it plane such that the absolute value of the difference of the distances (focal radii) from two given points (foci) is constant.

Hyperbolic trigonometric functions The hyperbolic sine of x, denoted sin x is defined to be $\sinh x = \dfrac{e^x - e^{-x}}{2}$ while the hyperbolic cosine cosh is defined to be $\cosh x = \dfrac{e^x + e^{-x}}{2}$. Note that each is the other's derivative. All of the other hyperbolic trig functions are defined in terms of these two in exactly the same way the other trig functions are defined in terms of the sine and cosine functions.

Hypergeometric distribution A probability distribution of a discrete variable generally associated with sampling from a finite population without replacement. An example may be that given a lot with 25 good units and five faulty. The probability that a sample of five will yield not more than one faulty item follows a hypergeometric distribution.

Hyperspace Multidimensional space beyond the three dimensions that we can easily represent.

Hypotenuse The longest side of a right triangle. It is always located directly across from the right angle.

Hypotenuse The longest side of a right triangle. It is opposite the right angle.

Hypothesis testing A process by which decisions are made concerning the values of parameters.

I

I The Latin lower case letter i represents the imaginary unit, the imaginary number that is the square root of minus one (–1).

I Case An automated systems development environment that provides numerous tools to create diagrams, forms, and reports; provides analysis, reporting, and code generation facilities; and seamlessly shares and integrates data across and between tools.

Icosa The prefix icosa- means twenty (20).

Icosad The noun and adjective icosad means a group of twenty (20) members.

Icosagon The noun and adjective icosagon means a plain figure with twenty (20) straight sides.

Icosagram The noun and adjective icosagram means an interconnected star figure with twenty (20) points formed by stellating a icosagon.

Icosahedron The noun and adjective icosahedron means a solid figure with twenty (20) planar faces.

Icosidi The prefix icosidi- means twenty-two (22). Icosidi- is an alternate form of the prefix icosikaidi-.

Icosiennea 0The prefix icosiennea- means twenty-nine (29). Icosiennea- is an alternate form of the prefix icosikaiennea-.

Icosihena The prefix icosihena- means twenty-one (21). Icosihena- is an alternate form of the prefix icosikaihena-.

Icosihepta The prefix icosihepta- means twenty-seven (27). Icosihepta- is an alternate form of the prefix icosikaihepta-.

Icosihexa The prefix icosihexa- means twenty-six (26). Icosihexa- is an alternate form of the prefix icosikaihexa-.

Icosikaidi The prefix icosikaidi- means twenty-two (22).

Icosikaidigon The noun and adjective icosikaidigon means a plain figure with twenty-two (22) straight sides.

Icosikaidigram The noun and adjective icosikaidigram means an interconnected star figure with twenty-two (22) points formed by stellating a icosikaidigon.

Icosikaidihedron The noun and adjective icosikaidihedron means a solid figure with twenty-two (22) planar faces.

Icosikaidyad The noun and adjective icosikaidyad means a group of twenty-two (22) members.

Icosikaiennea The prefix icosikaiennea- means twenty-nine (29).

Icosikaiennead The noun and adjective icosikaiennead means a group of twenty-nine (29) members.

Icosikaienneagon The noun and adjective icosikaienneagon means a plain figure with twenty-nine (29) straight sides.

Icosikaienneagram The noun and adjective icosikaienneagram means an interconnected star figure with twenty-nine (29) points formed by stellating a icosikaienneagon.

Icosikaienneahedron The noun and adjective icosikaienneahedron means a solid figure with twenty-nine (29) planar faces.

Icosikaihena The prefix icosikaihena- means twenty-one (21).

Icosikaihenad The noun and adjective icosikaihenad means a group of twenty-one (21) members.

Icosikaihenagon The noun and adjective icosikaihenagon means a plain figure with twenty-one (21) straight sides.

Icosikaihenagram The noun and adjective icosikaihenagram means an interconnected star figure with twenty-one (21) points formed by stellating a icosikaihenagon.

Icosikaihenahedron The noun and adjective icosikaihenahedron means a solid figure with twenty-one (21) planar faces.

Icosikaihepta The prefix icosikaihepta- means twenty-seven (27).

Icosikaiheptad The noun and adjective icosikaiheptad means a group of twenty-seven (27) members.

Icosikaiheptagon The noun and adjective icosikaiheptagon means a plain figure with twenty-seven (27) straight sides.

Icosikaiheptagram The noun and adjective icosikaiheptagram means an interconnected star figure with twenty-seven (27) points formed by stellating a icosikaiheptagon.

Icosikaiheptahedron The noun and adjective icosikaiheptahedron means a solid figure with twenty-seven (27) planar faces.

Icosikaihexa The prefix icosikaihexa- means twenty-six (26).

Icosikaihexad The noun and adjective icosikaihexad means a group of twenty-six (26) members.

Icosikaihexagon The noun and adjective icosikaihexagon means a plain figure with twenty-six (26) straight sides.

Icosikaihexagram The noun and adjective icosikaihexagram means an interconnected star figure with twenty-six (26) points formed by stellating a icosikaihexagon.

Icosikaihexahedron The noun and adjective icosikaihexahedron means a solid figure with twenty-six (26) planar faces.

Icosikaiocta The prefix icosikaiocta- means twenty-eight (28).

Icosikaioctad The noun and adjective icosikaioctad means a group of twenty-eight (28) members.

Icosikaioctagon The noun and adjective icosikaioctagon means a plain figure with twenty-eight (28) straight sides.

Icosikaioctagram The noun and adjective icosikaioctagram means an interconnected star figure with twenty-eight (28) points formed by stellating a icosikaioctagon.

Icosikaioctahedron The noun and adjective icosikaioctahedron means a solid figure with twenty-eight (28) planar faces.

Icosikaipenta The prefix icosikaipenta- means twenty-five (25).

Icosikaipentad The noun and adjective icosikaipentad means a group of twenty-five (25) members.

Icosikaipentagon The noun and adjective icosikaipentagon means a plain figure with twenty-five (25) straight sides.

Icosikaipentagram The noun and adjective icosikaipentagram means an interconnected star figure with twenty-five (25) points formed by stellating a icosikaipentagon.

Icosikaipentahedron The noun and adjective icosikaipentahedron means a solid figure with twenty-five (25) planar faces.

Icosikaitetra The prefix icosikaitetra- means twenty-four (24).

Icosikaitetrad The noun and adjective icosikaitetrad means a group of twenty-four (24) members.

Icosikaitetragon The noun and adjective icosikaitetragon means a plain figure with twenty-four (24) straight sides.

Icosikaitetragram The noun and adjective icosikaitetragram means an interconnected star figure with twenty-four (24) points formed by stellating a icosikaitetragon.

Icosikaitetrahedron The noun and adjective icosikaitetrahedron means a solid figure with twenty-four (24) planar faces.

Icosikaitri The prefix icosikaitri- means twenty-three (23).

Icosikaitriad The noun and adjective icosikaitriad means a group of twenty-three (23) members.

Icosikaitrigon The noun and adjective icosikaitrigon means a plain figure with twenty-three (23) straight sides.

Icosikaitrigram The noun and adjective icosikaitrigram means an interconnected star figure with twenty-three (23) points formed by stellating a icosikaitrigon.

Icosikaitrihedron The noun and adjective icosikaitrihedron means a solid figure with twenty-three (23) planar faces.

Icosiocta The prefix icosiocta- means twenty-eight (28). Icosiocta- is an alternate form of the prefix icosikaiocta-.

Icositetra The prefix icositetra- means twenty-four (24). Icositetra- is an alternate form of the prefix icosikaitetra-.

Icositri The prefix icositri- means twenty-three (23). Icositri- is an alternate form of the prefix icosikaitri-.

Ideal elements In Geometry it's often a convenience to introduce a point at infinity as an ideal (probably too good to be real) element. Completion of a metric space by incorporating ideal elements which are limits of Cauchy sequences results in a complete metric space.

Ideal subrings Subrings of a ring are its subsets which are rings in their own right. For a given ring R, its subring I is (an) ideal if for every $r \in R$ and $I \in I$ both $r \cdot i$ and $i \cdot r$ belong to I. For an ideal I it's possible to define a factor ring R/I.

Ideal The word ideal appears in Mathematics in at least a couple of contexts.

Idempotent matrix An idempotent matrix A has the property that

$A * A = A$

An idempotent matrix is not quite a projection matrix.

Simple facts about an idempotent matrix A:

The identity matrix is idempotent;

Every point of the form A*x is a fixed point of A, that is $A * (A * x) = A * x$;

If A is idempotent, and it is not the identity matrix, it must be singular;

The only eigenvalues of A are 0 and 1;

I–A is also idempotent;

rank(A) = trace(A);

I–2*A is involutory.

Identifier A candidate key that has been selected as the unique, identifying characteristic for an entity type.

Identities for negatives Fundamental identities that involve the basic trig functions of negative angles.

Identity An equation made up of trigonometric functions of an angle that is valid for all values of the angle Also called trigonometric identity.

Identity element A number which added to or multiplied times any number in a set yields that same number in the set.

Identity properties There are real numbers 0 and I such that for each real number a, a + 0, 0 + a = a and a x 1 = 1 x a = a.

Ill Conditioned linear system When solving a linear system A * x = b, the matrix A is said to be ill conditioned if small errors or perturbations in the coefficient matrix A or right hand side b correspond to large errors or perturbations in the solution x.

A numerical scale for ill conditioning is provided by the condition number.

Linear systems which are extremely ill conditioned may be impossible to solve accurately. A standard example of an ill conditioned matrix is the Hilbert Matrix, with $A_{i,j} = 1 / (I + J)$.

Imaginary axis An axis in the complex plane.

Imaginary number An imaginary number is a number that can be represented as the square root of a negative real number.

Imaginary unit The imaginary unit (symbol: i) is the imaginary number that is the square root of minus one (–1).

Improper fraction A fraction which is equal to one or more than one.

Incalculable The adjective incalculable means infinite, or more generally, a very large number.

Incenter The point of intersection of the angle bisectors of a triangle.

Inconsistent system of equations A system with no soiution.

Increasing function A function such that for any numbers a and b in the domain. if a > b, then f(a) > f(b).

Incremental commitment A strategy in systems analysis and design in which the project is reviewed after each phase and continuation of the project is rejustified in each of these reviews.

Indefinite integral The indefinite integral of a function f(x)is another function F(x) with the distinguishing feature that the derivative of F(x) is f(x). Not to be confused with the definite integral, which gives a number as an answer. Also called the antiderivative of f(x).

Indefinitely An unspecified amount, having no exact limits.

Independent events Two events A and B are independent if the probability that they happen at the same time is the product of the probabilities that each occurs individually; i.e., if P(A & B) = P(A)P(B). In other words, learning that one event occurs does not give any information about whether the other event occurred too: the conditional probability of A given B is the same as the unconditional probability of A, i.e., P(A/B) = P(A).

Independent events Two events not affected by each other. Two events A and B are independent if and only if P(A and then B) P(A) x P(B).

Independent variable If the value of a depends upon the value of b, or if a is usually defined in terms of b, b is called the independent variable, and it is graphed on the horizontal axis.

Independent variables Those variables controlled by the experimenter.

Index champion A number is an index champion if its index is larger than the index of any smaller number. Also called superabundant numbers. The first few index champions are 1, 2, 4, 6, 12, 24, 36, 48, 60, 120, 180, 240, 360, 720.

Index of summation The variable used in sigma notation.

Index plot An index plot plots each residual, leverage, or Cook's distance against the corresponding observation (row) number (i or index) in the dataset. In many cases, the row number corresponds to the order in which the data were collected. If this is the case, this would be similar to plotting the residuals (or another diagnostic quantity) against time. The index plot is a helpful diagnostic test for normal linear and particularly generalised linear models. Both outliers and influential points can be detected by the index plot. It is particularly useful when the data is in time order so that pattern in the residuals, etc. over time can be detected. If a residual index plot is showing a trend in time, then they are not independent (violation of a major assumption of linear regression).

Index The number represented by n in n?b.

Indifferent condition In a decision table, a condition whose value does not affect which actions are taken for two or more rules.

Indirect measurement A measurement which is found by using a formula or other strategy and not actually measuring something (e.g., finding the height of a tree without actually holding a ruler next to it).

Individual The noun and adjective individual means one (1) particular being or thing as distinguished from a group.

Inductive reasoning A type of type of mathematical reasoning which involves observing patterns and using those observations to make generalizations.

Inequality A mathematical sentence in which the value of the expressions on either side of the relation symbol are unequal. relation symbols include > (greater than), < (less than), >=greater than or equal to),<= less than or equal to), (e.g., x < y, 7 > 3).

Inertia The oddly named inertia of a (square) matrix is the numbers of negative, zero, and positive eigenvalues.

Sylvester's Law of Inertia states that if A and B are congruent matrices, then we cannot guarantee that they have the same eigenvalues, but they do have the same inertias.

This theorem allows us to determine if a symmetric matrix is positive definite. Because the matrix is symmetric, we can compute an LDL factorization:

A = L * D * L'

where L is unit lower triangular, and D is diagonal. This means that A is congruent to the diagonal matrix D. But the eigenvalues of D are easily determined, from which we can get the inertia of D. This is equal to the inertia of A; in particular, if D has only positive eigenvalues, then so does A, which is therefore positive definite.

Moreover, if such a factorization is cheap, as for a tridiagonal symmetric matrix, then we can search for eigenvalues by seeking diagonal shifts of the matrix that cause the number of negative eigenvalues to change by 1.

Inference A conclusion drawn from data.

Inferential statistics That branch of statistics that involves drawing inferences about parameters of the population(s) from which you have sampled.

Infinite geometric series A geometric series with an infinitely large number of terms.

Infinite sequence A sequence that does not have a last term.

Infinitesimal Infinitesimal is a variable whose limit is zero. Development by Abraham Robinson (1960) of the Nonstandard Analysis conferred new significance on infinitesimals and brought them closer to the vision of Leibniz (1646-1716) who introduced the dy/dx notation for the derivative and perceived infinitesimals more like small but constant quantities.

Infinitesimality The noun infinitesimality (symbol: 1/") means a quantity smaller than every real number greater than zero (0).

Infinitesimally The adverb infinitesimally means immeasurably small.

Infinity Infinity pops up in a variety of places. A function may grow to infinity when it's not bounded from above. In Geometry, it's often convenient to think of parallel lines as intersecting at a point at infinity. G. Cantor was the first to introduce and systematically study various kinds of infinities.

Influence A measure of the degree to which an individual data point can influence the obtained value of a regression coefficient.

Influential points Observations that actually dominate a regression analysis (due to high leverage, high residuals or their combination). The method of ordinary least squares gives equal weight to every observation. However, every observation does not have equal impact on the least squares results. The slope, for example, is influenced most by the observations having values of the independent variable farthest from the mean. An observation is influential if deleting it from the dataset would lead to a substantial change in the fit of the generalised linear model. High-leverage points have the potential to dominate a regression analysis but not necessarily exert an influence (i.e., a point may have high leverage but low influence as measured by Cook statistics). Cook statistics is used to determine the influence of a data point on the model.

Informal system The way a system actually works.

Information Data that have been processed and presented in a form suitable for human interpretation, often with the purpose of revealing trends or patterns.

Information center An organizational unit whose mission is to support users in exploiting information technology.

Information systems analysis and design The process of developing and maintaining an information system.

Initial point The beginning point of a vector.

Initial side Side of angle where angle measurement begins.

Initial transition An initial transition is a transition that has no prior state conjunction on it.

Initiator A line-segment or figure that begins as the beginning geometric shape for a fractal. The initiator is then replaced by the generator for the fractal.

Inner fences Points that are 1.5 times the H-spread above and below the appropriate hinge.

Inner product An inner product is a scalar-valued function of two vectors x and y, denoted (x,y), with the properties that:

(x, y) = (y, x), symmetry

(x, s * y) = s * (x, y) for any scalar s, linearity

(x, y + z) = (x, y) + (x, z) additivity

(x, x) >= 0, and equal to zero only if x = 0, positivity.

(x, A * y) = (A' * x, y).

A vector inner product (x,y) can be used to define a corresponding vector norm ||x||:

|| x || = sqrt (x, x).

If the inner product and norm are related in this way, then the Cauchy-Schwarz inequality relates them.

The inner product is sometimes referred to as the dot product (because it is often represented as x dot y), or as the scalar product (because its result is a scalar value).

Innumerable The adjective innumerable means infinite, or more generally, a very large number of individuals or discrete objects.

Innumeracy The noun innumeracy means an incapacity to understand and work with numbers.

Innumerate The adjective innumerate means lacking the capacity to understand and work with numbers.

Innumerous The adjective innumerous means infinite, or more generally, a very large number of individuals or discrete objects.

Input The number or value that is entered, for example, into a function machine. The number that goes into the machine is the input

Inside arc The shortest arc of a turn.

Inspections A testing technique in which participants examine program code for predictable language-specific errors.

Installation The organizational process of changing over from the current information system to a new one.

Intangible benefit A benefit derived from the creation of an information system that cannot be easily measured in dollars or with certainty.

Intangible cost A cost associated with an information system that cannot be easily measured in terms of dollars or with certainty.

Integer number The compound noun integer number means an integer.

Integer This is a very basic notion. To define it rigorously we may need a set of axioms, like those proposed by G.Peano. Usually, integers are numbers from the sequence ..., –2, –1, 0, 1, 2, ..., positive members of which are often called counting or natural numbers. Sometimes, 0 is also judged to be natural.

Integrable A function is integrable if its integral exists. Most of the standard functions are integrable.

Integrand The function inside the integral that is being integrated. Found between the ∫ and the dx.

Integrandstand A stand upon which to put the function inside the integral that is being integrated.

Integration testing The process of bringing together all of the modules that a program comprises for testing purposes. Modules are typically integrated in a top-down, incremental fashion.

Interaction If the effect of one factor depends on the level of another factor, the two factors involved are said to interact, and a contrast involving all these levels is called their interaction. Factors A and B interact if the effect of factor A is not independent of the level of factor B. For example, when there are two main effects on a response variable, if their combined effect is higher than the sum of their main effects due to a bonus (say, the effects of a kind of exercise and a kind of diet on blood lipid levels), they have an interaction (meaning a simple additive model is not sufficient to account for the observed data and a multiplicative term must be added). Briefly, interaction is a deviation from additivity. Also, there would be an interaction between the factors sex and treatment if the effect of treatment was not the same for males and females in a drug trial. Interaction is closely linked with effect modification in epidemiology .

Intercept In linear regression, the intercept is the mean value of the response variable when the explanatory variable takes the value of zero (the value of y when x=0).

Intercept The value of Y when X is 0.

Intercorrelation matrix A matrix (table) showing the pairwise correlations between all variables.

Interface Point of contact where a system meets its environment or where subsystems meet each other.

Internal documentation System documentation that is part of the program source code or is generated at compile time.

Internal validity The degree to which a study is logically sound and free of confounding variables.

Internet A large worldwide network of networks that use a common protocol to communicate with each other; a global computing network to support business-to-consumer electronic commerce.

Interpolation Making deductions from a model for values that lie between data points. Deductions for values beyond the data points is called extrapolation and the results are not valid.

Interquartile range (dQ) dQ is a measure of spread and is the counterpart of the standard deviation for skewed distributions. dQ is the distance between the upper and lower quartiles (Q_U-Q_L).

Interrelated components Dependence of one part of the system on one or more other system parts.

Intersection constraint An intersection constraint is used when a specialization object class has more than one generalization object class and shows that the specialization object class is the intersection of the generalization object classes. In other words, any object in all of the generalizations is also in the specialization.

Intersection of sets The intersection of two or more sets is the set of elements that all the sets have in common; in other words, all the elements contained in every one of the sets. The mathematical symbol for intersection is ∩.

Intersection The term intersect is used when lines, rays, line segments or figures meet, that is, they share a common point. The point they share is called the point of intersection. We say that these figures intersect.

Example: In the diagram below, line AB and line GH intersect at point D

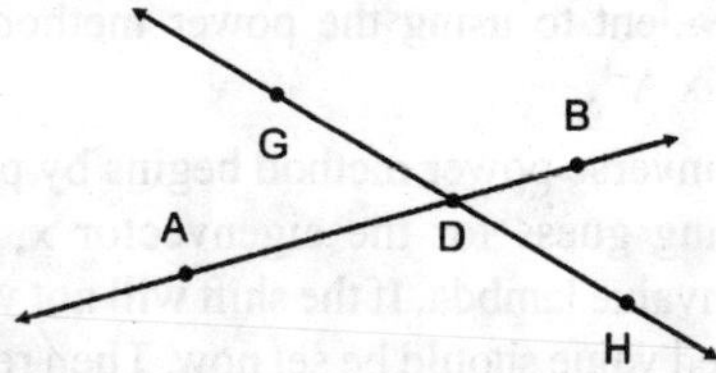

Example: In the diagram below, line 1 intersects the square in points M and N:

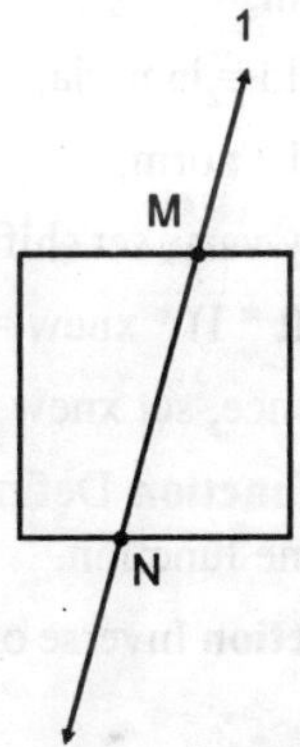

Example: In the diagram below, line 2 intersects the circle at point P:

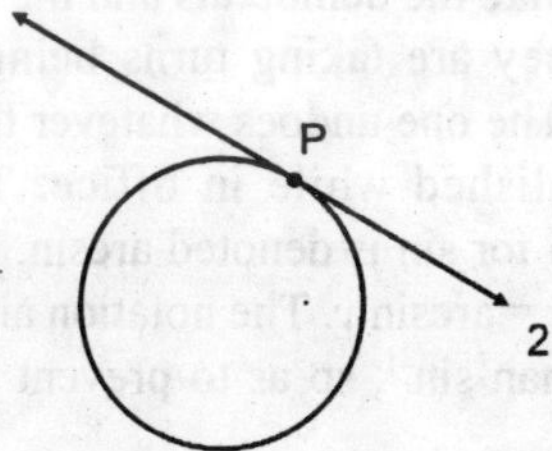

Interval estimate A range of values estimated to include the parameter.

Interval scale Scale on which equal intervals between objects represent equal differences. (differences are meaningful)

Interval variable (equivalent to continuous variable) A quantitative variable measured on a scale with constant intervals (like days, millilitres, kilograms, miles so that equal-sized differences on different parts of the scale are equivalent) where the zero point and unit of measurement are arbitrary. When temperature is measured on two scales, Fahrenheit and Centigrade, the zero points in these two scales do not correspond, and a 10% increase in Fahrenheit (from 50^0 to 55^0) is not a 10% increase in the corresponding Centigrade scale (10° to 12.8^o = 2.8%); these two measurements cannot be mixed or compared. For estimation of correlation coefficients, data should be interval type.

Intraclass correlation A measure of the degree of relationship between two variables. It is usually squared. The variables need not be more than nominal.

Intranet Internet-based communication to support business activities within a single organization.

Inverse A term which means turned upside down. For example the ratio top/bottom is expressed inversely as bottom/top.

Inverse cosine function Inverse of the restricted cosine function.

Inverse cotangent function Defined in terms of the restricted tangent function.

Inverse functions Functions and g are inverse functions if and only if f(g(x)) = g(f(x)) =x for all numbers x in the domains of both f and 9.

Inverse matrix The inverse matrix of a square matrix A, if it exists, is a matrix denoted A^{-1} with the property that

$A * A^{-1} = A^{-1} * A = I.$

If the inverse matrix exists, it is unique, and A is said to be nonsingular or invertible. Otherwise, A is singular.

If the inverse of A exists, then the solution of

$A * x = b$

can be immediately written down:

$x = A^{-1} * b$.

However, it's not a good idea to solve a linear system in this way. The inverse is relatively expensive to compute, and subject to greater inaccuracies than other solution methods. This is not to say that the inverse isn't useful. Orthogonal and unitary transformations are so popular in numerical linear algebra because their inverses come for free; (and their inverses are very well conditioned).

Simple facts:

if lambda is an eigenvalue of A, then 1/lambda is an eigenvalue of A^{-1};

if x is an eigenvector of A, x is also an eigenvector of A^{-1}.

LAPACK and LINPACK include routines for explicitly computing the inverse of a given matrix.

Inverse notation Notation used to express an angle in terms of the value of trigonometric functions.

Inverse of a relation Relation B is the inverse of relation A if it consists of the inverses of the individual ordered pairs in relation A.

Inverse operations Two operations that undo each other (e.g., addition and subtraction).

Inverse power method The inverse power method is a technique for solving the eigenvalue problem.

The inverse power method is related to the power method, but is more flexible. Through the use of shifts, it can be aimed to seek the eigenvalue that is closest to any particular target value. Moreover, once a rough estimate of an eigenvalue is made, the shift can be set to this value, which will increase the convergence rate of the algorithm.

The inverse power method only requires that the user be able to repeatedly solve linear systems of the form:

A * x = b

or, when shifts are used,

(A – shift * I) * x = b.

Thus, for instance, if A is a band matrix or a sparse matrix, the user can employ a storage method and solution algorithm appropriate to the particular form of the problem.

The inverse power method gets its name from the fact that, when the shift is zero, it is equivalent to using the power method on the matrix A^{-1}.

The inverse power method begins by picking a starting guess for the eigenvector x, and the eigenvalue lambda. If the shift will not vary then a fixed value should be set now. Then repeat the following steps as often as necessary:

Set norm = the norm of x;

Set x = x / norm;

Set lambda_Old = lambda;

Set lambda = 1 / norm;

If the shift may vary, set shift = lambda;

Solve (A - shift * I) * xnew = x.

If no convergence, set xnew = x and repeat.

Inverse secant function Defined in terms of the restricted cosine function.

Inverse sine function Inverse of the restricted sine function.

Inverse trigonometric function An inverse trigonometric function is the function that reverses the effect of the original trig function, kind of like the democrats and the republicans, when they are taking turns being elected to power. The one undoes whatever the other had accomplished while in office. The inverse function for sin is denoted arcsin. So if $y = \sin x$, then $x = \arcsin y$. The notation arcsin is used, rather than $\sin^{-1}$, so as to prevent people from confusing the inverse trig function with $\frac{1}{\sin x}$.

Inverse variation Two quantities x and y vary inversely if and only if there is a constant κ. $\kappa \neq 0$, such that $\Psi = K/\Xi$.

Inverse, additive A number when added to a given number yields zero.

Inverse, multiplicative A number when multiplied by a given number yields one.

Involutory Matrix An involutory matrix A has the property that

A * A = I.

Simple facts about an involutory matrix A:

The identity matrix is involutory;

A is invertible, and $A^{-1}=A$;

Irrational number A real number that is not rational, which is to say that it cannot be written as a fraction of two integers. Classic examples include π, e and $\sqrt{2}$. Every irrational number has a decimal representation that is non-repeating. There are tons of these numbers, actually more of these than there are of the rational numbers.

Irrational number A real number that is not a rational number, i.e. cannot be expressed as the quotient of two integers.

Irreducible matrix An irreducible matrix is a (square) matrix which is not reducible.

Definition One: A reducible matrix is one which can be rearranged into the following block form:

(P Q)

(0 R)

where P and R are square sub-blocks, and 0 represents a (nonempty) rectangular block of zero entries. The rearrangement can be done by simultaneous row and column permutations. An irreducible matrix is one which is not reducible.

Definition Two: A matrix is irreducible if and only if, for any row index i and column index j, there is always a nonnegative integer p (which may be 0) and a sequence of integers k1, ..., kp so that the product

$A_{i,k1} * A_{k1,k2} * ... * A_{kp,j}$ is nonzero.

Theorem: A nonnegative matrix A is irreducible if and only if, for any vector x>0 it is the case that A*x>0.

Definition Three: A matrix of order N is irreducible if, for any division of the integers between 1 and N into two disjoint sets K1 and K2, there is always a nonzero element $A_{I,J}$ with I in K1 and J in K2.

The concept of an irreducible matrix is mainly of interest in the analysis of the convergence of certain iterative schemes for linear equations. One key idea is the following: if the matrix A is irreducible and diagonally dominant, then A is nonsingular.

If you only know that the matrix is diagonally dominant, then Gershgorin's theorem would still not rule out an eigenvalue of zero, and hence singularity. It's the irreducibility that guarantees no zero eigenvalue here. The 1, -2, 1 tridiagonal matrix is an example where this theorem applies.

Here's an example of an irreducible matrix with zero main diagonal:

0 1 0 0

1 0 1 0

0 1 0 1

0 0 1 0

If you're familiar with graph theory, then a matrix A is irreducible if and only if the digraph with corresponding adjacency matrix is strongly connected. (Consider node I to be connected to node J if $A_{I,J}$ is nonzero. The digraph is strongly connected if you can get from any node to any other node following directed edges.)

A simple algorithm to determine if a matrix is irreducible can be guided by the idea of the strong connectivity of the corresponding adjacency matrix. First, we choose any node P, and ensure that it can be reached from at least one other node. (We're looking at the P column of the matrix.) If that is not the case, then the matrix is reducible, and we're done. Otherwise, we determine all the nodes we can reach from node P (that's the nonzero entries in row P). Now we take all those nodes, and see what new nodes we can reach. We repeat this operation until no new nodes are found. If any node wasn't reached, the graph is not strongly connected, and the matrix is reducible. Otherwise, the matrix is irreducible.

Irregular fractals Complex fractals whose dimension is often difficult to determine and in some cases is unknown.

Isosceles right triangle A 45° right triangle. There is only one right triangle that has two equal legs - the 45° right triangle.

Isosceles triangle A triangle that has two equal legs and two equal angles.

Item The things or objects that are the subject of a bar graph .

Iteration Repeating a set of rules or steps over and over. One step is called an iterate.

Iterative methods for eigen values A method is called iterative when it consists of a basic series of operations which are carried out over and over again, until the answer that is produced is no longer significantly changing, or some exceptional error occurs, or the limit on the number of steps is exceeded.

All eigenvalue problems are solved by iterative methods, except for the toy problems presented in textbooks. This is because the computation of eigenvalues is equivalent to finding the roots of a polynomial, and there is no explicit method for finding the roots of a general polynomial of degree 5 or higher.

The best known methods for the eigenvalue problem include:

the power method;

the inverse power method;

the Jacobi Algorithm;

the QR method.

Iterative methods for linear equations A method is called iterative when it consists of a basic series of operations which are carried out over and over again, until the answer that is produced is no longer significantly changing, or some exceptional error occurs, or the limit on the number of steps is exceeded.

A direct method is the opposite of an iterative method. A fixed number of operations are carried out once, at the end of which the solution is produced. Gauss elimination on a linear system is an example of such a direct method. Direct methods are the primary method for solving small or dense or nonsymmetric linear systems.

The most common reason for using an iterative method is that it can require far less storage than a direct method. An iterative method typically only sets aside storage for the original nonzero entries of the matrix; no fill in occurs. Standard direct methods must set aside an entry for every possible position in the matrix, though some reduction in this requirement is possible if the matrix is banded.

Secondly, each iteration typically takes much less time than a full direct solve; thus, it is possible, for some problems, that an iterative method will actually converge to an acceptable answer more quickly than a direct method.

An iterative method has numerous disadvantages. You will need a starting point, and if you pick a poor one that may slow down convergence. Your system matrix usually needs to satisfy extra conditions beyond merely being nonsingular. The rate of convergence may be extremely slow, although this can be helped by a suitable preconditioner. If you have a very large problem and are storing the matrix in a compact form, the programming and computational cost involved in storing and retrieving coefficient data can exceed that of the solution phase.

Iterative methods are generally only suitable for certain kinds of system matrices. The most common requirements are that the matrix be positive definite symmetric or strictly diagonally dominant.

Iterative methods for solving systems of linear equations include:

Gauss Seidel Iteration;

the Jacobi Algorithm;

the conjugate gradient method;

Successive Overrelaxation Method (SOR).

Iterative refinement An attempt to improve a computed solution x0 for the linear system of equations A*x=b.

The algorithm is occasionally effective when the coefficient matrix is ill conditioned, a problem that may become evident if the residual error is computed, and seen to be relatively large.

A single step of the algorithm involves computing the residual error:

r := b – A * x,

solving the linear system:

A * dx = r,

and adding the correction dx to the original solution x.

x := x + dx.

In order to achieve any improvement in accuracy, the residual error should be calculated in higher arithmetic precision. The rest of the calculation can be done in single precision, allowing the use of the already computed LU factorization of A, if available. However, to compute the residual error, the original matrix A is needed. It is preferable to figure out how to use the LU factors to compute the residual error, rather than keeping two copies of the matrix, one factored and one untouched.

If the residual error of the new solution is still too large, but somewhat better, then the procedure may be repeated as often as desired.

A subroutine SGEIM for carrying out iterative refinement using the LINPACK routines is described in the LINPACK User's Guide.

J

J The English lower case letter j is an alternate symbol for the imaginary unit used in engineering and technology.

Jacobi algorithm for eigenvalues For computing eigenvalues is an iterative method, which may be applied to a symmetric matrix A, to compute its eigenvalues and eigenvectors.

The method will produce an orthogonal matrix Q and an approximately diagonal matrix lambda such that

A = Q' * lambda * Q.

The steps of the iteration compute a sequence of orthogonal matrices Q1, Q2, ... which transform the original matrix A into matrices A1, A2, ... each of which has the same eigenvalues as A, but each of which is more diagonal than A was. Starting with A, we determine Q1, and produce A1, which is orthogonally similar to A:

A1 = Q1' * A * Q1.

We then determine a matrix Q2 that produces A2:

A2 = Q2' * A1 * Q2 = (Q1 * Q2)' * A * (Q1 * Q2).

The point of the algorithm is how we choose the matrices Q at each step. Q1 is chosen in such a way as to annihilate the (1,2) and (2,1) elements of A. That is, the (1,2) and (2,1) elements of A1 will be zero. Q2 will eliminate the (1,3) and (3,1) elements of A1. Thus A2 will have zeroes in the (1,3) and (3,1) positions. Unfortunately, the (1,2) and (2,1) positions of A2, which we just zeroed out on the previous step, will not remain zero, but will fill in again. However, the fill in values are generally smaller than the original values. As we can sweep through the entire matrix, we repeatedly annihilate the off-diagonal elements until they all have decreased below some tolerance.

It can be shown that the sum of the squares of the off diagonal elements always decreases to zero with the iteration. Thus, for some iteration step M, one can expect to have a matrix AM and a matrix

Q = Q1*Q2*...*QM

so that

AM = Q' * A * Q

where AM is essentially diagonal. At that point, we can rewrite this equation as

A * Q = lambda * Q

where lambda is the diagonal entries of AM, and is the eigenvalues of A, and the transformation matrix Q is the matrix of eigenvectors of A. Thus, if the off diagonal elements disappear as promised, we have approximately solved our eigenvalue problem.

So which orthogonal matrix Q zeroes out a specific pair of entries? The formula is fairly simple. To annihilate the arbitrary entries $A_{i,j}$ and $A_{j,i}$, the matrix Q is equal to the identity matrix, except for:

Q(I,I) = C Q(I,J) = S

Q(J,I) = – S Q(J,J) = C

where C and S are the cosine and sine of some rotation angle THETA. Thus, each matrix Q is a Givens rotation matrix. We can compute C and S directly (skipping the computation of THETA) by the following formula:

U = (A(J,J) – A(I,I)) / (2 * $A_{i,j}$)

T = sign (U) / (| U | + sqrt (U^2 + 1))

C = 1 / sqrt (T^2 + 1)

S = T / sqrt (T^2 + 1)

The Jacobi method is simple and easy to program, but is usually slower than the QR method.

JAD session leader The trained individual who plans and leads Joint Application Design sessions.

Jillion The slang noun and adjective jillion means an very large number. Jillion is a parody of the Chuquet number naming system.

Joint application design (JAD) A structured process in which users, managers, and analysts work together for several days in a series of intensive meetings to specify or review system requirements.

Joint probability The probability of event A and event B happening at the same time is expressed as P(A & B). For independent events A and B, P(A & B)=P(A)P(B). P(A & B) is also known as the probability of intersection of events A and B, from the Venn diagram description.

Jordan canonical form A matrix A is an upper bidiagonal matrix whose main diagonal contains the eigenvalues of A, and whose superdiagonal contains only zeroes or ones. The location of the ones in the superdiagonal is determined by the algebraic and geometric multiplicity of the eigenvalues.

Here is an example of a Jordan canonical form:

4 0 0 0 0

0 2 1 0 0

0 0 2 1 0

0 0 0 2 0

0 0 0 0 3

Every matrix is unitarily similar to its Jordan Canonical Form. That is, for any matrix A, there exists a unitary matrix U so that

A = U^* * J * U

where J has Jordan canonical form. This form can also be regarded as a matrix factorization.

If A is real, but has complex eigenvalues, the matrix J has complex entries.

The Jordan canonical form is of little interest to computational linear algebraists. Unless exact arithmetic is used, it is extremely sensitive to small errors, and the information it provides can be computed more reliably in other ways.

Julia set The set of all the points for a function of the form Z^2+C. The iterations will either approach zero, approach infinity, or get trapped.

Jump A function has a jump at a point where its limits from right and left both exist but do not coincide.

K The metric prefix symbol k– (metric prefix: kilo-) denotes a multiple of one thousand (10^3 = 10 raised to power +03 = 1,000) as defined by Le Système International d'Unités (SI).

Kappa Cohen's measure of agreement based on a contingency table.

Kendall's coefficient of concordance (W) A coefficient of agreement among two or more judges.

Kendall's tau A correlation for ranked data that relies on the number of inversions in the rank order of one variable when the other variable is ranked in order.

Kershaw matrix The Kershaw matrix is a simple 4 by 4 positive definite symmetric matrix, which is an example for which the symmetric version of the incomplete Cholesky factorization gives a negative pivot, causing the algorithm to break down.

The Kershaw matrix:

3 –2 0 2

–2 3 –2 0

0 –2 3 –2

2 0 –2 3

Ki The compound binary prefix symbol Ki- (binary prefix: kibi-) denotes a multiple of two to the tenth power (2^{10} = 1,024) as defined by the International Electrotechnical Commission (IEC).

Kibi The binary prefix kibi- (prefix symbol: Ki-) denotes a multiple of two to the tenth power (2^{10} = 1,024) as defined by the International Electrotechnical Commission (IEC).

Kibin The adjective and noun kibin means one thousand twenty-four (2^{10} = 1024).

Kibinal The adjective kibinal means based on the number one thousand twenty-four (2^{10} = 1024).

Kilia The prefix kilia- means one thousand (1000).

Kilo The metric prefix kilo- (prefix symbol: k-) denotes a multiple of one thousand (10^3 = 10 raised to power +03 = 1,000) as defined by Le Système International d'Unités (SI).

Kilobinary The adjective kilobinary means two to the tenth power (2^{10} = 1,024).

Kolmogorov-smirnov two-sample test A non-parametric test applicable to continuous frequency distributions. It is considered to be the equivalent of the á²-test for quantitative data and has greater power than the G-statistics or á²-test for goodness of fit especially when the sample size is small. It can be used to compare two independent groups. The test is based on differences between two cumulative relative frequency distributions (it compares the distributions not the parameters). Thus, the Kolmogorov-Smirnov test is also sensitive to differences in the general shapes of the distributions in the two samples such as differences in dispersion, skewness. Its interpretation is similar to that of the Wald-Wolfowitz runs test.

Kruskal-wallis test (One-way ANOVA by ranks) It is one of the non-parametric tests equivalent to one-way ANOVA that are used to compare multiple independent samples (another one is the Median test). This test assesses the hypothesis that the different samples in the comparison were drawn from the same distribution or from distributions with the same median. It can be used to analyse ordinal variables. It is an extension of the Mann-Whitney (U) test. The interpretation of the Kruskal-Wallis test is identical to that of one-way ANOVA, but is based on ranks rather than means.

Krylov matrix Given a square matrix A and some initial vector b, the first k elements of the sequence of Krylov vectors are:

$b, Ab, A^2b, A^3b, \ldots A^{k-1}b$;

the corresponding Krylov subspace is the subspace spanned by these vectors; the Krylov matrix is the matrix whose columns are formed by the Krylov vectors.

Kurtosis Kurtosis is a measure of whether the data are peaked or flat in its distribution relative to a normal distribution (whose kurtosis is zero). Positive kurtosis indicates a 'peaked' distribution and negative kurtosis indicates a 'flat' distribution (data sets with high kurtosis have a distinct peak near the mean and decline rapidly; data sets with low kurtosis tend to have a flat top near the mean rather than a sharp peak).

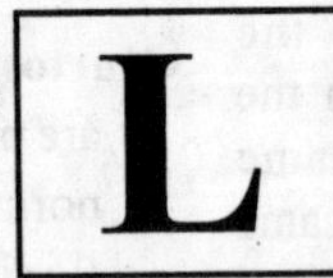

L Infinity matrix norm A matrix norm that is vector-bound to, and hence compatible with, the L Infinity vector norm.

Thus, the formal definition of the norm is

||A|| = max (|| A*x || / ||x||)

where the vector norm used on the right hand side is the L Infinity vector norm, and the maximum is taken over all nonzero vectors x.

However, it is easy to show that the L Infinity matrix norm has a simpler formula: ||A|| = the maximum, over all matrix rows, of the sum of the absolute values of the entries in the row.

L infinity vector norm A vector norm defined as

||x|| = max (1 <= I <= N) |x(i)|.

The L Infinity vector norm is an example of a Hölder norm.

L1 Matrix norm A matrix norm that is vector-bound to, and hence compatible with, the L1 vector norm.

Thus, the formal definition of the norm is

||A|| = max (|| A*x || / ||x||)

where the vector norm used on the right hand side is the L1 vector norm, and the maximum is taken over all nonzero vectors x.

However, it is easy to show that the L1 matrix norm has a simpler formula: ||A|| = the maximum, over all matrix columns, of the sum of the absolute values of the entries in the column.

L1 Vector norm A vector norm defined as

||x|| = sum (1 <= I <= N) |x(i)|.

The L1 vector norm is an example of a Holder norm.

L2 Matrix norm A matrix norm that is vector-bound to, and hence compatible with, the L2 vector norm.

Thus, the formal definition of the norm is

||A|| = max (|| A*x || / ||x||)

where the vector norm used on the right hand side is the L2 vector norm, and the maximum is taken over all nonzero vectors x.

The L2 matrix norm has another formulation: ||A|| = the square root of the maximum absolute value of the eigenvalues of A' * A.

The computation of the L2 norm is expensive, and so it is often simpler to use the easily-computed Frobenius matrix norm, which is not vector-bound to the L2 vector norm, but is compatible with it.

L2 Vector norm A vector norm defined as

||x|| = sqrt (sum (1 <= I <= N) $x(i)^2$).

The L2 vector norm is also known as the Euclidean vector norm or the root-mean-square vector norm.

The L2 vector norm is an example of a Holder norm.

Lakh The adjective and noun lakh means one hundred thousand (100,000 usually punctuated 1,00,000 in India.)

Lapack A set of linear algebra routines, intended as the replacement for LINPACK and EISPACK. It is a project sponsored by the Argonne National Laboratories and the Numerical Algorithms Group (NAG). The LAPACK routines are intended to achieve optimal performance on various machines by calling the level 3 BLAS, which use block methods to achieve high performance.

Large sample effect In large samples, even small or trivial differences can become statistically significant. This should be distinguished from biological/clinical importance.

Latin square design A design which varies the order of presentation of stimuli in such a way as to distribute sequence effects across the design.

Latus chord A segment whose endpoints are on a parabola. which passes through the focus, and is parallel to the directrix.

Law of cosines A relationship between the lengths of the three sides of a triangle and the cosine of one of the angles.

Law of excluded middle One of the two: either a statement is true or it is false. There's nothing in-between. An axiom of 2-valued logic.

Law of sines A relationship between the ratios of the sines of angles of a triangle and the side opposite those angles.

LDL factorization A symmetric matrix is a decomposition of the form:

A = L * D * L',

involving a unit lower triangular matrix L, and a diagonal matrix.

The LDL factorization is a special case of the LU Factorization, in which we give up the option of Pivoting in order to get a very simple factorization. If the matrix A has a zero pivot, the factorization is still valid; we just can't guarantee that we can solve linear systems.

If the matrix A is actually positive definite, then we can get the even stronger Cholesky factorization.

Leading coefficient The coefficient of the term of a polynomial with the highest degree.

Leading digits (most significant digits) Left-most digits of a number.

Least significant difference test (LSD) A technique in which we run t tests between pairs of means only if the analysis of variance was significant. Also known as the protected-t.

Least squares method A method of fitting a straight line or curve based one minimisation of the sum of squared differences (residuals) between the predicted and the observed points. Given the data points (x_i, y_i), it is possible to fit a straight line using a formula, which gives the y=a+bx. The gradient of the straight line b is given by $[?(x_i - m_x)(y_i - m_y)] / [(?(x - m_x))^2]$, where m_x and m_y are the means for x_i and y_i. The intercept a is obtained by $m_y - bm_x$.

Least The superlative adjective least means the smallest non-discrete quantity.

Leaves Horizontal axis of display containing the trailing digits.

Legal and contractual feasibility The process of assessing potential legal and contractual ramifications due to the construction of a system.

Legs of a right triangle The sides of a right triangle other than the hypotenuse.

Leptokurtic A distribution that has relatively more scores in the center and in the tails.

Less The comparative adjective less means a smaller non-discrete quantity.

Lesser The comparative adjective lesser means having a smaller size or quantity.

Level Even, flat, not having any part higher or lower than another part.

Level-0 diagram A data flow diagram that represents a system's major processes, data flows, and data stores at a high level of detail.

Level-n diagram A DFD that is the result of n nested decompositions of a series of subprocesses from a process on a level-0 diagram.

Leverage points In regression analysis, these are the observations that have an extreme value on one or more explanatory variable. The leverage values indicate whether or not X values for a given observation are outlying (far from the main body of the data). A high leverage value indicates that the particular observation is distant from the centre of the X observations. High-leverage points have the potential to dominate a regression analysis but not necessarily influential. If the residual of the same data point and Cook's distance are also high, then it is an influential point.

Leverage The degree to which an observation is unusual with respect to the predictor variables. Similar to an outlier.

Lexical object class A lexical object class is a class whose members have a one-to-one mapping to their representation.

Lightweight graphics The use of small simple images to allow a Web page to be more quickly displayed.

Like terms Terms that contain identical vari- ables with identical exponents.

Likelihood The probability of a set of observations given the value of some parameter or set of parameters. This function, which applies equally to continuous density and discrete mass functions, is the basis of maximum likelihood estimation.

Likelihood distance test (likelihood residual, deletion residual) This test is based on the change in deviance when one observation is excluded from the dataset. It uses the difference between the log-likelihood of the complete dataset and the log-likelihood when a particular observation is removed. A relatively large difference indicates that the observation involved is an outlier (poorly fitted by the model).

likelihood function (or just likelihood) A conditional probability function considered a function of its second argument with its first argument held fixed. For example, imagine pulling a numbered ball with the number k from a bag of n balls, numbered 1 to n. Then you could describe a likelihood function for the random variable N as the probability of getting k given that there are n balls : the likelihood will be 1/n for n greater or equal to k, and 0 for n smaller than k. Unlike a probability distribution function, this likelihood function will not sum up to 1 on the sample space.

Likelihood ratio chi square An alternative procedure for calculating the chi-square statistic—most commonly used in log-linear models

Limit In calculus, a limit is the number you approach as you plug values into a function, and the values get closer and closer to a given number.

Line The equation of a line has two general forms, the point-slope form $y-y_0 = m(x-x_0)$, where (x_0,y_0) is a point on the line and m is the slope of the line) and the slope-y-intercept form $y=mx+b$, where b is the y-coordinate where the line intersects the y-axis and m is the slope).

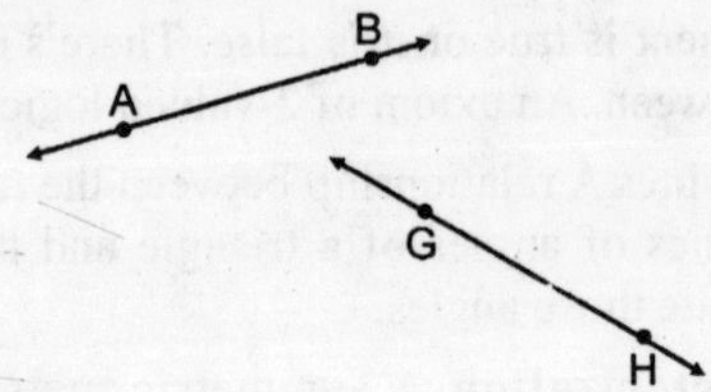

Line graph A graphical representation using points connected by line segments to show how something changes over time.

Line of best fit A line drawn on a scatter plot to estimate the relationship between two sets of data.

Line plot A graph using marks (e.g., x, ?) above a number on a number line to show the frequency of data.

Line segments A line segment is one of the basic terms in geometry. We may think of a line segment as a straight line that we might draw with a ruler on a piece of paper. A line segment does not extend forever, but has two distinct endpoints. We write the name of a line segment with endpoints A and B as line segment AB or as $\overline{AB}$. Note how there are no arrow heads on the line over AB such as when we denote a line or a ray.

Example: The following is a diagram of two line segments: line segment CD and line segment PN, or simply segment CD and segment PN.

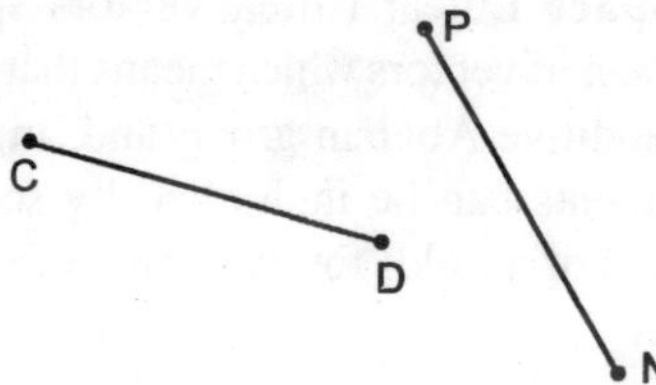

Line symmetry If a figure is divided by a line and both divisions are mirrors of each other, the figure has line symmetry. The line that divides the figure is the line of symmetry.

Linear An equation or graph is linear if the graph of an equation is a straight line.

Linear combination An equation that is formed by adding multiples of two equations together.

Linear contrast A linear combination where the sum of the squared coefficients (weights) is 0.

Linear dependence A set of M vectors, each of order N, is called linearly dependent if there is some linear combination of the vectors, with at least one nonzero coefficient, which equals the zero vector.

If the I-th vector is used as row I of an M by N matrix A, then this is equivalent to saying there is a nonzero vector C such that

$A * C = 0$

If no such combination is possible, then the vectors are linearly independent.

Simple facts:

if any of the vectors is the zero vector, then the set is linearly dependent;

if M is greater than N, the set is linearly dependent;

if M = N, the vectors are linearly independent if and only if the matrix A is nonsingular.

Linear equation An equation in a general form $Ax + By + C = 0$, where A, B and C are constants that are possibly zero.

Linear equation in two variables An equation of the form $y = mx + b$.

Linear expression A polynomial expression with the degree of polynomial being 1. It will be something like, $f(x)=2x^1+3$, but not x^2+2x+4.

Linear function A function is a linear function if and only if its domain is the set of all real numbers and its equation can be written in the form $y = mx + b$.

Linear interpolation A method of approximating values in a table using adjacent table values.

Linear least squares A linear least squares problem seeks an approximate solution x to a linear system $A * x = b$, where x is to be chosen in such a way as to minimize the L2 or Euclidean norm of the residual error $b - A * x$.

A least squares approach is usually tried when the standard algorithms for solving the linear system, such as Gauss Elimination, are not suitable, perhaps because the matrix is square, but singular;

the matrix is square, nonsingular, but with a high condition number;

the matrix is rectangular, with more rows than columns (probably overdetermined);

the matrix is rectangular, with more columns than rows (underdetermined);

In the most common case, there are more equations than unknowns, so that A is actually a matrix with rectangular order of M rows by N columns. Then the system may or may not be consistent, and A itself may perhaps not have the fullest possible rank.

Despite the fact that this problem can't be solved by the usual means, it is still the case that:

a unique exact solution x may exist;

many exact solutions might exist;

there might be no exact solutions, but clearly some vectors x are better than others, in the sense that they produce a smaller residual error b – A * x.

By choosing to use the L2 vector norm, we are using a least-squares approach. Two advantages are that in case 2, there will always exist a unique solution of minimal norm, and in case 3, there are good algorithms for finding the unique approximate solution which has a residual of minimum norm.

If the columns of the coefficient matrix are linearly independent, then a simple approach to solving the problem is to form and solve the normal equations, although this incurs a significant penalty in terms of the condition number.

Other, superior, methods of solving this problem are preferred, usually via the QR factorization or the pseudoinverse. Such methods can also handle the case where the matrix does not have maximal rank, or in which the number of columns is greater than the number of rows.

Linear logistic model A linear logistic model assumes that for each possible set of values for the independent (X) variables, there is a probability p that an event (success) occurs. Then the model is that Y is a linear combination of the values of the X variables: $Y = b_o + b_1*X_1 + b_2*X_2 + b_3*X_3 + \dots b_k*X_k$, where Y is the logit transformation of the probability p. Logistic in statistical usage stems from logit and has nothing to do with the military use of the word which means the provision of material.

Linear polynomial A polynomial of degree 1.

Linear programming A branch of mathematics based on the maximum-minimum property.

Linear regression models In the context of linear statistical modelling, 'linear' means linear in the parameters (coefficients), not the explanatory variables. The explanatory variables can be transformed (say, x^2), but the model will still be linear if the coefficients remain linear. When the overall function (Y) remains a sum of terms that are each an X variable multiplied by a coefficient, the function Y is said to be linear in the coefficients. A non-linear model is different in that it has a non-constant slope.

Linear regression Regression in which the relationship is linear.

Linear relationship A situation in which the best-fitting regression line is a straight line.

Linear space Linear (often Vector) space is a collection of vectors which means that the space is an additive Abelian group and, in addition, its elements can be multiplied by scalars, i.e. elements of a field; for example, a field of real numbers.

Linear term of a quadratic equation The term bx in an equation of the form y = ax^2+ bx + c.

Linear transformation A transformation involving addition, subtraction, multiplication, or division of or by a constant.

Linear velocity Defined in terms of arc length and time.

Linearity of regression The assumption that the best fitting line for a bivariate set of data in linear (straight).

Link function A particular function of the expected value of the response variable used in modelling as a linear combination of the explanatory variables. For example, logistic regression uses a logit link function rather than the raw expected values of the response variable; and Poisson regression uses the log link function. The response variable values are predicted from a linear combination of explanatory variables, which are connected to the response variable via one of these link functions. In the case of the general linear model for a single response variable (a special case of the generalised linear model), the response variable has a normal distribution and the link function is a simple identity function (i.e., the linear combination of values for the predictor variable(s) is not transformed).

Linkage disequilibrium Also called gametic association, which is more appropriate. Increased probability for two or more alleles to be on the same chromosome at the population level. In a population at equilibrium, haplotype frequency is obtained by multiplying the allele frequencies x2. When there is linkage disequilibrium, the observed frequency (say by family analysis or sperm typing) is different from the expected frequency. The difference gives the D value (D for difference), which can be tested for significant difference from 0 by a 2x2 table analysis (for two alleles). The D value can be negative or positive. Linkage disequilibrium can derive from population admixture, tight linkage or elapse of insufficient time for the population to reach equilibrium. A classic example in immunogenetics is the HLA-A1-B8-DR3 haplotype which shows significant linkage disequilibrium extending over 6.5 Mb.

Linpack A standard linear algebra package for factoring matrices, computing matrix determinants, condition numbers, and inverses, and for solving linear systems. Additional capabilities include least squares solutions, QR and singular value decomposition.

Many matrix storage modes are allowed, including dense, banded, symmetric, positive definite, symmetric banded, but LINPACK does not handle sparse matrices, nor does it employ iterative methods.

Here are the routines available for single precision computations. There is a related set available for complex matrices, with names which begin with C instead of S.

SCHDC computes the Cholesky Factorization of a positive definite matrix in general storage.

SCHDD downdates a Cholesky Factorization.

SCHEX updates the Cholesky Factorization of a permuted matrix.

SCHUD updates a Cholesky Factorization.

SGBCO factors a general band matrix and estimates its condition number.

SGBDI computes the determinant of a matrix factored by SGBCO or SGBFA.

SGBFA factors a general band matrix.

SGBSL solves a linear system involving a general band matrix.

SGECO factors a general matrix and estimates its condition number.

SGEDI gets determinant or inverse of a matrix factored by SGECO or SGEFA.

SGEFA factors a general matrix.

SGESL solves a linear system factored by SGECO or SGEFA.

SGTSL solves a linear system, unfactored tridiagonal matrix.

SPBCO factors a positive definite band matrix and estimates its condition number.

SPBDI computes the determinant or inverse of a matrix factored by SPBCO or SPBFA.

SPBFA factors a positive definite band matrix.

SPBSL solves a linear system factored by SPBCO or SPBFA.

SPOCO factors a positive definite matrix and estimates its condition number.

SPODI computes the determinant or inverse of a matrix factored by SPOCO or SPOFA.

SPOFA factors a positive definite matrix.

SPOSL solves a linear system factored by SPOCO or SPOFA.

SPPCO factors a positive definite packed matrix and estimates its condition number.

SPPDI computes the determinant or inverse of a matrix factored by SPPCO or SPPFA.

SPPFA factors a positive definite packed matrix.

SPPSL solves a linear system factored by SPPCO or SPPFA.

SPTSL solves a linear system for a positive definite tridiagonal matrix.

SQRDC computes the QR factorization of a general matrix.

SQRSL solves an overdetermined system, given QR factorization.

SSICO factors a symmetric indefinite matrix and estimates its condition number.

SSIDI computes the determinant or inverse of a matrix factored by SSICO or SSIFA.

SSIFA factors a symmetric indefinite matrix.

SSISL solves a linear system factored by SSIFA or SSICO.

SSPCO factors a symmetric indefinite packed matrix and estimates its condition number.

SSPDI computes the determinant or inverse of a matrix factored by SSPCO or SSPFA.

SSPFA factors a symmetric indefinite packed matrix.

SSPSL solves a linear system factored by SSPFA or SSPCO.

SSVDC computes the singular value decomposition of a general matrix.

STRCO estimates the condition number of a triangular matrix.

STRDI finds the inverse or determinant of a triangular matrix.

STRSL solves a triangular linear system.

Literal equation An equation in which con- stants are represented by letters.

Little The positive adjective little means having a relatively small size or quantity.

LOD score Stands for the logarithm of odds. It is a statistical measure of the likelihood that two genetic markers occur together on the same chromosome and are inherited as a single unit of DNA (co-segregation). The LOD score serves as a test of the null hypothesis of free recombination versus the alternative hypothesis of linkage. Determination of LOD scores requires pedigree analysis and a score of >3 is traditionally taken as evidence for linkage. Linkage is between two genetic loci but not alleles. An example is the linkage between the hemochromatosis gene (HFE) and HLA–A. This means that within the same family all affected subjects will have the same HLA–A allele but not necessarily a particular one, i.e., there will be no recombination between HFE and HLA–A. LOD score has nothing to do with linkage disequilibrium.

Log 0 is undefined. If we need to use a log transformation but some data values are 0, the usual way to get round this problem is to add a small positive quantity (such as 1/2) to all the values before taking logs.

Log linear models Models for handling multiple categorical variables, such as a contingency table with three or more variables.

Log transformation This transformation pulls smaller data values apart and brings the larger data values together and closer to the smaller data values (shrinkage effect). Thus, it is mostly used to shrink highly positively skewed data.

Logarithm The exponent of the power to which a base number must be raised to equal a given number. An example: 2 is the logarithm of 100 to the base 10. One can look at this way: 10 * 10 = 100, which is the same as 10^2, and 2 is the exponent referred to above.

Logarithmic equation An equation involving logarithms of expressions containing variables.

Logarithmic function For all positive real numbers x and b, b ??1, there is a real number v such that v = logbx if and only if x = b^x.

Logistic (binary) regression A statistical analysis most frequently models the relationship between a dichotomous (binary) outcome variable (such as diseased or healthy; dead or alive; relapsed or not relapsed), and a set of explanatory variables of any kind (such as age, HLA type, blood pressure, kind of treatment, disease stage etc). It can also be used when the outcome variable is polytomous (several categories of the prognosis; including ordinal response 'ordinal logistic regression' or 'proportional odds ratio model'), and when there are several outcome variables (multinomial logistic regression - a special class of loglinear models). Analysis of data from case-control studies via logistic regression can proceed in the same way as cohort studies.

Logistic regression A variant of standard regression used when the dependent variable is a dichotomy, such as success/failure.

Logit transformation The logit (or logistic) transformation Y of a probability p of an event is the logarithm of the ratio between the probability that the event occurs (p) and that the event does not occur (1–p): Y = ln (p/(1–p)). Thus it is a transformation of a binary (dichotomous) response variable. The logit transformation of p is also known as the log odds of p, since it is the logarithm of the odds. There are other link functions for binary response variables.

Loglinear model Multinomial data (from contingency tables) can be fitted by using a generalised linear model with a Poisson response distribution and a log link function. The resulting models for counts in the cells of a contingency table are known as loglinear models in which the logarithm of the expected value of a count variable is modelled as a linear function of parameters. Loglinear models try to model all important associations among all variables. In this respect, they are related to ANOVA models for quantitative data. Loglinear modelling allows more than two discrete variables to be analysed and interactions between any combination of them to be identified. Associations between variables in log-linear models are analogous to interactions in ANOVA models. The aim is to find a small model, which achieves a reasonable fit (small residual). Data sets with a binary response (outcome) variable and a set of explanatory variables that are all categorical can be modelled either by logistic regression or by loglinear modelling.

In loglinear modelling, the counts in the cells of the contingency table are treated as values of the response variable rather than either of the categorical variables defining the rows and columns. A loglinear model relates the distribution of the counts to the values of the explanatory variables (rows and columns), and tests for the presence of an interaction between them. Whether the interaction term is necessary is tested by fitting two models, one without and one with the interaction term, and using the difference between the regression deviance values for the two models (as well as the difference in df). If the SP obtained by the ?²-test is small, the interaction term should be in the model. This corresponds to a significant difference in Fisher's exact test or the ?²-test. For a 2x2 table, the difference between the regression deviances in the two models is the same as the residual deviance in the model with no interaction term. In loglinear modelling, it makes no sense if any main effect is omitted.

In loglinear modelling (as in other GLMs), the saturated model, which includes all interactions, fits the data exactly, and the fitted values are exactly equal to the observed values (the residual deviance is zero). The residual deviance of any other model can be used to test how worse it is compared to the saturated model. A small SP arising from a high residual deviance would mean that it does not fit better than the saturated model (the alternative model is rejected). If the associated SP is not small, then the model is not significantly worse than the saturated model (i.e., the exact fit).

An important constraint on the choice of models to fit, i.e., which terms and which interactions to include, is that whether any marginal totals (row or column) for combinations of terms are fixed in advance. If this applies to any combination of terms, their interaction must be included (the main effects and this particular interaction make up the null model for such data). When only the overall sample size is fixed in advance, there is no constraint on the models that can be fit.

When there are more than two variables, whether any interaction can be omitted from the model can be tested by fitting a model containing all the interactions of a particular order (say, second-order interactions). Then, new models omitting each interaction are fitted. Each new model's regression deviance is compared with the regression deviance of the model containing all interactions. If the resulting difference in regression deviances (and df) results in a large SP, that interaction can be omitted (it is not significantly different but has fewer terms; or it does not significantly contribute to the model).

The loglinear model can be used to find a conditional probability involving the factors in the contingency table for one of the factors chosen as a response variable. When the factor chosen as a response variable is binomial, logistic regression can be used to analyse the same data (a binomial response variable and categorical explanatory variables). Logistic regression substitutes the loglinear model with equal results as long as the fitted loglinear model includes the main effects of the response variable and a saturated model for all the explanatory variables.

Log-rank test (of equality across strata) A non-parametric test of significance for the difference between two survival curves (for categorical data). It is a special application of the Mantel-Haenszel test. It can be adjusted for confounders (not preferable to Cox proportional hazard regression which is a semi-parametric model), or performed for trend.

Longitudinal data Data collected over a period of time as in cohort studies. These data are usually analysed by using survival analysis techniques.

Lower CASE CASE tools designed to support the systems implementation and operation phase of the systems development life cycle.

Lowest term fraction A fraction which can not be reduced to a lower term.

Lowest terms (simplest form) A fraction is expressed in lowest terms when its numerator and denominator have no common factors other than 1 or – 1.

LU factorization A matrix is a decomposition of the form:

A = P * L * U,

Involving a permutation matrix P, a unit lower triangular matrix L, and an upper triangular matrix U.

P records the pivoting operations carried out in Gauss elimination, L the row multipliers used during elimination and U contains the pivot values and other information.

The factors are typically computed via Gauss elimination. Once the factors are computed, they may be used to solve one or more linear systems A*x=b;

solve one or more transposed linear systems A'*x=b;

compute the inverse;

compute the determinant.

The LU factorization is generally computed only for a nonsingular square matrix, but the LU factorization exists and is useful even if the matrix is singular, or rectangular.

While the LU factorization is usually quite satisfactory, it is possible to prefer a factorization of the form

A = P * L * U * Q,

by using Gaussian elimination with complete pivoting, in which case Q is a permutation matrix selecting the variable (or column) to be eliminated at each step. Another rival factorization is the QR factorization

A = Q *R,

which is slightly more expensive, but has better stability and accuracy properties.

LAPACK and LINPACK include routines to compute the LU factorization of a matrix stored in a variety of formats. Note that the P, L and U factors themselves are scrambled and compressed in the data storage used by these routines, so that it's difficult to determine their actual values.

M The metric prefix symbol M- (metric prefix: mega-) denotes a multiple of one million (10^6 = 10 raised to power +06 = 1,000,000) as defined by Le Système International d'Unités (SI).

Magnitude of a vector The length of the directional line segment.

Magnitude of effect A measure of the degree to which variability among observations can be attributed to treatments.

Main diagonal of a square matrix The diagonal from the upper left to the lower right.

Main diagonal The diagonal cells of a matrix from upper left to lower right.

Main effect The effect of one independent variable averaged across the levels of the other independent variable(s).

Maintenance Changes made to a system to fix or enhance its functionality.

Major axis The longer axis of an ellipse.

Majority The noun majority means more than one half of the members of a group. For example, a majority of 25 is 13 to 25 and a majority of 26 is 14 to 26.

Mann-whitney (U) test A non-parametric test for comparing the distribution of a continuous variable between two independent groups. It is analogous to the independent two-sample t-test, so that it can be used when the data are ordinal or not normally distributed. The Wilcoxon signed ranks T-test for independent samples is another non-parametric alternative to the t-test in this context (for paired samples, Wilcoxon matched pairs signed rank test should be used).

Many The noun and positive adjective many means a relatively large number of individuals or discrete objects.

The adjectives many/more/most are used with count nouns and collection nouns (integers), while the adjectives much/more/most are used with mass nouns (non-discrete quantities).

Example: Many students feel they spend too much money on books.

Map "Map" is another word used for a function. It could be used as in " This is a map from the reals to the reals," translation," This is a function that takes a real number and turns it into another real number."

Mapping An association from the elements of the domain to the elements of the range.

Marginal distribution Given two jointly distributed random variables X and Y, the marginal distribution of X is simply the probability distribution of X ignoring information about Y.

Marginal probability The probability of an event, ignoring any information about other events. The marginal probability of A is written P(A). Contrast with conditional probability.

Marginal totals Totals for the levels of one variable summed across the levels of the other variable.

Mark twain The archaic expression mark twain was used by Nineteenth Century American rivermen to denote a water depth of two (2) fathoms (approximately 3.7 meters.) American writer Samuel Langhorne Clemens (1835-1910) used Mark Twain as his pseudonym.

Matched samples An experimental design in which the same subject is observed under more than one treatment.

Matched samples t test A t test comparing the means of matched (or repeated) samples.

Mathematical induction A method of mathematical proof.

Matrices Rectangular arrays of numbers arranged in rows and columns.

Matrix algebra Algebra in which you work with matrices of elements or variables instead of individual elements.

Matrix exponential A square matrix A is a matrix B(A,t) = exp (A * t), which has properties similar to those of the exponential function of a scalar argument.

In particular: B(A,0) = I, the identity matrix; d B(A,t)/dt = A * B(A,t);

Lambda is an eigenvalue of A if and only if exp (Lambda * t) is an eigenvalue of B(A,t).

B(A,t) is never singular, for any values of A or t;

B(A,t) = sum (I = 0 to Infinity) (A * t)^I / I!;

Inverse (B(A,t)) = B(A,–t);

If A = M * J * Inverse(M) is the Jordan canonical form of A, then B(A,t) = M * exp (J*t) * Inverse(M). If J is diagonal, then exp(J*t) is a diagonal matrix whose entries are the exponentials of J(I,I)*t;

If the matrix A can be is orthogonally diagonalizable, so that

A = Q * Lambda * Q',

then, using the power series definition of B(A,t) results in:

B(A,t) = Q * sum (I = 0 to Infinity) (Lambda * t)^I / I! * Q';

or

B(A,t) = Q * exp (Lambda, t) * Q',

where exp (Lambda, t) is the diagonal matrix whose I-th diagonal entry is exp(lambda(i)), the exponential of the I-th eigenvalue of A.

In the general case where A cannot be orthogonally diagonalized, or where the eigenvalues and eigenvectors cannot be reliably computed, the computation of the matrix exponential is difficult.

Matrix factorization The process of rewriting a matrix A as a product of factors with certain special properties. Matrix factorization is a key technique in solving linear systems, determining eigenvalues, and many other tasks.

Useful matrix factorizations include:

The Cholesky Factorization incomplete Cholesky factorization

Incomplete LU Factorization

Jordan_Canonical_Form

The LDL Factorization

The LU Factorization

The Polar Decomposition

The QR Factorization

Row Echelon Form

The Schur Decomposition

The Singular Value Decomposition

Matrix matrices A rectangular array of numbers arranged in Tows and columns.

Matrix multiplication The computation of the matrix-vector product A * x or the matrix-matrix product A * B, where A and B are (possibly rectangular) matrices, and x is a vector.

A matrix product is only meaningful when the factors are conformable. This is a condition on the dimensions of the factors. If we are computing A * x or A * B, then the column order of A must equal the order of x or the row order of B.

To multiply the L by M array A times the M by N array B, the product C will have L rows and N columns, and a typical entry $C_{i,j}$ is:

$C_{i,j}$ = sum (K = 1 to M) $A_{i,k}$ * $B_{k,j}$.

Matrix multiplication is not commutative. A*B and B*A are not even of the same order unless both matrices are square. Even if they are square, the two products will general have different numerical values.

Matrix multiplication is associative: (A*B)* C=A*(B*C). While the result is not affected by the order of operations, the amount of work can change, if the matrices are of different orders. For products of a large number of matrices, techniques of dynamic programming are required to determine the most efficient order in which to carry out the operations.

As a computational task, matrix multiplication is relatively expensive. To multiply two matrices of order N takes roughly $2*N^3$ floating point operations, which is more expensive than Gauss elimination of a matrix of order N. On vector and parallel machines, it is important to write a multiplication algorithm carefully, to take advantage of the potential for speedup.

Matrix order The order of a square matrix is the number of rows and columns. Thus a matrix of order 5 is a square matrix with 5 rows and 5 columns.

The order of a rectangular matrix is described by giving both the number of rows and the number of columns. Thus a rectangular matrix might have order 5 by 4.

The order of a matrix refers to a property of the mathematical object, which does not depend on how the information is stored in a computer. The actual numbers representing the matrix may be stored in a rectangular two dimensional array, whose row and column lengths are equal to or greater than the mathematical orders, a rectangular array of lesser size (for band matrix storage, say), or even in a collection of several separate one dimensional arrays.

Matrix properties Any features of a matrix which may be of use in choosing a storage scheme or algorithm, or analyzing convergence or error properties, but which are not immediately evident from the simple arrangement of zeroes in the matrix, or any symmetries among the nonzero elements.

Matrix properties include:

diagonal dominance;

invertibility;

irreducibility;

M matrix;

normality;

orthogonality;

positivity;

positive definiteness;

property A;

rank;

stochasticness;

Matrix rank The rank of a matrix is a measure of the linear independence of its rows and columns.

The row rank of a matrix of order M by N is the number of linearly independent rows in the matrix, while the column rank is the number of linearly independent columns. The row rank will be between 0 and M, the column rank between 0 and N. If the row rank is equal to M, the matrix is said to have maximal or full row rank; there are corresponding terms for column rank.

For a square matrix, of order N, the row and column ranks will be equal. A square matrix is nonsingular if and only if it has maximal row rank; in other words, if no row is a linear combination of the other rows, and similarly for columns. A square matrix with full rank has an inverse, a nonzero determinant, and Gauss elimination with pivoting can be used to solve linear systems involving the matrix.

Every singular matrix is almost nonsingular. That is, using any matrix norm you like, and no matter how small you specify the (positive) tolerance epsilon, there is a nonsingular matrix closer than epsilon to the singular matrix. Another way to look at this is to realize that it is always possible, by making tiny changes to the entries of a singular matrix, to turn it into a nonsingular matrix. (Consider the zero matrix; add epsilons to the diagonal entries, and it's nonsingular.) Thus, given the roundoff implicit in computation, the determination of matrix rank is not a reliable process.

If the rank of a matrix is actually desired, a reasonable method is to compute the QR factorization. Very small diagonal terms in the R factor may indicate linear dependence of the corresponding columns of the matrix. The singular value decomposition will also give this sort of information.

Matrix splitting A matrix splitting is a decomposition of the system matrix:

A = M – N

in order to analyze the behavior of an iterative method for solving the linear system A*x = b. The M matrix is the multiplier of the next iterate, and the N matrix is the multiplier of the current iterate.

For instance, consider the Jacobi iteration. If we decompose the matrix into its strictly lower triangular, diagonal, and strictly upper triangular parts, we can write:

A = L + D + U

The Jacobi iteration can be written as:

D * xnew = – (L + U) * x – b

Hence, in the language of matrix splittings,

M = D

N = (L + U).

The matrix splitting gives us a convenient way of expressing the iteration matrix, which is the multiplier of the current iterate in the (explicit) formula for the next one. In terms of the matrix splitting, this iteration matrix always has the form M^{-1} * N. For instance, for the Jacobi iteration, the explicit formula for xnew is:

xnew = – D^{-1} * (L + U) * x – D^{-1} * b

and so the iteration matrix is D^{-1} * (L + U).

Matrix square root The square root of a matrix A is a matrix X with the property that

A = X * X.

For a matrix to have a square root, it must be symmetric. If a matrix is positive semidefinite, it is guaranteed to have a square root. The square root can be determined from the Cholesky factorization as follows:

A = L * L'

Now determine the Singular Value Decomposition of L

L = U * D * V'

and write

A = L * L'
= U * D * V' * (U * D * V')'
= U * D * V' * V * D * U'
= U * D * D * U'
= U * D * U' * U * D * U'
= X * X

where X is the desired matrix square root of A. In particular:

X = U * D * U'

Similarly, if B^3=A, we say that B is the cube root or third root of A, and for any integer N, B^N=A means that B is called an N-th root of A.

Matrix storage Matrix storage formats are the schemes for storing the values of a mathematical matrix into memory locations in a computer, or, in some cases, for writing this information to an external file.

A special storage format may be chosen because of the overall matrix structure of zero entries, or because of the matrix symmetry involving nonzero entries.

Common matrix storage formats include: Band Matrix Storage

General Matrix Storage

HBSMC Finite Element Matrix Storage

HBSMC Sparse Matrix Storage

Skyline Matrix Storage

Sparse Matrix Storage

Symmetric Matrix Storage

In addition, many linear programming problems are stored in files that are formatted according to the MPS Storage Format.

Matrix structure Matrix structure The classification of matrices according to the pattern of its zero entries.

If a matrix has a known structure, it may be possible to use a specialized storage scheme that takes less space, or an algorithm that can execute more quickly.

A matrix about which nothing is known, or which exhibits no special pattern, may be called full or dense or general.

Matrix structure patterns that can occur include:

band;

bidiagonal;

block;

diagonal;

Hessenberg;

rectangular;

sparse;

square;

trapezoidal;

triangular;

tridiagonal.

Maxima The greatest value in a data set or the greatest value of a function.

Maximum likelihood This method is a general method of finding estimated values of parameters. It yields values for the unknown parameters, which maximise the probability of obtaining the observed values. The estimation process involves considering the observed data values as constants and the parameter to be estimated as a variable, and then using differentiation to find the value of the parameter that maximises the likelihood function. First a likelihood function is set up which expresses the probability of the observed data as a function of the unknown parameters. The maximum likelihood estimators of these parameters are chosen to be those values, which maximise this function. The resulting estimators are those, which agree most closely with the observed data. This method works best for large samples, where it tends to produce estimators with the smallest possible variance. The maximum likelihood estimators are often biased in small samples .Another method for point estimation is the method of moments.

Maximum value The Y-coordinate of the vertex of the graph of a quadratic function that opens downward.

McNemar's test A special form of the Chi-squared test used in the analysis of paired (not independent) proportions. This non-parametric test compares two correlated dichotomous responses and finds its most frequent use in situations where the same sample is used to find out the agreement (concordance) of two diagnostic tests or difference (discordance) between two treatments. If the pairs of data points are the measurements on two matched people (such as affected and unaffected siblings) in a case-control study, or two measurements on the same person, the appropriate test for equality of proportions is the McNemar's test. It can be used to assess the outcome of two treatments applied to the same individuals or the significance of the agreement between two detection methods of a physical sign. If there are more than two periods of data collection (such as pretest, posttest and follow-up), Cochran's Q test should be used.

Mean (or average) A measure of location for a batch of data values; the sum of all data values divided by the number of elements in the distribution. Its accompanying measure of spread is usually the standard deviation. The mean is one of several indices of central tendency that statisticians use to indicate the point on the scale of measures where the population is centered.

The mean is the average of the scores in the population. Numerically, it equals the sum of the scores divided by the number of scores. It is of interest that the mean is the one value which, if substituted for every score in a population, would yield the same sum as the original scores, and hence it would yield the same mean.

This illustration shows a population of 9 persons. It also shows the height of each person as well as the formula and calculations of the mean of the population. As is customary, we have used the Greek letter mu to represent the mean of a population.

54 77 67 68 46 64 62 56 38

Height (inches)

$$\mu = \frac{\sum X}{N} = \frac{54+77+67+68+46+64+62+58+38}{9} = 59.11''$$

$\mu = 59.11''$

0 10 20 30 40 50 60 70 80 90 100 110

Scale of Measures (inches)

Mean absolute deviation The average absolute deviation from the mean for a set of numbers.

Mean proportional The geometric mean between two numbers.

Mean square In an ANOVA, the term Mean Square refers to an estimate of the population variance based on the variability among a given set of measures.

In a one way ANOVA, the Within Mean Square is an estimate of the population variance based on the average of all s-square within the several samples.

In a one way ANOVA the Between Mean Square is an estimate of the population variance based on the s-square of the sample means multiplied by n (the size of the samples).

Mean time between failures (MTBF) A measurement of error occurrences that can be tracked over time to indicate the quality of a system.

Measurement The assignment of numbers to objects.

Measurement data Data obtained by measuring objects or events.

Measurement type The data may be measured in the following scales: nominal, ordinal, interval or ratio scales (known as Stevens' typology). The scale of the measurements may be (other than nominal scale measurements) either continuous or discrete, and either bounded or unbounded.

Measures of association These measures include the Phi coefficient of association, Cramer's contingency coefficient (C) and V, Kendall's tau-b and (Stuart's) tau-c, Somers' D (a modification of Kendall's tau-b), Yule's Q, gamma, Spearman's rank correlation coefficient (rho), Pearson's correlation coefficient, lambda (symmetric and asymmetric), uncertainty coefficients (symmetric and asymmetric), Guttman's coefficient of predictability (lambda)/ Goodman-Kruskal's lambda, Goodman-Kruskal's gamma and Goodman-Kruskal's tau (concentration coefficient).

Measures of central tendency Numbers which tend to cluster around the middle of a set of values. three such numbers are mean, median, and mode.

Measures of location Another term for measures of central tendency.

Measuring fractions The fractions used with English units of measurement when measuring distance.

Mebi The binary prefix mebi- (prefix symbol: Mi-) denotes a multiple of two to the twentieth power (2^{20} = 1,048,576) as defined by the International Electrotechnical Commission (IEC).

Mebin The adjective and noun mebin means two to the twentieth power (2^{20} = 1,048,576).

Median Another measure of location just like the mean. The value that divides the frequency distribution in half when all data values are listed in order. It is insensitive to small numbers of extreme scores in a distribution. Therefore, it is the preferred measure of central tendency for a skewed distribution (in which the mean would be biased) and is usually paired with the interquartile range (dQ) as the accompanying measure of spread.

Median location The location of the median in an ordered series.

Median test This is a crude version of the Kruskal-Wallis ANOVA in that it assesses the difference in samples in terms of a contingency table. The number of cases in each sample that fall above or below the common median is counted and the Chi-square value for the resulting 2xk samples contingency table is calculated. Under the null hypothesis (all samples come from populations with identical medians), approximately 50% of all cases in each sample are expected to fall above (or below) the common median. The median test is particularly useful when the scale contains artificial limits, and many cases fall at either extreme of the scale. In this case, the median test is the most appropriate method for comparing samples.

Mega The metric prefix mega- (prefix symbol: M-) denotes a multiple of one million (10^6 = 10 raised to power +06 = 1,000,000) as defined by Le Système International d'Unités (SI).

Megabinary The adjective megabinary means two to the twentieth power (2^{20} = 1,048,576). Membership condition A membership condition is a constraint imposed on the object class by the ORM. To be a member of an object class, an object must satisfy the membership conditions for that class. Over time, an object may be a member of different object classes. This is referred to as class migration.

Mersenne number A number which is one less than a power of two; 2^n – 1 for a positive integer n.

Mersenne numbers are named for Father Marin Mersenne, a 17th century French mathematician, philosopher and theologian who published a remarkable (but incorrect) statement about which of these numbers are prime.

Mersenne prime A Mersenne number which is prime. In order for 2^n – 1 to be prime, n itself must be prime.

Only 40 Mersenne primes are known; they include the very largest known primes.

Mesokurtic A distribution with a neutral degree of kurtosis.

Meta analysis A systematic approach yielding an overall answer by analysing a set of studies that address a related question. This approach is best suited to questions, which remain unanswered after a series of studies. Meta-analysis provides a weighted average of the measure of effect (such as odds ratio). The rationale is to increase the power by analysing the sets of data. The selection of studies to include in a meta-analysis study is the main problem with this approach. Funnel Plot is an informal method to assess the effect of publication bias in this context.

Metric completeness A metric space is complete if all Cauchy sequences of its elements converge.

Metric measurement Measurements which use metric units of measurement.

Metric space A space X is metric if there is defined a real non-negative function (metric, or metric function) of two variables d(A, B). The function is known as the distance between the two points. It's characterized by the following properties. For $A \in X$, $B \in X$, and $C \in X$

d(A, B) = 0 if and only if A = B. (the distance is 0 if and only if the points coincide)

d(A, B) = d(B, A) (the distance from A to B is the same as the distance from B to A)

$d(A,B)+d(B,C) \geq d(A,C)$(the sum of two sides of a triangle is never less than the third side.)

There are many examples of distance functions.

Metropolis-hastings algorithms These algorithms are a class of Markov chains which are commonly used to perform large scale calculations and simulations in Physics and Statistics.

Mi The compound binary prefix symbol Mi- (binary prefix: mebi-) denotes a multiple of two to the twentieth power (2^{20} = 1,048,576) as defined by the International Electrotechnical Commission (IEC).

Micro The metric prefix micro- (prefix symbol: µ- {Greek lower case letter mu}) denotes a multiple of one one-millionth (10^{-6} = 10 raised to power –06 = 0.000 001) as defined by Le Système International d'Unités (SI).

Midpoint formula If the coordinates of point A are (x1,y1) and the coordinates of point B are (x2,y2), then the coordinates of the midpoint of segment AB are {(x1+x2)/2, (y1+y2)/2}.

Midpoint The point on a segment that is equidistant from the endpoints.

Mil The noun mil means divided by one thousand (1/1,000 = 0.001). In thickness measurement, a mil is equal to one one-thousandth of an inch or 25.400 micrometers. In angular measurement, a mil is equal to one sixty-four-hundredth (1/6,400) of a circle (or roughly one one-thousandth of a radian.)

Mile The noun mile literally means one thousand (1,000) paces. An international mile is equal to 5,280 feet or 1,609.344 meters. A nautical mile is equal to 1,852 meters or approximately 6,076 feet.

Mill The noun mill means divided by one thousand (1/1,000 = 0.001). In currency, a mill is equal to one one-thousandth of a dollar ($0.001).

Mille The numeric prefix mille- means one thousand (1,000). This prefix is derived from the Latin cardinal number mille (m or M).

Millenary 1. The noun and adjective millenary means a group of one thousand (1000).

2. The noun and adjective millenary means a period of one thousand (1000) years.

3. The noun and adjective millenary means a one thousand (1000) year anniversary.

4. The adjective and noun millenary means one thousandth (1000th) in rank or order.

Millennial 1. The adjective and noun millennial means recurring every one thousand (1000) years.

2. The adjective and noun millennial means lasting one thousand (1000) years.

3. The noun and adjective millennial means a one thousand (1000) year anniversary.

Millennium The noun and adjective millennium means a period of one thousand (1000) years.

Millesimal The adjective millesimal means based on the number one thousand (1000). The SI (Le Système International d'Unités) metric system employs millesimal prefixes.

Milli 1. The numeric prefix milli- means one thousand (1,000). This prefix is derived from the Latin cardinal number mille (m or M).

2. The metric prefix milli- (prefix symbol: m-) denotes a multiple of one one-thousandth (10^{-3} = 10 raised to power –03 = 0.001) as defined by Le Système International d'Unités (SI).

Milliard The noun and adjective milliard means multiplied by ten to the ninth power (10^9 = 10 raised to power +09 = 1,000,000,000) in the modified Chuquet number naming system.

Million The noun and adjective million means multiplied by one million (1,000,000).

Millionth 1. As part of the name of an ordinal number, the noun and adjective millionth means a multiple of one million (1,000,000).

2. As part of the name of a rational fraction, the noun and adjective millionth means divided by one million (1/1,000,000 = 0.000 001).

Millipede The noun millipede literally means having one thousand (1,000) feet. A millipede is an insect of the class Diplopoda.

Minima Look, we just explained maxima. Do we have to do everything around here?

Minimum value The y-coordinate of the vertex of the graph of a quadratic function that opens upward.

Minor axis The shorter axis of an ellipse.

Minor of an element The determinant that remains after the row and column containing the element have been deleted.

Minority The noun minority means less than one half of the members of a group. For example, a minority of 25 is 0 to 12 and a minority of 26 is also 0 to 12. Minority can also mean an under-represented portion of a population.

Minus 1. The preposition minus (symbol: " {minus sign}) denotes either the operation of subtraction or the operation of unary negation.

2. The adjective and noun minus (prefix symbol: " {minus sign}) means negative or less than zero (<0).

The symbol " {minus sign} is pronounced as minus. The minus sign is used as a prefix to denote a negative number. As a mathematical operator, the minus sign represents either the operation of subtraction or the operation of unary negation. The minus sign is also used to denote a negative polarity or direction.

Minute The noun minute means divided by sixty (1/60). In time measurement, a minute is one sixtieth (1/60) of an hour or one one-thousand-four-hundred-fortieth (1/1,440) of a day. In angular measurement, a minute (suffix symbol: -2 {prime}) is one sixtieth (1/60) of a degree (suffix symbol: -° {degree symbol}) or one twenty-one-thousand-six-hundredth (1/21,600) of a circle.

Miter To cut material at an angle.

Mixed model designs Anova designs with one or more between subjects factors and one or more repeated measures factors. Also refers to designs with both fixed and random independent variables.

Mixed numbers Numbers that have both whole numbers and decimals, such as 4.567.

Modality The term used to refer to the number of major peaks in a distribution.

Mode The mode is one of several measures of central tendency that statisticians use to indicate the point (or points) on the scale of measures where the population is centered. It is the score in the population that occurs most frequently. Please notice that the mode is not the frequency of the most numerous score. It is the value of that score itself.

Also notice that if there are two (or more) different scores that occur with equal frequency and that frequency is higher than the frequency of any of the other scores, the population is described as multi-modal.

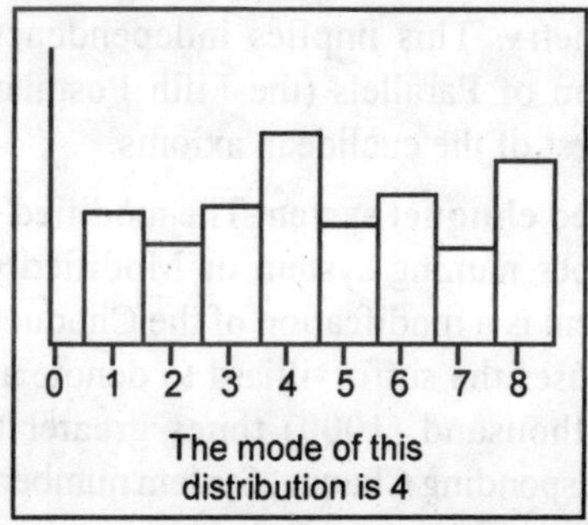

The mode of this distribution is 4

Model building The traditional approach to statistical model building is to find the most parsimonious model that still explains the data. The more variables included in a model (overfitting), the more likely it becomes mathematically unstable, the greater the estimated standard errors become, and the more dependent the model becomes on the observed data. Choosing the most adequate and minimal number of explanatory variables helps to find out the main sources of influence on the response variable, and increases the predictive ability of the model. Ideally, there should be more than 10 observations for each variable in the model. The usual procedures used in variable selection in regression analysis are: univariate analysis of each variable (using ?² test), stepwise method (backward or forward elimination of variables; using the deviance difference), and best subsets selection. Once the essential main effects are chosen, interactions should be considered next. As in all model building situations in biostatistics, biological considerations should play a role in variable selection.

Model Kepler and Newton laws are models of the real world. These are very good models due to their predictive power. Mathematical biologists derive equations (create models) of population growth and spread of epidemics. Mathematical economist create models that describe some aspects of and help study real world economy. Modeling being so important, there is of course the Model Theory. It's a general theory of interpretations of axiomatic systems. A hyperbolic (non-euclidean) geometry can be modeled in a circle of the regular euclidean geometry. This implies independence of the Axiom of Parallels (the Fifth Postulate) from the rest of the euclidean axioms.

Modified chuquet system The modified Chuquet number naming system or Modified Chuquet System is a modifcation of the Chuquet System that uses the suffix -illiard to denote a number one thousand (1000) times greater than the corresponding Chuquet System number with the -illion suffix.

Example: In the Modified Chuquet System, one billiard = $10^{(6\times2)+3} = 10^{12+3} = 10^{15} = 10$ raised to power +15 = 1,000,000,000,000,000.

Modify interaction A modify interaction is an interaction that update some data in object.

Modular arithmetic A method for finding remainders where all the possible numbers (the numbers less than the divisor) are put in a circle, and then by counting around the circle the number of times of the number being divided, the remainder will be the final number landed on.

Modularity Dividing a system into chunks or modules of equal size.

Modulus A unit of measure. For example, when measuring days, a modulus could be 24 for the number of hours in a day. 75 hours would be divided by 24 to give 3 remainder 3, or 3 days and 3 hours.

Modulus of a complex number Same as absolute value of a complex number.

Monad The noun and adjective monad means a group with only one (1) member.

Monarch The noun monarch means one (1) person who rules alone.

Monarchy The noun monarchy means rule by only one (1) person.

Mono The prefix mono- means one (1).

Monogamic The adjective monogamic means having a single (1) marriage or exclusive sexual relationship between only two (2) individuals.

Monogamous The adjective monogamous means having a single (1) marriage or exclusive sexual relationship between only two (2) individuals.

Monogamously The adverb monogamously means maintaining a single (1) marriage or exclusive sexual relationship between only two (2) individuals.

Monogamy The noun monogamy means a single (1) marriage or exclusive sexual relationship between only two (2) individuals.

Monogram 1. The noun and verb monogram means a graphic design composed of the first (1st) character of one or more (most commonly three (3)) words or parts of a name.

2. The noun monogram means a word consisting of only one (1) character.

Monograph The noun and adjective monograph means a scholarly book or treatise on one (1) specific subject.

Monomial A constant (a number), a variable, or a product of constants and/or variables.

Monotonic relationship A relationship represented by a regression line that is continually increasing (or decreasing), but perhaps not in a straight line.

Monte carlo trial Studying a complex relationship difficult to solve by mathematical analysis by means of computer simulations.

Monthly The adjective monthly means recurring once (1) each month.

More The comparative adjective more means a larger number or quantity.

Most The superlative adjective most means the largest number or quantity.

$MS_{between\ groups}$ (MS_{group}) Variability among group means.

MS_{within} (MS_{error}) Variability among subjects in the same treatment group.

Much The noun and positive adjective much means a relatively large non-discrete quantity.

Multi The prefix multi- means having more than one (>1).

Multicategory data A situation in which data can be sorted into more that two categories.

Multicolinearity In multiple regression, two or more X variables are colinear if they show strong linear relationships. This makes estimation of regression coefficients impossible. It can also produce unexpectedly large estimated standard errors for the coefficients of the X variables involved. This is why an exploratory analysis of the data should be first done to see if any collinearity among explanatory variables exists. Multicolinearity is suggested by non-significant results in individual tests on the regression coefficients for important explanatory (predictor) variables. Multicolinearity may make the determination of the main predictor variable having an effect on the outcome difficult.

Multinomial distribution The distribution in which each of a number of independent trials results in one of two or more mutually exclusive outcomes.

Multiperfect number (MPFN) A number whose index is an integer; equivalently, a number such that the sum of its divisors is a multiple of the number. If the index is two, the number is perfect; if it is greater than two, the number is proper multiperfect. Often multiperfect is used to mean proper multiperfect. MPFNs have also been called multiply perfect, pluperfect, k-fold perfect or k-ply perfect. In French, multiparfait; in German mehrfach vollkommen.

The only MPFN with index one is one itself. The first few MPFNs are 1, 6, 28, 120, 496, 672, 8128, 30240, 32760, 523776, 2178540, ...

Multiple 1. The adjective multiple means having more than one (>1).

2. The noun multiple means the product of a given number times an integer.

Multiple comparison techniques Techniques for making comparisons between two or more group means subsequent to an analysis of variance.

Multiple correlation coefficient ($R_{0.123..p}$) The correlation between one variable (Y) and a set of p predictors.

Multiple regression To quantify the relationship between several independent (explanatory) variables and a dependent (outcome) variable. The coefficients (?, $?_1$ to $?_i$) are estimated by the least squares method, which is equivalent to maximum likelihood estimation. A multiple regression model is built upon three major assumptions:

1. The response variable is normally distributed,

2. The residual variance does not vary for small and large fitted values (constant variance),

3. The observations (explanatory variables) are independent.

Multiple regression is the prototype for general linear models because the response variable should be normally distributed and there is no link function, whereas, simple linear regression is a special case for generalised linear models. The extension of multiple regression to multivariate data analysis is called canonical correlation .

Multiple regression correlation coefficient (R^2 – R-squared) R^2 is a measure of the proportion of variability explained by, or due to the regression (linear relationship) in a sample of data (equal to the percentage of variance accounted for). It is a measure of the effect of X in reducing the uncertainty in predicting Y. The R^2 statistic is given by R^2 = s.s. (regression) / s.s. (total) [i.e., ESS/TSS] where s.s. is the sum of squares. It is also called the coefficient of determination. When all observations fall on the fitted regression line, then s.s. (residual) = 0; and ESS=TSS; thus, $R^2 = 1$. When the fitted regression line is horizontal (s.s = 0), then s.s. (regression) = 0 and $R^2 = 0$. The adjusted R^2_a is different in that it takes into account the sample size and the number of explanatory variables. The fit of a model should never be judged from the R^2 value even if the statistical significance is achieved (for example, a high value may arise when the relationship between the two variables is non-linear; or a small R^2 may be significant due to large sample effect). The square root of the R^2 is the correlation coefficient (r) and has the same sign (minus or plus) as the slope. The r value does not have a clear-cut interpretation as the R^2 in linear regression. The R^2 and r are simply descriptive measures of the degree of linear association between X and Y in the sample observations .

Multiples The product of multiplying a number by a whole number. For example, multiples of 5 are 10, 15, 20, or any number that can be evenly divided by 5.

Multiplicand A quantity multiplied by another. A product is obtained by multiplying two or more multiplicands.

Multiplication rule The probability of two or more statistically independent events all occurring is equal to the product of their individual probabilities. If the probability of having trisomy 21 is a, and trisomy 6 is b, assuming independence of these two events, for a baby the probability of having both trisomies is (a x b). One of the most critical errors of judgement in the use of independence assumption relates to a court case in the UK (Watkins, 2000).

Multiplication The operation by which the product of two quantities is calculated. To multiply a number b by c is to add b to itself c times.

Multiplicative A function f on the positive integers is called multiplicative if f(rs) = f(r) f(s) whenever r and s are relatively prime. It follows that f is determined by the values it takes on prime powers: if n = 2^a 3^b 5^c ..., then f(n) = f(2^a) f(3^b) f(5^c) ...

Multiplicative identity matrix A square matrix with I's on the main diagonal (upper left to lower right) and other entries 0.

Multiplicative inverse Two numbers are multiplicative inverses of each other if their product is 1 (e.g., since 4 x 1/4 = 1, 1/4 and 4 are multiplicative inverses).

Multiplicative law of probability The rule giving the probability of the joint occurrence of independent events.

Multiplicity Indicates how many objects participate in a given relationship.

Multivalued attribute An attribute that may take on more than one value for each entity instance.

Multivariate analysis Methods to deal with more than one related 'outcome/dependent variable' (like two outcome measures from the same individual) simultaneously with adjustment for multiple confounding variables (covariates). When there is more than one dependent variable, it is inappropriate to do a series of univariate tests. Hotelling's T^2 test is used when there are two groups (like cases and controls) with multiple dependent measures (may be more than two), and multivariate analysis of variance (MANOVA) is used for more than two groups. Wilks' lambda is the most common parameter used to quantitate the significance of the result of a multivariate procedure (others are Roy's largest criterion, Hotelling-Lawley trace, and Pillai-Bartlett trace). Unfortunately, the word 'multivariate' is most frequently used instead of 'multivariable' analysis (which means multiple independent/explanatory variables but one outcome/dependent variable). Canonical correlation, cluster analysis, factor analysis (FA) and principal component analysis (PCA) are types of multivariate analysis. PCA aims at reducing a large set of variables to a small set that still contains most of the information in the large set. While PCA summarises or approximates the data using fewer dimensions (to visualise it, for example), FA provides an explanatory model for the correlations among the data.

Multivariate analysis of variance (MANOVA) An extension of Hotelling's T^2 test to more than two groups with related 'multiple' outcome measures. Groups are compared on all variables simultaneously as a global test (rather than one-by-one as ANOVA does). .

Multivariate normal distribution A generalization of the normal distribution to the joint distribution of two or more variables.

Multivariate outliers Observations that are outliers in some multivariate space.

Multivariate procedures Procedures that deal with two or more dependent variables simultaneously.

Mutual exclusion constraint The mutual exclusion constraint indicates that the specialization object classes are pairwise disjoint. In other words, no object can be a member of more than one specialization object class.

Mutual independence A collection of events is mutually independent if for any subset of the collection, the joint probability of all events occurring is equal to the product of the joint probabilities of the individual events. Think of the result of a series of coin-flips. This is a stronger condition than pairwise independence.

Mutually exclusive events Two events A and B are mutually exclusive if and only if they have no outcomes in common.

Mutually exclusive Two events are mutually exclusive when the occurrence of one precludes the occurrence of the other.

Myria The prefix myria- means ten thousand (10000).

Myriad The adjective and noun myriad derives from the Greek cardinal number myriad meaning ten thousand ($10^4 = 10{,}000$). Myriad can more generally mean a large number.

Myriamyriad The adjective and noun myriamyriad means one hundred million ($10^8 = 100{,}000{,}000$).

N factorial (!) N x (N–1) x (N–2) x (N–3) x...x 1

N The metric prefix symbol n– (metric prefix: nano-) denotes a multiple of ten to the negative ninth power (10^{-9} = 10 raised to power –09 = 0.000 000 001) as defined by Le Système International d'Unités (SI).

Nano The metric prefix nano- (prefix symbol: n-) denotes a multiple of ten to the negative ninth power (10^{-9} = 10 raised to power –09 = 0.000 000 001) as defined by Le Système International d'Unités (SI).

Napier's constant Napier's Constant, also known as the natural logarithmic base, (symbol: e) is the base of the natural, hyperbolic, or Naperian logarithms. Napier's Constant is a transcendental number approximately 2.718 281 828 459. Scottish mathematician John Napier (1550-1617) published the first description of logarithms in 1614.

Narrowly The adverb narrowly means only slightly more than a required minimum.

Nary The archaic adjective nary means not any or zero (0). Nary is most commonly used with an indefinite article, as in Nary a pence could be had.

Natural (raw) residuals The difference between the observed (Y_i) and fitted values (Y_i-hat) of the response variable in regression. To obtain the standardised residual, this value is divided by its estimated standard deviation. Other residual types are Pearson residual, deviance residual and deletion residual. In all cases, a relatively large residual indicates an observation poorly fitted by the model. The one that is most closely related to maximum likelihood estimation (i.e., generalised linear models) is the standardised deviance residual (which is equal to the standardised Pearson residual in normal linear models).

Natural logarithm The natural logarithm, usually denoted ln, is the logarithm to the base e. Also, a method of birth control used at tree farms.

Natural logarithmic base The natural logarithmic base, also known as Napier's Constant, (symbol: e) is the base of the natural, hyperbolic, or Naperian logarithms. The natural logarithmic base is a transcendental number approximately 2.718 281 828 459. Scottish mathematician John Napier (1550-1617) published the first description of logarithms in 1614.

Natural logarithms Base-e logarithms.

Natural number A natural number is a finite positive integer, e.g., 1, 2, 3, 4, 5, etc.

Nea The nonstandard prefix nea- denotes a multiple of ten to the twenty-seventh power (10^{27} = 10 raised to power +27 = 1,000,000,000,000,000, 000,000,000,000).

Nearly The adverb nearly means slightly less than the specified amount.

Needs The unknowns of a formula, or the measurements needed to do a calculation.

Negative 1. The adjective and noun negative means less than zero (<0). A negative quantity is denoted with the prefix symbol " {minus sign}. Negative has the opposite polarity, direction, or meaning of positive. In physics, the electric charge of an electron is (arbitrarily) designated as negative.

2. The adjective and noun negative is a logical value. Negative and affirmative or negative and positive are sometimes used as binary logic pairs.

Negative angle Results from clockwise rotation.

Negative exponents For each nonzero number a and for each positive integer n, a^–n = 1/a^n.

Negative numbers Numbers less than zero. In graphing, numbers to the left of zero. Negative numbers are represented by placing a minus sign (–) in front of the number.

Negative predictive value Probability of a true negative as in a person identified healthy by a test is really free from the disease .

Negative reciprocals Two numbers having a product of – 1.

Negative relationship A relationship in which increases in one variable are associated with decreases in the other.

Negative square root function The function with the equation y= –?x.

Negative square root The negative of the principal square root.

Negatively skewed A distribution that trails off to the left.

Neighborhood (Set theoretical) Topology starts with a definition of open sets which are often and interchangeably called neighborhoods. By definition, the empty set is open and so are finite intersections and arbitrary unions of open sets. A set is a topological space if some of its subsets are declared to be open subject to these conditions. An open set is a neighborhood of all its points. A function f:A–>B from one topological space into another is continuous at a point a $\in$ A if for every neighborhood V of f(a) there exists a neighborhood U of a such that f(U) $\subset$ V. This is equivalent to saying that inverse images of sets open in B are open in A. Complements of open sets are closed by definition.

Network A set of points (vertices) connected by arcs.

Newman keuls test A multiple comparison procedure for making pairwise comparisons among means.

Nibble In computer terminology, the noun nibble means a set of four (4) binary units having two to the fourth power (2^4 = 16) possible states. A single sexdecimal digit can represent the status of a nibble. A nibble is equal to four (4) bits, two (2) crumbs, or half (1/2) a byte.

Nine The noun, pronoun, and adjective nine (9) is the standard name for the cardinal number that follows eight (8). Nine corresponds with the ordinal number ninth (9th).

Nineteen The noun, pronoun, and adjective nineteen (19) is the standard name for the cardinal number that follows eighteen (18). Nineteen corresponds with the ordinal number nineteenth (19th).

Nineteenth hole The compound noun nineteenth (19th) hole means the club house of a golf course.

Nineties 1. The plural noun and adjective nineties (90s) means an integer multiple of ninety (90).

2. The plural noun and adjective nineties (90's) means the range of numbers at least ninety (90) but less than one hundred (100). When referring to dates, nineties means a span of ten years in which the tens (10s) digit of the year number is nine (9).

Ninety The noun, pronoun, and adjective ninety (90) is the standard name for the cardinal number that follows eighty-nine (89). Ninety corresponds with the ordinal number ninetieth (90th).

Ninety-eight The noun and adjective ninety-eight (98) is the standard name for the cardinal number that follows ninety-seven (97). Ninety-eight corresponds with the ordinal number ninety-eighth (98th).

Ninety-eighth The compound adjective, adverb, and noun ninety-eighth (98th) is the standard name for the ordinal number that follows ninety-seventh (97th). Ninety-eighth corresponds with the cardinal number ninety-eight (98).

Ninety-five The noun and adjective ninety-five (95) is the standard name for the cardinal number that follows ninety-four (94). Ninety-five corresponds with the ordinal number ninety-fifth (95th).

Ninety-four The noun and adjective ninety-four (94) is the standard name for the cardinal number that follows ninety-three (93). Ninety-four corresponds with the ordinal number ninety-fourth (94th).

Ninety-nine The noun and adjective ninety-nine (99) is the standard name for the cardinal number that follows ninety-eight (98). Ninety-nine corresponds with the ordinal number ninety-ninth (99th).

Ninety-one The noun and adjective ninety-one (91) is the standard name for the cardinal number that follows ninety (90). Ninety-one corresponds with the ordinal number ninety-first (91st).

Ninety-second The compound adjective, adverb, and noun ninety-second (92nd or 92^{d}) is the standard name for the ordinal number that follows ninety-first (91st). Ninety-second corresponds with the cardinal number ninety-two (92).

Ninety-seven The noun and adjective ninety-seven (97) is the standard name for the cardinal number that follows ninety-six (96). Ninety-seven corresponds with the ordinal number ninety-seventh (97th).

Ninety-six The noun and adjective ninety-six (96) is the standard name for the cardinal number that follows ninety-five (95). Ninety-six corresponds with the ordinal number ninety-sixth (96th).

Ninety-three The noun and adjective ninety-three (93) is the standard name for the cardinal number that follows ninety-two (92). Ninety-three corresponds with the ordinal number ninety-third (93rd or 93^{d}).

Ninety-two The noun and adjective ninety-two (92) is the standard name for the cardinal number that follows ninety-one (91). Ninety-two corresponds with the ordinal number ninety-second (92nd or 92^{d}).

Ninth 1. The adjective, adverb, and noun ninth (9th) is the standard name for the ordinal number that follows eighth (8th). Ninth corresponds with the cardinal number nine (9).

2. The adjective and noun ninth (1/9) means divided by nine (9).

Nix The slang noun nix means zero (0) or nothing. Nix derives from the German word nichts meaning nothing.

No one The compound pronoun no one means no person or zero (0).

No The adjective, adverb, and noun no is a logical value. No and yes are often used as a binary logic pair.

Nobody The pronoun nobody means no person or zero (0). Nobody is a contraction of no body.

Nominal scale Numbers used only to distinguish among objects.

Nominal variable A qualitative variable defined by mutually exclusive unordered categories such as blood groups, races, sex etc.

Non equivalent groups design A design in which the experimental groups differ on one or more important variables at the start of the experiment.

Nona The numeric prefix nona- means nine (9). This prefix is derived from the Latin ordinal number nonus (VIIII).

Nonaconta The prefix nonaconta- means ninety (90).

Nonacta The prefix nonacta- means nine hundred (900).

Nonadeca The prefix nonadeca- means nineteen (19).

Nonagenarian The noun and adjective nonagenarian means between ninety (90) and one hundred (100) years of age.

Nonagesimal The adjective nonagesimal means based on the number ninety (90).

Naginta The numeric prefix nonaginta- means ninety (90). This prefix is derived from the Latin cardinal number nonaginta (LXXXX).

Nonagintennary The noun and adjective nonagintennary means a group of ninety (90) or a period of ninety (90) years.

Nonagintennial 1. The adjective and noun nonagintennial means recurring every ninety (90) years.

2. The adjective and noun nonagintennial means lasting ninety (90) years.

3. The noun and adjective nonagintennial means a ninety (90) year anniversary.

Nonagintennium The noun and adjective nonagintennium means a period of ninety (90) years.

Nonagintillion The noun and adjective nonagintillion means multiplied by 10^{273} in the American number naming system, but means multiplied by 10^{540} in the Chuquet number naming system.

Nonagon A nine-sided polygon. The sum of the angles of a nonagon is 1260 degrees.

Examples: A regular nonagon:

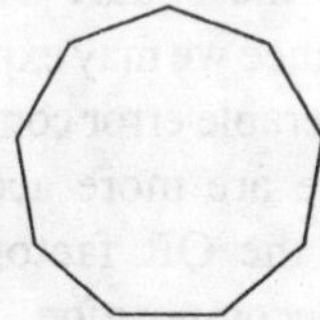

An irregular nonagon:

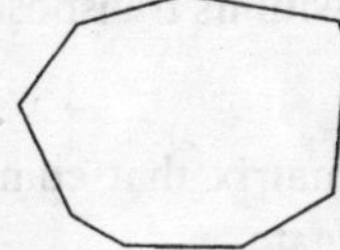

Nonagon The noun and adjective nonagon means a plane figure with nine (9) straight sides.

Nonaheptaconta The prefix nonaheptaconta- means seventy-nine (79).

Nonahexaconta The prefix nonahexaconta- means sixty-nine (69).

Nonaicosa The prefix nonaicosa- means twenty-nine (29).

Nonalia The prefix nonalia- means nine thousand (9000).

Nonanonaconta The prefix nonanonaconta- means ninety-nine (99).

Nonaoctaconta The prefix nonaoctaconta- means eighty-nine (89).

Nonapentaconta The prefix nonapentaconta- means fifty-nine (59).

Nonary 1. The adjective and noun nonary means ninth (9th) in order or importance.

2. The adjective nonary means having nine (9) parts.

Nonatetraconta The prefix nonatetraconta- means forty-nine (49).

Nonatriaconta The prefix nonatriaconta- means thirty-nine (39).

Noncentrality parameter The degree to which the mean of the sampling distribution of the test statistic departs from its mean when the null hypothesis is true.

Noncuple The verb, adjective, adverb, and noun noncuple means multiply by nine (9).

Nondirectional test A test that rejects extreme outcomes in either tail of the distribution. A two-tailed test.

None The pronoun none means not any or zero (0). None is a contraction of no one.

Nonequivalent transrormation A transformation that results in an equation not equivalent to the original.

Nonillion The noun and adjective nonillion means multiplied by 10^{30} in the American number naming system, but means multiplied by 10^{54} in the Chuquet number naming system.

Nonlinear regression Regression analysis in which the fitted (predicted) value of the response variable is a nonlinear function of one or more X variables. GraphPad Guide to Nonlinear Regression, Introduction to Nonlinear Regression, A GraphPad Practical Guide to Curve Fitting, Prophet Guide on Nonlinear Regression.

Nonlinear system A system of two or more equations at least one of which is nonlinear.

Nonnegative The adjective nonnegative means greater than or equal to zero (e"0).

Nonparametric methods (distribution free methods) Statistical methods to analyse data frgm populations, which do not assume a particular population distribution. Mann-Whitney U test, Kruskal-Wallis test and Wilcoxon's (T) test are examples. Such tests do not assume a distribution of the data specified by certain parameters (such as mean or variance). For example, one of the assumptions of the Student's t-test and ANOVA is normal distribution of the data. If this is not valid, a non-parametric equivalent must be used. If a wrong choice of test has been made, it does not matter very much if the sample size is large (a non-parametric test can be used where a parametric test might have been used but a parametric test can only be used when the assumptions are met). For a small sample size, non-parametric tests tend to give a larger P value. In general, parametric tests are more robust, more complicated to compute and have greater power efficiency. Parametric tests compare parameters such as the mean in t-test and variance in F-test as opposed to non-parametric tests that compare distributions. Nonparametric methods are most appropriate when the sample sizes are small. In large (e.g., n > 100) data sets, it makes little sense to use nonparametric statistics.

Nonparametric tests Statistical tests that do not rely on parameter estimation or precise distributional assumptions.

Nonpositive The adjective nonpositive means less than or equal to zero (d"0).

Nonuple The verb, adjective, adverb, and noun nonuple means multiply by nine (9).

Norm Another name for the magnitude of a vector.

Normal distribution (Gaussian distribution) Is a model for values on a continuous scale. A normal distribution can be completely described by two parameters: mean (m) and variance (v^2). It is shown as x ~ N(m, v^2). The distribution is symmetrical with mean, mode, and median all equal at x. In the special case of m = 1 and v^2 = 1, it is called the standard normal distribution.

Normal equations The normal equations for an M by N rectangular linear system

A * x = b

are computed by multiplying both sides by the transpose matrix:

A' * A * x = A' * b.

In the case where N < M, and the N columns of A are linearly independent, the matrix

B = A' * A

will be an invertible symmetric N by N matrix, and the normal equations can be solved by Gauss elimination. The matrix B is sometimes called the Gram matrix.

However, this method of producing an approximate solution of an overdetermined (and usually inconsistent) linear system is usually not recommended. The matrix B has a condition number that is the square of the condition number of A. Hence we may expect our solution to have a considerable error component. We can do better. There are more accurate methods which employ the QR factorization or the singular value decomposition.

Normal matrix A normal matrix A is a real matrix that commutes with its transpose:

A * A' = A' * A;

or a complex matrix that commutes with its transpose conjugate:

$A * A^H = A^H * A$.

Why define such an odd property? Because it's easy to check, and it turns out that a matrix is normal if and only if it is unitarily similar to a diagonal matrix. That's the same as saying the matrix is not only diagonalizable (similar to a diagonal matrix), but has a complete set of orthonormal eigenvectors.

A matrix is automatically normal if it is:

real and antisymmetric;

circulant;

diagonal;

Hermitian;

real and orthogonal;

Skew Hermitian;

real and symmetric;

unitary;

Note that an upper or lower triangular matrix is not normal, (unless it is actually diagonal!); such a matrix may have a complete set of eigenvectors (for example, if the eigenvalues are distinct), but the eigenvectors cannot be orthonormal.

Normal probability plot of the residuals A diagnostic test for the assumption of normal distribution of residuals in linear regression models. Each residual is plotted against its expected value under normality. A plot that is nearly linear suggests normal distribution of the residuals. A plot that obviously departs from linearity suggests that the error distribution is not normal.

Normality in arrays The assumption that the Y values in any array of X are normally distributed.

Normality Usually refers to the assumption behind most parametric tests that the data are normally distributed in the population.

Normalization The process of converting complex data structures into simple, stable data structures.

Notes Notes are explanatory text added to diagrams. They have no semantic significance to the model.

Nothing The pronoun and noun nothing means no thing or zero (0). Nothing is a contraction of no thing.

Nought Nought is an alternate spelling of the pronoun naught which means zero (0).

November The month name November literally means the ninth (9th) month. The Roman year began on the first (1st) day of March.

Novemcentenary 1. The noun and adjective novemcentenary means a group of nine hundred (900).

2. The noun and adjective novemcentenary means a period of nine hundred (900) years.

3. The noun and adjective novemcentenary means a nine hundred (900) year anniversary.

4. The adjective and noun novemcentenary means nine hundredth (900th) in rank or order.

Novemcentennial 1. The adjective and noun novemcentennial means recurring every nine hundred (900) years.

2. The adjective and noun novemcentennial means lasting nine hundred (900) years.

3. The noun and adjective novemcentennial means a nine hundred (900) year anniversary.

Novemcentennium The noun and adjective novemcentennium means a period of nine hundred (900) years.

Novemdecennary The noun and adjective novemdecennary means a group of nineteen (19) or a period of nineteen (19) years.

Novemdecennial 1. The adjective and noun novemdecennial means recurring every nineteen (19) years.

2. The adjective and noun novemdecennial means lasting nineteen (19) years.

3. The noun and adjective novemdecennial means a nineteen (19) year anniversary.

Novemdecennium The noun and adjective novemdecennium means a period of nineteen (19) years.

Novemdecentenary 1. The noun and adjective novemdecentenary means a group of one thousand nine hundred (1900).

2. The noun and adjective novemdecentenary means a period of one thousand nine hundred (1900) years.

3. The noun and adjective novemdecentenary means a one thousand nine hundred (1900) year anniversary.

4. The adjective and noun novemdecentenary means one thousand nine hundredth (1900th) in rank or order.

Novemdecentennial 1. The adjective and noun novemdecentennial means recurring every one thousand nine hundred (1900) years.

2. The adjective and noun novemdecentennial means lasting one thousand nine hundred (1900) years.

3. The noun and adjective novemdecentennial means a one thousand nine hundred (1900) year anniversary.

Novemdecentennium The noun and adjective novemdecentennium means a period of one thousand nine hundred (1900) years.

Novemdeci The numeric prefix novemdeci- means nineteen (19). This prefix is taken from novemdecim, a more regular form of the Latin cardinal number undeviginti (XVIIII) which literally means one (1) from twenty (20).

Novemdecillion The noun and adjective novemdecillion means multiplied by 10^{60} in the American number naming system, but means multiplied by 10^{114} in the Chuquet number naming system.

Novemmillenary The noun and adjective novemmillenary means a group of nine thousand (9000).

Novemmillennial The adjective and noun novemmillennial means recurring every nine thousand (9000) years.

Novemmillennium The noun and adjective novemmillennium means a period of nine thousand (9000) years.

Novemnonaginta The numeric prefix novemnonaginta- means ninety-nine (99). This prefix is derived from the Latin cardinal number novem et nonaginta (LXXXXVIIII).

Novemnonagintennary The noun and adjective novemnonagintennary means a group of ninety-nine (99) or a period of ninety-nine (99) years.

Novemnonagintennial The adjective and noun novemnonagintennial means recurring every ninety-nine (99) years.

Novemnonagintennium The noun and adjective novemnonagintennium means a period of ninety-nine (99) years.

Novemnonagintillion The noun and adjective novemnonagintillion means multiplied by 10^{300} in the American number naming system, but means multiplied by 10^{594} in the Chuquet number naming system.

Novemoctoginta The numeric prefix novemoctoginta- means eighty-nine (89). This prefix is derived from the Latin cardinal number novem et octoginta (LXXXVIIII).

Novemoctogintennary The noun and adjective novemoctogintennary means a group of eighty-nine (89) or a period of eighty-nine (89) years.

Novemoctogintennial The adjective and noun novemoctogintennial means recurring every eighty-nine (89) years.

Novemoctogintennium The noun and adjective novemoctogintennium means a period of eighty-nine (89) years.

Novemoctogintillion The noun and adjective novemoctogintillion means multiplied by 10^{270} in the American number naming system, but means multiplied by 10^{534} in the Chuquet number naming system.

Novemquadraginta The numeric prefix novemquadraginta- means forty-nine (49). This prefix is derived from the Latin cardinal number novem et quadraginta (XXXXVIIII).

Novemquadragintennary The noun and adjective novemquadragintennary means a group of forty-nine (49) or a period of forty-nine (49) years.

Novemquadragintennial The adjective and noun novemquadragintennial means recurring every forty-nine (49) years.

Novemquadragintennium The noun and adjective novemquadragintennium means a period of forty-nine (49) years.

Novemquadragintillion The noun and adjective novemquadragintillion means multiplied by 10^{150} in the American number naming system, but means multiplied by 10^{294} in the Chuquet number naming system.

Novemquinquaginta The numeric prefix novemquinquaginta- means fifty-nine (59). This prefix is derived from the Latin cardinal number novem et quinquaginta (LVIIII).

Novemquinquagintennary The noun and adjective novemquinquagintennary means a group of fifty-nine (59) or a period of fifty-nine (59) years.

Novemquinquagintennial The adjective and noun novemquinquagintennial means recurring every fifty-nine (59) years.

Novemquinquagintennium The noun and adjective novemquinquagintennium means a period of fifty-nine (59) years.

Novemquinquagintillion The noun and adjective novemquinquagintillion means multiplied by 10^{180} in the American number naming system, but means multiplied by 10^{354} in the Chuquet number naming system.

Novemseptuaginta The numeric prefix novemseptuaginta- means seventy-nine (79). This prefix is derived from the Latin cardinal number novem et septuaginta (LXXVIIII).

Novemseptuagintennary The noun and adjective novemseptuagintennary means a group of seventy-nine (79) or a period of seventy-nine (79) years.

Novemseptuagintennial The adjective and noun novemseptuagintennial means recurring every seventy-nine (79) years.

Novemseptuagintennium The noun and adjective novemseptuagintennium means a period of seventy-nine (79) years.

Novemseptuagintillion The noun and adjective novemseptuagintillion means multiplied by 10^{240} in the American number naming system, but means multiplied by 10^{474} in the Chuquet number naming system.

Novemsexaginta The numeric prefix novemsexaginta- means sixty-nine (69). This prefix is derived from the Latin cardinal number novem et sexaginta (LXVIIII).

Novemsexagintennary The noun and adjective novemsexagintennary means a group of sixty-nine (69) or a period of sixty-nine (69) years.

Novemsexagintennial The adjective and noun novemsexagintennial means recurring every sixty-nine (69) years.

Novemsexagintennium The noun and adjective novemsexagintennium means a period of sixty-nine (69) years.

Novemsexagintillion The noun and adjective novemsexagintillion means multiplied by 10^{210} in the American number naming system, but means multiplied by 10^{414} in the Chuquet number naming system.

Novemtriginta The numeric prefix novemtriginta- means thirty-nine (39). This prefix is derived from the Latin cardinal number novem et triginta (XXXVIIII).

Novemtrigintacentenary The noun and adjective novemtrigintacentenary means a group of three thousand nine hundred (3900).

Novemtrigintacentennial The adjective and noun novemtrigintacentennial means recurring every three thousand nine hundred (3900) years.

Novemtrigintacentennium The noun and adjective novemtrigintacentennium means a period of three thousand nine hundred (3900) years.

Novemtrigintennary The noun and adjective novemtrigintennary means a group of thirty-nine (39) or a period of thirty-nine (39) years.

Novemtrigintennial The adjective and noun novemtrigintennial means recurring every thirty-nine (39) years.

Novemtrigintennium The noun and adjective novemtrigintennium means a period of thirty-nine (39) years.

Novemtrigintillion The noun and adjective novemtrigintillion means multiplied by 10^{120} in the American number naming system, but means multiplied by 10^{234} in the Chuquet number naming system.

Novemvigintennary The noun and adjective novemvigintennary means a group of twenty-nine (29) or a period of twenty-nine (29) years.

Novemvigintennial 1. The adjective and noun novemvigintennial means recurring every twenty-nine (29) years.

2. The adjective and noun novemvigintennial means lasting twenty-nine (29) years.

3. The noun and adjective novemvigintennial means a twenty-nine (29) year anniversary.

Novemvigintennium The noun and adjective novemvigintennium means a period of twenty-nine (29) years.

Novemviginti The numeric prefix novemviginti- means twenty-nine (29). This prefix is derived from the Latin cardinal number novem et viginti (XXVIIII).

Novemviginticentenary 1. The noun and adjective novemviginticentenary means a group of two thousand nine hundred (2900).

2. The noun and adjective novemviginticentenary means a period of two thousand nine hundred (2900) years.

3. The noun and adjective novemviginticentenary means a two thousand nine hundred (2900) year anniversary.

4. The adjective and noun novemviginticentenary means two thousand nine hundredth (2900th) in rank or order.

Novemviginticentennial 1. The adjective and noun novemviginticentennial means recurring every two thousand nine hundred (2900) years.

2. The adjective and noun novemviginticentennial means lasting two thousand nine hundred (2900) years.

3. The noun and adjective novemviginticentennial means a two thousand nine hundred (2900) year anniversary.

Novemviginticentennium The noun and adjective ovemviginticentennium means a period of two thousand nine hundred (2900) years.

Novemvigintillion The noun and adjective novemvigintillion means multiplied by 10^{90} in the American number naming system, but means multiplied by 10^{174} in the Chuquet number naming system.

Novena The noun novena means a nine (9) day period of prayer.

Novennary The noun and adjective novennary means a group of nine (9) or a period of nine (9) years.

Novennial 1. The adjective and noun novennial means recurring every nine (9) years.

Novennium The noun and adjective novennium means a period of nine (9) years.

Nth root theorem An extension of De Moivre's theorem involving roots of complex numbers.

Null 1. The noun and adjective null means zero (0), empty, or invalid.

2. The verb null means to set to zero (0).

Null hypothesis The null hypothesis is a term that statisticians often use to indicate the statistical hypothesis tested. The purpose of most statistical tests, is to determine if the obtained results provide a reason to reject the hypothesis that they are merely a product of chance factors. For example, in an experiment in which two groups of randomly selected subjects have received different treatments and have yielded different means, it is always necessary to ask if the difference between the obtained means is among the differences that would be expected to occure by chance whenever two groups are randomly selected. In this example, the hypothesis tested is that the two samples are from populations with the same mean. Another way to say this is to assert that the investigator tests the null hypothesis that the difference between the means of the populations from which the samples were drawn, is zero. If the difference between the means of the samples is among those that would occur rarely by chance when the null hypothesis is true, the null hypothesis is rejected and the investigator describes the results as statistically significant.

Null model A model in which all parameters except the intercept are 0. It is also called the intercept-only model. The null model in linear regression is that the slope is 0, so that the predicted value of Y is the mean of Y for all values of X. The F test for the linear regression tests whether the slope is significantly different from 0, which is equivalent to testing whether the fit using non-zero slope is significantly better than the null model with 0 slope.

Null space The null space of a matrix A is the set of all null vectors x such that

A * x = 0.

Simple facts:

The null space is a linear space;

A square matrix is nonsingular if and only if its null space is exactly the zero vector;

If x is a solution of A * x = b, then so is (x + y), where y is any vector in the null space of A; linear solutions are only guaranteed to exist and to be unique when the null space is 0;

The eigenspace corresponding to an eigenvalue lambda is the null space of (A - lambda * I).

Null value A special field value, distinct from 0, blank, or any other value, that indicates that the value for the field is missing or otherwise unknown.

Null vector A null vector of a matrix A is a non-zero vector x with the property that A * x = 0.

Facts about a null vector:

A nonzero multiple of a null vector is also a null vector;

The sum of two null vectors is a null vector;

The set of all null vectors of a matrix forms the null space of the matrix;

If x solves A * x = b, and y is a null vector of A, then x + y is another solution, that is, A * (x + y) = b;

A square matrix A is singular if and only if it has a null vector;

For a square matrix A, a null vector is an eigenvector for the eigenvalue 0.

Number The noun and adjective number means an abstract dimensionless quantity.

Number needed to treat The reciprocal of the reduction in absolute risk between treated and control groups in a clinical trial. It is interpreted as the number of patients who need to be treated to prevent one adverse event.

Numberless The adjective numberless means a very large number of individuals or discrete objects.

Numeracy The noun numeracy means the capacity to understand and work with numbers.

Numeral The noun and adjective numeral means a single symbol or series of symbols used to represent a number. There are many modern and historical systems of numerals and number notation. In modern positional number notation systems, the Indio-Arabic numerals 0 through 9 followed by the English letters A through Z are used to represent the integers zero (0) through thirty-five (35). Roman numerals and Greek numerals are also commonly used to represent numbers.

Numerate The adjective numerate means having the capacity to understand and work with numbers.

Numerator The number above the fraction bar that indicates the number of parts of the whole there are in a rational number.

Numeric The adjective and noun numeric means pertaining to numbers.

Numerical The adjective numerical means of or relating to numbers.

Numerous The adjective numerous means a relatively large number of individuals or discrete objects.

Object An object is any unique thing such as a person, place, or any physical or conceptual thing.

Object class A set of objects that share a common structure and a common behavior.

Object class cardinality constraint An object-class cardinality constraint restricts the number of objects in an object class.

Object diagram A graph of instances that are compatible with a given class diagram.

Observed frequencies The cell frequencies that were actually observed—as distinguished from expected frequencies.

Obtuse angles An obtuse angle is an angle measuring between 90 and 180 degrees.

Example: The following angles are all obtuse.

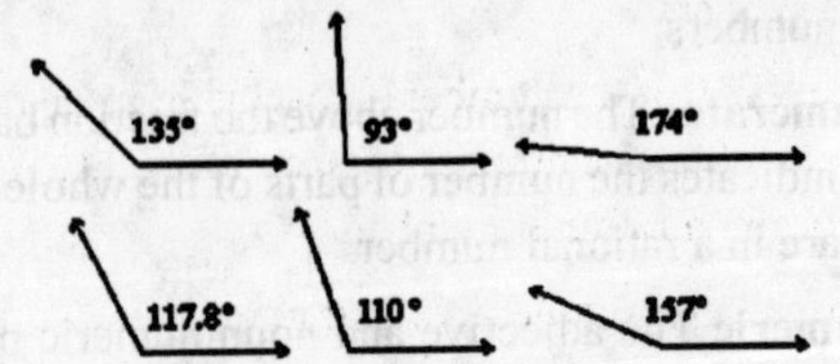

Obtuse triangle A triangle having an obtuse angle. One of the angles of the triangle measures more than 90 degrees.

Examples:

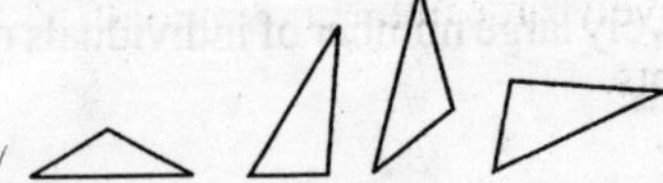

Oct The numeric prefix oct- means eight (8). This prefix is derived from the Latin cardinal number octo (VIII).

Octa The prefix octa- means eight (8).

Octaconta The prefix octaconta- means eighty (80).

Octacontad The noun and adjective octacontad means a group of eighty (80) members.

Octacontadi The prefix octacontadi- means eighty-two (82). Octacontadi- is an alternate form of the prefix octacontakaidi-.

Octacontaennea The prefix octacontaennea- means eighty-nine (89). Octacontaennea- is an alternate form of the prefix octacontakaiennea-.

Octacontagon The noun and adjective octacontagon means a plain figure with eighty (80) straight sides.

Octacontagram The noun and adjective octacontagram means an interconnected star figure with eighty (80) points formed by stellating a octacontagon.

Octacontahedron The noun and adjective octacontahedron means a solid figure with eighty (80) planar faces.

Octacontahena The prefix octacontahena- means eighty-one (81). Octacontahena- is an alternate form of the prefix octacontakaihena-.

Octacontahepta The prefix octacontahepta- means eighty-seven (87). Octacontahepta- is an alternate form of the prefix octacontakaihepta-.

Octacontahexa The prefix octacontahexa- means eighty-six (86). Octacontahexa- is an alternate form of the prefix octacontakaihexa-.

Octacontakaidi The prefix octacontakaidi- means eighty-two (82).

Octacontakaidigon The noun and adjective octacontakaidigon means a plain figure with eighty-two (82) straight sides.

Octacontakaidigram The noun and adjective octacontakaidigram means an interconnected star figure with eighty-two (82) points formed by stellating a octacontakaidigon.

Octacontakaidihedron The noun and adjective octacontakaidihedron means a solid figure with eighty-two (82) planar faces.

Octacontakaidyad The noun and adjective octacontakaidyad means a group of eighty-two (82) members.

Octacontakaiennea The prefix octacontakaiennea- means eighty-nine (89).

Octacontakaiennead The noun and adjective octacontakaiennead means a group of eighty-nine (89) members.

Octacontakaienneagon The noun and adjective octacontakaienneagon means a plain figure with eighty-nine (89) straight sides.

Octacontakaienneagram The noun and adjective octacontakaienneagram means an interconnected star figure with eighty-nine (89) points formed by stellating a octacontakaienneagon.

Octacontakaienneahedron The noun and adjective octacontakaienneahedron means a solid figure with eighty-nine (89) planar faces.

Octacontakaihena The prefix octacontakaihena- means eighty-one (81).

Octacontakaihenad The noun and adjective octacontakaihenad means a group of eighty-one (81) members.

Octacontakaihenagon The noun and adjective octacontakaihenagon means a plain figure with eighty-one (81) straight sides.

Octacontakaihenagram The noun and adjective octacontakaihenagram means an interconnected star figure with eighty-one (81) points formed by stellating a octacontakaihenagon.

Octacontakaihenahedron The noun and adjective octacontakaihenahedron means a solid figure with eighty-one (81) planar faces.

Octacontakaihepta The prefix octacontakaihepta- means eighty-seven (87).

Octacontakaiheptad The noun and adjective octacontakaiheptad means a group of eighty-seven (87) members.

Octacontakaiheptagon The noun and adjective octacontakaiheptagon means a plain figure with eighty-seven (87) straight sides.

Octacontakaiheptagram The noun and adjective octacontakaiheptagram means an interconnected star figure with eighty-seven (87) points formed by stellating a octacontakaiheptagon.

Octacontakaiheptahedron The noun and adjective octacontakaiheptahedron means a solid figure with eighty-seven (87) planar faces.

Octacontakaihexa The prefix octacontakaihexa- means eighty-six (86).

Octacontakaihexad The noun and adjective octacontakaihexad means a group of eighty-six (86) members.

Octacontakaihexagon The noun and adjective octacontakaihexagon means a plain figure with eighty-six (86) straight sides.

Octacontakaihexagram The noun and adjective octacontakaihexagram means an interconnected star figure with eighty-six (86) points formed by stellating a octacontakaihexagon.

Octacontakaihexahedron The noun and adjective octacontakaihexahedron means a solid figure with eighty-six (86) planar faces.

Octacontakaiocta The prefix octacontakaiocta- means eighty-eight (88).

Octacontakaioctad The noun and adjective octacontakaioctad means a group of eighty-eight (88) members.

Octacontakaioctagon The noun and adjective octacontakaioctagon means a plain figure with eighty-eight (88) straight sides.

Octacontakaioctagram The noun and adjective octacontakaioctagram means an interconnected star figure with eighty-eight (88) points formed by stellating a octacontakaioctagon.

Octacontakaioctahedron The noun and adjective octacontakaioctahedron means a solid figure with eighty-eight (88) planar faces.

Octacontakaipenta The prefix octacontakaipenta- means eighty-five (85).

Octacontakaipentad The noun and adjective octacontakaipentad means a group of eighty-five (85) members.

Octacontakaipentagon The noun and adjective octacontakaipentagon means a plain figure with eighty-five (85) straight sides.

Octacontakaipentagram The noun and adjective octacontakaipentagram means an interconnected star figure with eighty-five (85) points formed by stellating a octacontakaipentagon.

Octacontakaipentahedron The noun and adjective octacontakaipentahedron means a solid figure with eighty-five (85) planar faces.

Octacontakaitetra The prefix octacontakaitetra- means eighty-four (84).

Octacontakaitetrad The noun and adjective octacontakaitetrad means a group of eighty-four (84) members.

Octacontakaitetragon The noun and adjective octacontakaitetragon means a plain figure with eighty-four (84) straight sides.

Octacontakaitetragram The noun and adjective octacontakaitetragram means an interconnected star figure with eighty-four (84) points formed by stellating a octacontakaitetragon.

Octacontakaitetrahedron The noun and adjective octacontakaitetrahedron means a solid figure with eighty-four (84) planar faces.

Octacontakaitri The prefix octacontakaitri- means eighty-three (83).

Octacontakaitriad The noun and adjective octacontakaitriad means a group of eighty-three (83) members.

Octacontakaitrigon The noun and adjective octacontakaitrigon means a plain figure with eighty-three (83) straight sides.

Octacontakaitrigram The noun and adjective octacontakaitrigram means an interconnected star figure with eighty-three (83) points formed by stellating a octacontakaitrigon.

Octacontakaitrihedron The noun and adjective octacontakaitrihedron means a solid figure with eighty-three (83) planar faces.

Octacontapenta The prefix octacontapenta- means eighty-five (85). Octacontapenta- is an alternate form of the prefix octacontakaipenta-.

Octacontatetra The prefix octacontatetra- means eighty-four (84). Octacontatetra- is an alternate form of the prefix octacontakaitetra-.

Octacontatri The prefix octacontatri- means eighty-three (83). Octacontatri- is an alternate form of the prefix octacontakaitri-.

Octacta The prefix octacta- means eight hundred (800).

Octad The noun and adjective octad means a group of eight (8) members.

Octadeca The prefix octadeca- means eighteen (18).

Octagon An eight-sided polygon. The sum of the angles of an octagon is 1080 degrees.

Examples: A regular octagon:

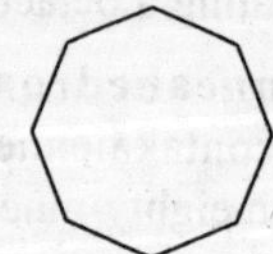

An irregular octagon:

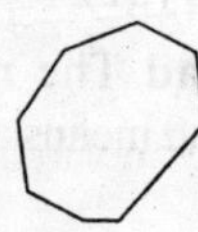

Octagram 1. The noun and adjective octagram means an interconnected star figure with eight (8) points formed by stellating a octagon.

2. The noun octagram means a word or character sequence consisting of eight (8) characters.

Octahedron The noun and adjective octahedron means a solid figure with eight (8) planar faces.

Octaheptaconta The prefix octaheptaconta- means seventy-eight (78).

Octahexaconta The prefix octahexaconta- means sixty-eight (68).

Octaicosa The prefix octaicosa- means twenty-eight (28).

Octakaideca The prefix octakaideca- means eighteen (18).

Octakaidecad The noun and adjective octakaidecad means a group of eighteen (18) members.

Octakaidecagon The noun and adjective octakaidecagon means a plain figure with eighteen (18) straight sides.

Octakaidecagram The noun and adjective octakaidecagram means an interconnected star figure with eighteen (18) points formed by stellating a octakaidecagon.

Octakaidecahedron The noun and adjective octakaidecahedron means a solid figure with eighteen (18) planar faces.

Octal The adjective octal means based on the number eight (8).

Octalia The prefix octalia- means eight thousand (8000).

Octameter The noun octameter means a verse consisting of eight (8) metrical feet.

Octanonaconta The prefix octanonaconta- means ninety-eight (98).

Octaoctaconta The prefix octaoctaconta- means eighty-eight (88).

Octapentaconta The prefix octapentaconta- means fifty-eight (58).

Octatetraconta The prefix octatetraconta- means forty-eight (48).

Octatriaconta The prefix octatriaconta- means thirty-eight (38).

Octave 1. The noun octave means a set of eight (8).

2. The noun octave means a religious observance of eight (8) days beginning with a feast day.

3. The noun octave means a stanza of eight (8) lines.

4. The noun and adjective octave means a frequency ratio of two (2).

Octavo The noun octavo (8vo) means a book with leaves printed on one eighth (1/8) of the printer sheet.

Octennary The noun and adjective octennary means a group of eight (8) or a period of eight (8) years.

Octennial 1. The adjective and noun octennial means recurring every eight (8) years.

2. The adjective and noun octennial means lasting eight (8) years.

3. The noun and adjective octennial means an eight (8) year anniversary.

Octennium The noun and adjective octennium means a period of eight (8) years.

Octilateral The adjective octilateral means having eight (8) sides.

Octillion The noun and adjective octillion means multiplied by 10^{27} in the American number naming system, but means multiplied by 10^{48} in the Chuquet number naming system.

October The month name October literally means the eighth (8th) month. The Roman year began on the first (1st) day of March.

Octocentenary The noun and adjective octocentenary means a group of eight hundred (800).

Octocentennial The adjective and noun octocentennial means recurring every eight hundred (800) years.

Octocentennium The noun and adjective octocentennium means a period of eight hundred (800) years.

Octodecennary The noun and adjective octodecennary means a group of eighteen (18) or a period of eighteen (18) years.

Octodecennial The adjective and noun octodecennial means recurring every eighteen (18) years.

Octodecennium The noun and adjective octodecennium means a period of eighteen (18) years.

Octodecentenary The noun and adjective octodecentenary means a group of one thousand eight hundred (1800).

Octodecentennial 1. The adjective and noun octodecentennial means recurring every one thousand eight hundred (1800) years.

2. The adjective and noun octodecentennial means lasting one thousand eight hundred (1800) years.

3. The noun and adjective octodecentennial means a one thousand eight hundred (1800) year anniversary.

Octodecentennium The noun and adjective octodecentennium means a period of one thousand eight hundred (1800) years.

Octodeci The numeric prefix octodeci- means eighteen (18). This prefix is taken from octodecim, a more regular form of the Latin cardinal number duodeviginti (XVIII) which literally means two (2) from twenty (20).

Octodecicycle The noun, adjective, and verb octodecicycle means having eighteen (18) wheels.

Octodecillion The noun and adjective octodecillion means multiplied by 10^{57} in the American number naming system, but means multiplied by 10^{108} in the Chuquet number naming system.

Octogenarian The noun and adjective octogenarian means between eighty (80) and ninety (90) years of age.

Octogesimal The adjective octogesimal means based on the number eighty (80).

Octoginta The numeric prefix octoginta- means eighty (80). This prefix is derived from the Latin cardinal number octoginta (LXXX).

Octogintennary The noun and adjective octogintennary means a group of eighty (80) or a period of eighty (80) years.

Octogintennial 1. The adjective and noun octogintennial means recurring every eighty (80) years.

2. The adjective and noun octogintennial means lasting eighty (80) years.

3. The noun and adjective octogintennial means an eighty (80) year anniversary.

Octogintennium The noun and adjective octogintennium means a period of eighty (80) years.

Octogintillion The noun and adjective octogintillion means multiplied by 10^{243} in the American number naming system, but means multiplied by 10^{480} in the Chuquet number naming system.

Octohedron The noun octohedron means a solid figure with eight (8) faces.

Octomillenary The noun and adjective octomillenary means a group of eight thousand (8000).

Octomillennial The adjective and noun octomillennial means recurring every eight thousand (8000) years.

Octomillennium The noun and adjective octomillennium means a period of eight thousand (8000) years.

Octonary The adjective and noun octonary means eighth (8th) in order or importance.

Octononaginta The numeric prefix octononaginta- means ninety-eight (98). This prefix is derived from the Latin cardinal number octo et nonaginta (LXXXXVIII).

Octononagintennary The noun and adjective octononagintennary means a group of ninety-eight (98) or a period of ninety-eight (98) years.

Octononagintennial The adjective and noun octononagintennial means recurring every ninety-eight (98) years.

Octononagintennium The noun and adjective octononagintennium means a period of ninety-eight (98) years.

Octononagintillion The noun and adjective octononagintillion means multiplied by 10^{297} in the American number naming system, but means multiplied by 10^{588} in the Chuquet number naming system.

Octooctoginta The numeric prefix octooctoginta- means eighty-eight (88). This prefix is derived from the Latin cardinal number octo et octoginta (LXXXVIII).

Octooctogintennary The noun and adjective octooctogintennary means a group of eighty-eight (88) or a period of eighty-eight (88) years.

Octooctogintennial 1. The adjective and noun octooctogintennial means recurring every eighty-eight (88) years.

2. The adjective and noun octooctogintennial means lasting eighty-eight (88) years.

3. The noun and adjective octooctogintennial means an eighty-eight (88) year anniversary.

Octooctogintennium The noun and adjective octooctogintennium means a period of eighty-eight (88) years.

Octooctogintillion The noun and adjective octooctogintillion means multiplied by 10^{267} in the American number naming system, but means multiplied by 10^{528} in the Chuquet number naming system.

Octoquadraginta The numeric prefix octoquadraginta means forty-eight (48). This prefix is derived from the Latin cardinal number octo et quadraginta (XXXXVIII).

Octoquadragintennary The noun and adjective octoquadragintennary means a group of forty-eight (48) or a period of forty-eight (48) years.

Octoquadragintennial The adjective and noun octoquadragintennial means recurring every forty-eight (48) years.

Octoquadragintillion The noun and adjective octoquadragintillion means multiplied by 10^{147} in the American number naming system, but means multiplied by 10^{288} in the Chuquet number naming system.

Octoquinquaginta The numeric prefix octoquinquaginta- means fifty-eight (58). This prefix is derived from the Latin cardinal number octo et quinquaginta (LVIII).

Octoquinquagintennary The noun and adjective octoquinquagintennary means a group of fifty-eight (58) or a period of fifty-eight (58) years.

Octoquinquagintennial The adjective and noun octoquinquagintennial means recurring every fifty-eight (58) years.

Octoquinquagintennium The noun and adjective octoquinquagintennium means a period of fifty-eight (58) years.

Octoquinquagintillion The noun and adjective octoquinquagintillion means multiplied by 10^{177} in the American number naming system, but means multiplied by 10^{348} in the Chuquet number naming system.

Octoseptuaginta The numeric prefix octoseptuaginta- means seventy-eight (78). This prefix is derived from the Latin cardinal number octo et septuaginta (LXXVIII).

Octoseptuagintennary The noun and adjective octoseptuagintennary means a group of seventy-eight (78) or a period of seventy-eight (78) years.

Octoseptuagintennial The adjective and noun octoseptuagintennial means recurring every seventy-eight (78) years.

Octoseptuagintennium The noun and adjective octoseptuagintennium means a period of seventy-eight (78) years.

Octoseptuagintillion The noun and adjective octoseptuagintillion means multiplied by 10^{237} in the American number naming system, but means multiplied by 10^{468} in the Chuquet number naming system.

Octosexaginta The numeric prefix octosexaginta- means sixty-eight (68). This prefix is derived from the Latin cardinal number octo et sexaginta (LXVIII).

Octosexagintennary The noun and adjective octosexagintennary means a group of sixty-eight (68) or a period of sixty-eight (68) years.

Octosexagintennial The adjective and noun octosexagintennial means recurring every sixty-eight (68) years.

Octosexagintennium The noun and adjective octosexagintennium means a period of sixty-eight (68) years.

Octosexagintillion The noun and adjective octosexagintillion means multiplied by 10^{207} in the American number naming system, but means multiplied by 10^{408} in the Chuquet number naming system.

Octotriginta The numeric prefix octotriginta- means thirty-eight (38). This prefix is derived from the Latin cardinal number octo et triginta (XXXVIII).

Octotrigintacentenary The noun and adjective octotrigintacentenary means a group of three thousand eight hundred (3800).

Octotrigintacentennial The adjective and noun octotrigintacentennial means recurring every three thousand eight hundred (3800) years.

Octotrigintacentennium The noun and adjective octotrigintacentennium means a period of three thousand eight hundred (3800) years.

Octotrigintennary The noun and adjective octotrigintennary means a group of thirty-eight (38) or a period of thirty-eight (38) years.

Octotrigintennial The adjective and noun octotrigintennial means recurring every thirty-eight (38) years.

Octotrigintennium The noun and adjective octotrigintennium means a period of thirty-eight (38) years.

Octotrigintillion The noun and adjective octotrigintillion means multiplied by 10^{117} in the American number naming system, but means multiplied by 10^{228} in the Chuquet number naming system.

Octovigintennary The noun and adjective octovigintennary means a group of twenty-eight (28) or a period of twenty-eight (28) years.

Octovigintennial The adjective and noun octovigintennial means recurring every twenty-eight (28) years.

Octovigintennium The noun and adjective octovigintennium means a period of twenty-eight (28) years.

Octoviginti The numeric prefix octoviginti- means twenty-eight (28). This prefix is derived from the Latin cardinal number octo et viginti (XXVIII).

Octoviginticentenary The noun and adjective octoviginticentenary means a group of two thousand eight hundred (2800).

Octoviginticentennial 1. The adjective and noun octoviginticentennial means recurring every two thousand eight hundred (2800) years.

2. The adjective and noun octoviginticentennial means lasting two thousand eight hundred (2800) years.

3. The noun and adjective octoviginticentennial means a two thousand eight hundred (2800) year anniversary.

Octoviginticentennium The noun and adjective octoviginticentennium means a period of two thousand eight hundred (2800) years.

Octovigintillion The noun and adjective octovigintillion means multiplied by 10^{87} in the American number naming system, but means multiplied by 10^{168} in the Chuquet number naming system. .

Octuplet The noun and adjective octuplet means one of eight (8) siblings born together or one of eight (8) very similar things.

Odd even identities Fundamental identities that involve the basic trig functions of negative angles. Also called identities for negatives.

Odd function A function is odd of $f(-x) = -f(x)$.

Odds multiplier In logistic regression, $x = \log_e$ (odds ratio), thus exp x = odds ratio. For a continuous (explanatory) variable, exp x is called the odds multiplier and corresponds to the odds ratio for a unit increase in the explanatory variable. The odds multiplier of the coefficient is the odds ratio for its level relative to the reference level. If x increases from a to b by c, the odds multiplier becomes exp (cx). The resulting value shows the proportional change in the odds associated with x = b relative to x = a. It follows that for binary variables where x can only get values of 0 and 1, exp x = odds ratio.

Odds ratio (OR) Also known as relative odds and approximate relative risk. It is the ratio of the odds of the risk factor in a diseased group and in a non-diseased (control) group (the ratio of the frequency of presence / absence of the marker in cases to the frequency of presence / absence of the marker in controls). The interpretation of the OR is that the risk factor increases the odds of the disease 'OR' times. OR is used in retrospective case-control studies (relative risk (RR) is the ratio of proportions in two groups which can be estimated in a prospective -cohort- study). These two and relative hazard (or hazard ratio) are measures of the strength/magnitude of an association. As opposed to the P value, these do not change with the sample size. OR and RR are considered interchangeable when certain assumptions are met, especially for large samples and rare diseases. Odds ratio is calculated as ad/bc where a,b,c,d are the entries in a 2x2 contingency table (hence the alternative definition as the cross-product ratio). In logistic regression, the coefficient L corresponds to the $\log_e$ of the odds ratio. There are statistical methods to test the homogeneity of odds ratios .

Odds The odds of a success is defined as the ratio of the probability of a success to the probability of a failure (p/(1–p)). If a team has a probability of 0.6 of winning the championship, the odds for winning is 0.6/(1–0.6) = 3:2. Similarly, the odd in a case-control study is the frequency of the presence of the marker divided by the frequency of absence of the marker (in cases or controls separately). The link function logit (or logistic) is the $\log_e$ of the odds.

Off diagonal elements The elements of a matrix that are not on the main diagonal.

Offset box An imaginary box which is used when drawing thumbnail sketches to calculate offsets.

Offset To change the direction of a line by making a turn and correcting back to the original direction by making another turn. Also, offset is the term used to identify such a turn.

Olympiad The noun olympiad means a period of four (4) years.

Omega squared A less biased measure of the magnitude of effect than .

Omnibus test If the chi-square test has more than one degree of freedom (larger than 2x2 table), it is called an 'omnibus' test, which evaluates the significance of an overall hypothesis containing multiple sub-hypotheses (these multiple sub-hypotheses then need to be tested using follow up tests).

Once The adverb once means one (1) time.

One dimensional A figure that has length but no width or height.

One eighth The compound noun, pronoun, and adjective one eighth (1/8 = 0.125) is the standard name for the rational fraction equal to one (1) divided by eight (8). One eighth equals two to the negative third power (2^{-3} = 1/8 = 0.125). One eighth is the third Eye of Horus fraction.

One fifth The compound noun, pronoun, and adjective one fifth (1/5 = 0.200) is the standard name for the rational fraction equal to one (1) divided by five (5).

One fourth The compound noun, pronoun, and adjective one fourth (1/4 = 0.250) is the standard name for the rational fraction equal to one (1) divided by four (4). One fourth equals two to the negative second power (2^{-2} = 1/4 = 0.250). One fourth is the second Eye of Horus fraction.

One half The compound noun, pronoun, and adjective one half (1/2 = 0.500) is the standard name for the rational fraction equal to one (1) divided by two (2). One half equals two to the negative first power (2^{-1} = 1/2 = 0.500). One half is the first Eye of Horus fraction.

One hundred million The compound noun and adjective one hundred million (100,000,000) is the standard name for the cardinal number that equals ten to the eighth power (10^8 = 10 raised to power +08 = 100,000,000).

One hundred The compound noun, pronoun, and adjective one hundred (100) is the standard name for the cardinal number that follows ninety-nine (99). One hundred equals ten to the second power (10^2 = 10 raised to power +02 = 100). One hundred corresponds with the ordinal number one hundredth (100th).

One hundred thousand The compound noun and adjective one hundred thousand (100,000) is the standard name for the cardinal number that equals ten to the fifth power (10^5 = 10 raised to power +05 = 100,000).

One hundredth The compound adjective, adverb, and noun one hundredth (100th) is the standard name for the ordinal number that follows ninety-ninth (99th). One hundredth corresponds with the cardinal number one hundred (100).

One million The compound noun and adjective one million (1,000,000) is the standard name for the cardinal number that equals ten to the sixth power (10^6 = 10 raised to power +06 = 1,000,000).

One ninth The compound noun, pronoun, and adjective one ninth (1/9) is the standard name for the rational fraction equal to one (1) divided by nine (9).

One one-hundred-millionth The compound noun and adjective one one-hundred-millionth (1/100,000,000 = 0.000 000 010) is the standard name for the rational fraction that equals ten to the negative eighth power (10^{-8} = 10 raised to power –08 = 0.000 000 010).

One one-hundredth The compound noun and adjective one one-hundredth (1/100 = 0.010) is the standard name for the rational fraction that equals ten to the negative second power (10^{-2} = 10 raised to power –02 = 0.010).

One one-hundred-thousandth The compound noun and adjective one one-hundred-thousandth (1/100,000 = 0.000 010) is the standard name for the rational fraction that equals ten to the negative fifth power (10^{-5} = 10 raised to power –05 = 0.000 010).

One one-millionth The compound noun and adjective one one-millionth (1/1,000,000 = 0.000 001) is the standard name for the rational fraction that equals ten to the negative sixth power (10^{-6} = 10 raised to power –06 = 0.000 001).

One one-thousandth The compound noun and adjective one one-thousandth (1/1,000 = 0.001) is the standard name for the rational fraction that equals ten to the negative third power (10^{-3} = 10 raised to power –03 = 0.001).

One quarter The compound noun, pronoun, and adjective one quarter is an alternate name for the rational fraction one fourth (1/4 = 0.250).

One seventh The compound noun, pronoun, and adjective one seventh (1/7) is the standard name for the rational fraction equal to one (1) divided by seven (7).

One sixteenth The compound noun, pronoun, and adjective one sixteenth (1/16 = 0.062 500) is the standard name for the rational fraction equal to one (1) divided by sixteen (16). One sixteenth equals two to the negative fourth power (2^{-4} = 1/16 = 0.062 500). One sixteenth is the fourth Eye of Horus fraction.

One sixth The compound noun, pronoun, and adjective one sixth (1/6) is the standard name for the rational fraction equal to one (1) divided by six (6).

One sixty-fourth The compound noun, pronoun, and adjective one sixty-fourth (1/64 = 0.015 625) is the standard name for the rational fraction equal to one (1) divided by sixty-four (64). One sixty-fourth equals two to the negative sixth power ($2^{-6} = 1/64 = 0.015\ 625$). One sixty-fourth is the sixth Eye of Horus fraction.

One tailed test The test of a given statistical hypothesis in which only a value of the statistic that is, for example, sufficiently large will lead to rejection of the hypothesis tested.

The statistical tables for F, for [S-squared x over Sigma squared x] and for Chi. Sq. provide the critical values for a one-tailed test in which the entire alpha region is assigned to the upper end of the sampling distribution when the hypothesis tested is true.

One ten-millionth The compound noun and adjective one ten-millionth (1/10,000,000 = 0.000 000 100) is the standard name for the rational fraction that equals ten to the negative seventh power (10^{-7} = 10 raised to power –07 = 0.000 000 100).

One tenth The compound noun and adjective one tenth (1/10 = 0.100) is the standard name for the rational fraction that equals ten to the negative first power (10^{-1} = 10 raised to power –01 = 0.100).

One ten-thousandth The compound noun and adjective one ten-thousandth (1/10,000 = 0.000 100) is the standard name for the rational fraction that equals ten to the negative fourth power (10^{-4} = 10 raised to power –04 = 0.000 100).

One The noun, pronoun, and adjective one (1) is the standard name for the cardinal number that represents unity. One corresponds with the ordinal number first (1st). .

One third The compound noun, pronoun, and adjective one third (1/3) is the standard name for the rational fraction equal to one (1) divided by three (3).

One thirty-second The compound noun, pronoun, and adjective one thirty-second (1/32 = 0.031 250) is the standard name for the rational fraction equal to one (1) divided by thirty-two (32). One thirty-second equals two to the negative fifth power ($2^{-5} = 1/32 = 0.031\ 250$). One thirty-second is the fifth Eye of Horus fraction.

One thousand The compound noun and adjective one thousand (1,000) is the standard name for the cardinal number that equals ten to the third power (10^3 = 10 raised to power +03 = 1,000).

One time cost A cost associated with project start-up and development, or system start-up.

One to one A characteristic of functions where each element in the domain is pairs with one and only one element in the range and vice versa.

One way ANOVA An analysis of variance where the groups are defined on only one independent variable.

One-hundred-millionth The compound adjective and noun one-hundred-millionth (1/100,000,000 = 0.000 000 010) means divided by one hundred million (100,000,000).

One-hundredth The compound adjective and noun one-hundredth (1/100 = 0.010) means divided by one hundred (100).

One-hundred-thousandth The compound adjective and noun one-hundred-thousandth (1/100,000 = 0.000 010) means divided by one hundred thousand (100,000).

One-millionth The compound adjective and noun one-millionth (1/1,000,000 = 0.000 001) means divided by one million (1,000,000).

Ones The plural noun and adjective ones (1s) means an integer multiple of one (1).

One-thousandth The compound adjective and noun one-thousandth (1/1,000 = 0.001) means divided by one thousand (1,000).

One-way anova A comparison of several groups of observations, all of which are independent and subjected to different levels of a single treatment (such as cells exposed to different dosage of a growth factor). It may be that different groups were exposed to the same treatment (different cell types exposed to a new agent). The main interest focuses on the differences among the means of groups.

Only 1. The adjective and adverb only means one (1) alone.

2. The adjective only means exactly the specified amount or number.

Open ended questions Questions in interviews and on questionnaires that have no prespecified answers.

Open interval Open interval is a piece of a straight line that does not contain the endpoints. On the number line, open intervals (a,b) are defined as {x: a < x < b}. Open intervals are open sets.

Operation A function or a service that is provided by all the instances of a class.

Operational feasibility The process of assessing the degree to which a proposed system solves business problems or takes advantage of business opportunities.

Opposite side The side of a right triangle that is opposite the reference angle.

Opposites Two numbers whose sum is zero. Also called additive inverses.

Optical illusion A drawing or object that appears to have an effect that it does not really have, such as when a flat painting seems to have three-dimensional depth.

Order effect The effect on performance attributable to the order in which treatments were administered.

Order of magnitude The expression order of magnitude (properly decimal order of magnitude) means an exponential range of ten to one (10:1). Two orders of magnitude means an exponential range of one hundred to one (100:1), etc.

Ordered pair A pair of numbers that identifies the location of a particular point on a graph.

Ordered triples (X, Y, Z) The set that is the solution of an equation in three variables.

Ordinal interaction An interaction in which the group differences do not reverse their sign.

Ordinal number An ordinal number is an adjective that describes the order or numerical position of an object, e.g., first, second, third, fourth, fifth, sixth, seventh, etc.

Ordinal scale A scale on which numbers are used only to place objects in order.

Ordinal variable An ordered (ranked) qualitative/categorical variable. The degree of HLA matching (one, two, three or four antigen matching in two loci) in transplant pairs, or HLA sharing in parents (one-to-four shared antigens) are ordinal variables although the increments do not have to be equal in magnitude. An ordinal variable may be a categorised quantitative variable. When two groups are compared for an ordinal variable, it is inappropriate to use ordinary Chi-squared test but the trend test or its equivalents must be used.

Ordinate Vertical (Y) axis.

Origin (on a graph) The point of intersection of the coordinate axes. It is designated (0. 0).

Origin In the Cartesian coordinate plane, the origin is the point at which the horizontal and vertical axes intersect, at zero (0,0).

Original The adjective original means first (1st).

Orthocenter The point of intersection of the altitudes of a triangle.

Orthogonal contrasts A set of contrasts that are independent of one another.

Orthonormal vectors A set of orthonormal vectors is a collection of M vectors X(I), each of order N, each of length 1 in the Euclidean norm:

X(I)' * X(I) = (X(I), X(I)) = 1,

and pairwise orthogonal, so that if I and J are distinct:

X(I)' * X(J) = (X(I), X(J)) = 0.

Given any set of vectors, Gram Schmidt othogonalization or the QR factorization can produce an orthonormal set of vectors, possibly fewer in number, that form a basis for the linear space that is spanned by the original set.

Outcome Any one of the possible results of an experiment.

Outcome of an event One of the possible occurrances in a probability situation.

Outcome space The outcome space is the set of all possible outcomes of a given experiment.

Outer product An outer product of two vectors x of dimension M and y of dimension N is a matrix A of order M by N whose entries are defined by:

A = x * y',

or, entrywise:

$A_{i,j} = x_i * y_j$

A matrix defined as the outer product of two vectors will usually have rank equal to 1. The only other possibility is that the matrix might actually have rank 0.

Outlier An extreme observations that is well separated from the remainder of the data. In regression analysis, not all outlying values will have an influence on the fitted function. Those outlying with regard to their X values (high leverage), and those with Y values that are not consistent with the regression relation for the other values (high residual) are expected to be influential. The test the influence of such values, the Cook statistics is used.

Output The number or value that comes out from a process. For example, in a function machine, a number goes in, something is done to it, and the resulting number is the output.

Outside arc The longest arc of a turn.

Outsourcing The practice of turning-over responsibility of some to all of an organization's information systems applications and operations to an outside firm.

Overdetermined system An overdetermined linear system A * x = b is, loosely speaking, a set of M linear equations in N variables, with M > N.

Actually, it is possible for many of the equations to be redundant, so that what looks formally like an overdetermined system could actually be determined or underdetermined, if we'd just throw away the redundant equations. However, just recognizing and removing the redundancy is a difficult task, except for small problems and exact arithmetic. Thus, we will simply use the term overdetermined for any situation in which there are more equations than variables.

Except in special artificial cases, an overdetermined linear system has no solution. In such a case, we might be interested in an approximate solution which satisfies as many equations exactly as possible, which we might find simply by using pivoting to choose the equations to be satisfied, or a solution x which minimizes the residual norm A*x-b, which we might find using the QR factorization.

Overdispersion Dispersion is a measure of the extent to which data are spread about an average. Overdispersion is the situation that occurs most frequently in Poisson and binomial regression when variance is much higher than the mean (normally, it should be the same). It is evident with a high (>2) residual mean deviance (which should normally be around one) and the presence of too many outliers. The reasons for overdispersion may be outliers, misspecification of the model, variation between the response probabilities and correlation between the binary responses. It distorts standard error and confidence interval estimations. In the analysis, overdispersion may be taken into account by estimating a dispersion parameter.

Overfitting In a multivariable model, having more variables than can be justified from sample size. The statistical rule of thumb is to have at least ten subjects for each variable investigated.

Overmatching When cases and controls are matched by an unnecessary non-confounding variable, this is called overmatching and causes underestimation of an association. For example, matching for a variable strongly correlated with the exposure but not with the outcome will cause loss of efficiency. Another kind of overmatching is matching for a variable which is in the causal chain or closely linked to a factor in the causal chain. This will also obscure evidence for a true association. Finally, if a too strict matching scheme is used, it will be very hard to find controls for the study.

P

p level The probability that a particular result would occur by chance if H_0 is true. The exact probability of a Type I error.

P The metric prefix symbol P- (metric prefix: peta-) denotes a multiple of ten to the fifteenth power (10^{15} = 10 raised to power +15 = 1,000,000,000,000,000) as defined by Le Système International d'Unités (SI).

P value (SP = significance probability) The P value gives the probability that the null hypothesis is correct; therefore, if it is a small value (like <0.05), null hypothesis is rejected. More technically, it is the probability of the observed data or more extreme outcome would have occurred by chance, i.e., departure from the null hypothesis when the null hypothesis is true. In a genetic association study, the P value represents the probability of error in accepting the alternative hypothesis (or rejecting the null hypothesis) for the presence of an association. For example, the P level of 0.05 (i.e., 1/20) indicates that assuming there was no relation between those variables whatsoever (the null hypothesis is correct), and we were repeating experiments like ours one after another, we could expect that approximately in every 20 replications of the experiment, there would be one in which the relation between the variables in question would be equal to or more extreme than what has been found. In the interpretation of a P value, it is important to know the accompanying measure of association and the biological/clinical significance of the significant difference. However small, a P value does not indicate the size of an effect (odds ratio/relative risk/hazard ratio do). A P value >0.05 does not necessarily mean lack of association. It does so only if there is enough power to detect an association. Most statistical nonsignificance is due to lack of power to detect an association (poor experimental design). Both 'p' and 'P' are used to indicate significance probability but the international standard is P (capital italic).

Pairwise independence A pairwise independent collection of random variables is a set of random variables any two of which are independent.

Palindrome Words, numbers and phrases that can be read the same backwords as forwards. Some examples include: mom, racecar, 34543, or the phrase never odd or even.

Parabola A certain type of curve. An equation of the form $y=Ax^2 + Bx + C$ will always give a parabola. Most common example is $y=x^2$ when you have a curve passing through the origin in the shape of an upward opening cup.

Paradox The Greek philosopher Zeno (circa 460 BC) is famous for several paradoxes. In one Achilles couldn't catch a tortoise because he had first to reach the point where the tortoise started.

Meanwhile, the tortoise would move to another point, etc. B. Russell has discovered paradoxes of infinity. There are semantic paradoxes and self-referential ones.

Parallel lines Two lines in the same plane which never intersect are called parallel lines. We say that two line segments are parallel if the lines that they lie on are parallel. If line 1 is parallel to line 2, we write this as line 1 || line 2

When two line segments DC and AB lie on parallel lines, we write this as

Segment DC || segment AB.]

Example: Lines 1 and 2 below are parallel.

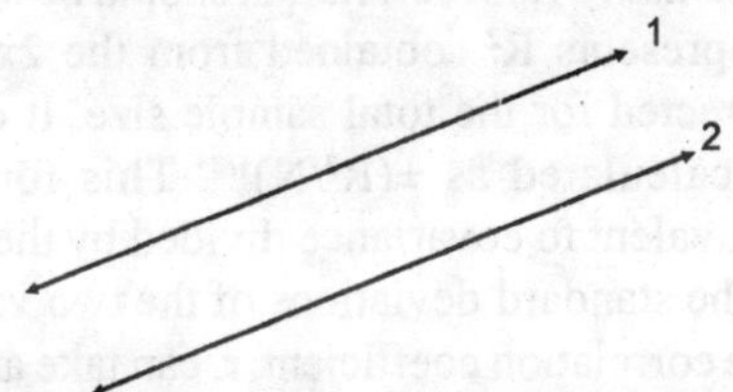

Example: The opposite sides of the rectangle below are parallel. The lines passing through them never meet.

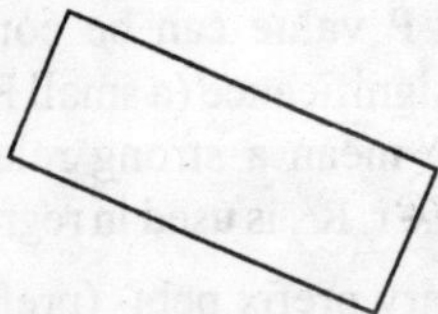

Parallelogram A four-sided polygon with two pairs of parallel sides. The sum of the angles of a parallelogram is 360 degrees.

Examples:

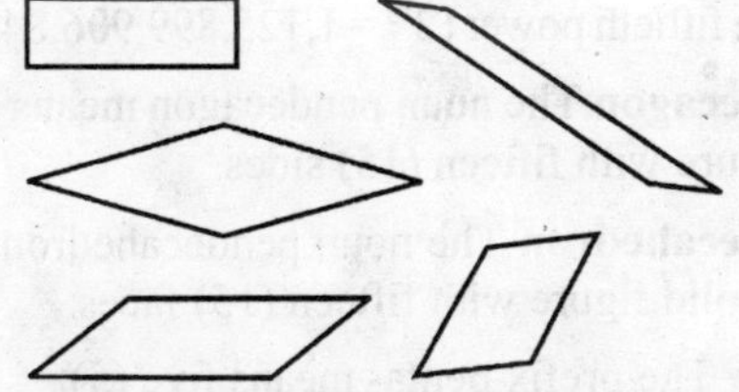

Parallelogram rule A process used to add together two nonparallel vectors.

Parameter A parameter is a measurement on a population that characterizes one of its features. An example of a parameter is the mode. The mode is the value in the population that occurs most frequently. Other examples of parameters are a population's mean (or average) and its variance.

$\mu_x = \frac{\sum x}{N}$ is a parameter

so is $\sigma_x^2 = \frac{\sum (X-\mu_x)^2}{N}$

Parametric tests Statistical tests that involve assumptions about, or estimation of, population parameters.

Parity The word parity applies to situations where two items or their properties may be juxtaposed as being opposites (in a certain context) of each other. Integers are of either odd or even parity when they are, respectively, odd or even. The convenience is in being able to say two numbers of different parities without having to explicitly mention which is which.

Partial correlation ($r_{01.2}$) The correlation between the dependent and independent variables with the effects of one or more additional variables removed from both sides of the equation.

Partial sum of a series The sum of a finite number of terms of the series.

Partialing To hold constant the effect of one variable when looking at the effects of two or more other variables.

Participation constraint A participation constraint defines the number of times an object in an object class can participate in a connected relationship set.

Participatory design (PD) A systems development approach that originated in northern Europe in which users and the improvement of their work lives are the central focus.

Particular The adjective and noun particular means of, relating to, or being a single being or thing.

Partition constraint The partition constraint indicates that every object in a generalization object class must also be a member of at least one of the specialization object classes, and that the specialization object classes are pairwise disjoint.

Partition To divide up a sum of squares—usually the $SS_{treatment}$.

Parts per billion In North American use, the compound noun parts per billion (symbol: ppb) means a relative portion represented as a ratio to ten to the negative ninth power (10^{-9} = 10 raised to power –09 = 0.000 000 001) of the whole, usually by mass.

Parts per million The compound noun parts per million (symbol: ppm) means a relative portion represented as a ratio to one one-millionth (1/ 1,000,000 = 0.000 001) of the whole, usually by mass.

Parts per quadrillion In North American use, the compound noun parts per quadrillion (symbol: ppq) means a relative portion represented as a ratio to ten to the negative fifteenth power (10^{-15} = 10 raised to power –15 = 0.000 000 000 000 001) of the whole, usually by mass.

Parts per thousand The compound noun parts per thousand (symbol: ppt) means a relative portion represented as a ratio to one one-thousandth (1/ 1,000 = 0. 001) of the whole, usually by mass.

Parts per trillion In North American use, the compound noun parts per trillion (symbol: ppt) means a relative portion represented as a ratio to ten to the negative twelfth power (10^{-12} = 10 raised to power –12 = 0.000 000 000 001) of the whole, usually by mass.

Pascal's triangle A triangular arrangement of numbers in which each number is the sum of the two numbers above it in the preceding row.

Path marker A path marker is used to apply a real-time constraint to a part of a state net. They are shown on the diagram as identifiers in braces ({}) placed along directed arrows.

Patterns Regularities in situations such as those in nature, events, shapes, designs, and sets of numbers (e.g., spirals on pineapples, geometric designs in quilts, the number sequence 3, 6, 9, 12, . . .).

Pearson product-moment correlation coefficient (r) The most common correlation coefficient.

Pearson's chi square The traditional chi-square statistic—as opposed to the likelihood ratio chi-square.

Pearson's correlation coefficient (r) A measure of the strength of the 'linear' relationship between two quantitative variables. A major assumption is the normal distribution of variables. If this is not met (for example, due to outliers), the non-parametric equivalent Spearman's rank correlation should be used. The r represents R^2 obtained from the 2x2 table, corrected for the total sample size. It can then be calculated as $\pm(R^2/N)^{1/2}$. This formula is equivalent to covariance divided by the prodict of the standard deviations of the two variables. The correlation coefficient, r, can take any value between -1 and +1; 0 meaning no linear relationship (there may still be a strong non-linear relationship). It is the absolute value of r showing the strength of relationship. An associated P value can be computed for the statistical significance (a small P value does not necessarily mean a strong relationship). The square of the r, R^2, is used in regression analysis.

Pebi The binary prefix pebi- (prefix symbol: Pi-) denotes a multiple of two to the fiftieth power (2^{50} = 1,125,899,906,842,624) as defined by the International Electrotechnical Commission (IEC).

Pebin The adjective and noun pebin means two to the fiftieth power (2^{50} = 1,125,899,906,842,624).

Pendecagon The noun pendecagon means a plane figure with fifteen (15) sides.

Pendecahedron The noun pendecahedron means a solid figure with fifteen (15) faces.

Penta The prefix penta- means five (5).

Pentacle The noun and adjective pentacle means a pentagram inscribed in a circle.

Pentaconta The prefix pentaconta- means fifty (50).

Pentacontad The noun and adjective pentacontad means a group of fifty (50) members.

Pentacontadi The prefix pentacontadi- means fifty-two (52). Pentacontadi- is an alternate form of the prefix pentacontakaidi-.

Pentacontaennea The prefix pentacontaennea- means fifty-nine (59). Pentacontaennea- is an alternate form of the prefix pentacontakaiennea-.

Pentacontagon The noun and adjective pentacontagon means a plain figure with fifty (50) straight sides.

Pentacontagram The noun and adjective pentacontagram means an interconnected star figure with fifty (50) points formed by stellating a pentacontagon.

Pentacontahedron The noun and adjective pentacontahedron means a solid figure with fifty (50) planar faces.

Pentacontahena The prefix pentacontahena- means fifty-one (51). Pentacontahena- is an alternate form of the prefix pentacontakaihena-.

Pentacontahepta The prefix pentacontahepta- means fifty-seven (57). Pentacontahepta- is an alternate form of the prefix pentacontakaihepta-.

Pentacontahexa The prefix pentacontahexa- means fifty-six (56). Pentacontahexa- is an alternate form of the prefix pentacontakaihexa-.

Pentacontakaidi The prefix pentacontakaidi- means fifty-two (52).

Pentacontakaidigon The noun and adjective pentacontakaidigon means a plain figure with fifty-two (52) straight sides.

Pentacontakaidigram The noun and adjective pentacontakaidigram means an interconnected star figure with fifty-two (52) points formed by stellating a pentacontakaidigon.

Pentacontakaidihedron The noun and adjective pentacontakaidihedron means a solid figure with fifty-two (52) planar faces.

Pentacontakaidyad The noun and adjective pentacontakaidyad means a group of fifty-two (52) members.

Pentacontakaiennea The prefix pentacontakaiennea- means fifty-nine (59).

Pentacontakaiennead The noun and adjective pentacontakaiennead means a group of fifty-nine (59) members.

Pentacontakaienneagon The noun and adjective pentacontakaienneagon means a plain figure with fifty-nine (59) straight sides.

Pentacontakaienneagram The noun and adjective pentacontakaienneagram means an interconnected star figure with fifty-nine (59) points formed by stellating a pentacontakaienneagon.

Pentacontakaienneahedron The noun and adjective pentacontakaienneahedron means a solid figure with fifty-nine (59) planar faces.

Pentacontakaihena The prefix pentacontakaihena- means fifty-one (51).

Pentacontakaihenad The noun and adjective pentacontakaihenad means a group of fifty-one (51) members.

Pentacontakaihenagon The noun and adjective pentacontakaihenagon means a plain figure with fifty-one (51) straight sides.

Pentacontakaihenagram The noun and adjective pentacontakaihenagram means an interconnected star figure with fifty-one (51) points formed by stellating a pentacontakaihenagon.

Pentacontakaihenahedron The noun and adjective pentacontakaihenahedron means a solid figure with fifty-one (51) planar faces.

Pentacontakaihepta The prefix pentacontakaihepta- means fifty-seven (57).

Pentacontakaiheptad The noun and adjective pentacontakaiheptad means a group of fifty-seven (57) members.

Pentacontakaiheptagon The noun and adjective pentacontakaiheptagon means a plain figure with fifty-seven (57) straight sides.

Pentacontakaiheptagram The noun and adjective pentacontakaiheptagram means an interconnected star figure with fifty-seven (57) points formed by stellating a pentacontakaiheptagon.

Pentacontakaiheptahedron The noun and adjective pentacontakaiheptahedron means a solid figure with fifty-seven (57) planar faces.

Pentacontakaihexa The prefix pentacontakaihexa- means fifty-six (56).

Pentacontakaihexad The noun and adjective pentacontakaihexad means a group of fifty-six (56) members.

Pentacontakaihexagon The noun and adjective pentacontakaihexagon means a plain figure with fifty-six (56) straight sides.

Pentacontakaihexagram The noun and adjective pentacontakaihexagram means an interconnected star figure with fifty-six (56) points formed by stellating a pentacontakaihexagon.

Pentacontakaihexahedron The noun and adjective pentacontakaihexahedron means a solid figure with fifty-six (56) planar faces.

Pentacontakaiocta The prefix pentacontakaiocta- means fifty-eight (58).

Pentacontakaioctad The noun and adjective pentacontakaioctad means a group of fifty-eight (58) members.

Pentacontakaioctagon The noun and adjective pentacontakaioctagon means a plain figure with fifty-eight (58) straight sides.

Pentacontakaioctagram The noun and adjective pentacontakaioctagram means an interconnected star figure with fifty-eight (58) points formed by stellating a pentacontakaioctagon.

Pentacontakaioctahedron The noun and adjective pentacontakaioctahedron means a solid figure with fifty-eight (58) planar faces.

Pentacontakaipenta The prefix pentacontakaipenta- means fifty-five (55).

Pentacontakaipentad The noun and adjective pentacontakaipentad means a group of fifty-five (55) members.

Pentacontakaipentagon The noun and adjective pentacontakaipentagon means a plain figure with fifty-five (55) straight sides.

Pentacontakaipentagram The noun and adjective pentacontakaipentagram means an interconnected star figure with fifty-five (55) points formed by stellating a pentacontakaipentagon.

Pentacontakaipentahedron The noun and adjective pentacontakaipentahedron means a solid figure with fifty-five (55) planar faces.

Pentacontakaitetra The prefix pentacontakaitetra- means fifty-four (54).

Pentacontakaitetrad The noun and adjective pentacontakaitetrad means a group of fifty-four (54) members.

Pentacontakaitetragon The noun and adjective pentacontakaitetragon means a plain figure with fifty-four (54) straight sides.

Pentacontakaitetragram The noun and adjective pentacontakaitetragram means an interconnected star figure with fifty-four (54) points formed by stellating a pentacontakaitetragon.

Pentacontakaitetrahedron The noun and adjective pentacontakaitetrahedron means a solid figure with fifty-four (54) planar faces.

Pentacontakaitri The prefix pentacontakaitri- means fifty-three (53).

Pentacontakaitriad The noun and adjective pentacontakaitriad means a group of fifty-three (53) members.

Pentacontakaitrigon The noun and adjective pentacontakaitrigon means a plain figure with fifty-three (53) straight sides.

Pentacontakaitrigram The noun and adjective pentacontakaitrigram means an interconnected star figure with fifty-three (53) points formed by stellating a pentacontakaitrigon.

Pentacontakaitrihedron The noun and adjective pentacontakaitrihedron means a solid figure with fifty-three (53) planar faces.

Pentacontaocta The prefix pentacontaocta- means fifty-eight (58). Pentacontaocta- is an alternate form of the prefix pentacontakaiocta-.

Pentacontapenta The prefix pentacontapenta- means fifty-five (55). Pentacontapenta- is an alternate form of the prefix pentacontakaipenta-.

Pentacontatetra The prefix pentacontatetra- means fifty-four (54). Pentacontatetra- is an alternate form of the prefix pentacontakaitetra-.

Pentacontatri The prefix pentacontatri- means fifty-three (53). Pentacontatri- is an alternate form of the prefix pentacontakaitri-.

Pentacta The prefix pentacta- means five hundred (500).

Pentad The noun and adjective pentad means a group of five (5) members.

Pentadeca The prefix pentadeca- means fifteen (15).

Pentagon A five-sided polygon. The sum of the angles of a pentagon is 540 degrees.

Examples: A regular pentagon:

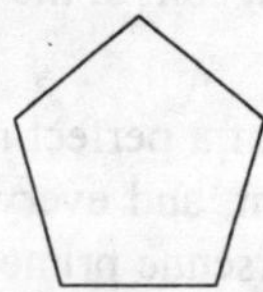

An irregular pentagon:

Pentagram 1. The noun and adjective pentagram means an interconnected star figure with five (5) points formed by stellating a pentagon.

2. The noun pentagram means a word or character sequence consisting of five (5) characters.

Pentahedron The noun and adjective pentahedron means a solid figure with five (5) planar faces.

Pentaheptaconta The prefix pentaheptaconta- means seventy-five (75).

Pentahexaconta The prefix pentahexaconta- means sixty-five (65).

Pentaicosa The prefix pentaicosa- means twenty-five (25).

Pentakaideca The prefix pentakaideca- means fifteen (15).

Pentakaidecad The noun and adjective pentakaidecad means a group of fifteen (15) members.

Pentakaidecagon The noun and adjective pentakaidecagon means a plain figure with fifteen (15) straight sides.

Pentakaidecagram The noun and adjective pentakaidecagram means an interconnected star figure with fifteen (15) points formed by stellating a pentakaidecagon.

Pentakaidecahedron The noun and adjective pentakaidecahedron means a solid figure with fifteen (15) planar faces.

Pentalia The prefix pentalia- means five thousand (5000).

Pentameter The noun pentameter means a verse consisting of five (5) metrical feet.

Pentangle The noun pentangle means a plane figure with five (5) vertices.

Pentanonaconta The prefix pentanonaconta- means ninety-five (95).

Pentapentaconta The prefix pentapentaconta- means fifty-five (55).

Pentathlon The noun pentathlon means an athletic competition consisting of five (5) separate events.

Pentatriaconta The prefix pentatriaconta- means thirty-five (35).

Pentecost The noun Pentecost means the fiftieth (50th) day of the Christian Easter Season.

Penult The noun penult means the next to the last member of a series.

Penultimate The adjective penultimate means the next to the last member of a series.

Per centum The compound noun per centum is the original Latin form of percent.

Per hundred The compound noun per hundred (suffix symbol: –% {percent sign}) means a relative portion represented as a ratio to one one-hundredth (1/100 = 0.010) of the whole.

Per hundred thousand The compound noun per hundred thousand means a relative portion represented as a ratio to one one-hundred-thousandth (1/100,000 = 0.000 010) of the whole.

Per mil The compound noun per mil (suffix symbol: –‰ {per mille sign}) means a relative portion represented as a ratio to one one-thousandth (1/1,000 = 0.001) of the whole. Small portions may be represented per mil rather than percent.

Per mille The compound noun per mille is the original Latin form of per mil.

–‰

The suffix symbol –‰ {per mille sign} denotes a quantity as per mil which means a relative portion represented as a ratio to one one-thousandth (1/1,000 = 0.001) of the whole.

Per million The compound noun per million means a relative portion represented as a ratio to one one-millionth (1/1,000,000 = 0.000 001) of the whole.

Per ten thousand The compound noun per ten thousand means a relative portion represented as a ratio to one ten-thousandth (1/10,000 = 0.000 100) of the whole.

Per thousand The compound noun per thousand (suffix symbol: –‰ {per mille sign}) means a relative portion represented as a ratio to one one-thousandth (1/1,000 = 0.001) of the whole.

Percent The noun percent (suffix symbol: –% {percent sign}) means a relative portion represented as a ratio to one one-hundredth (1/100 = 0.010) of the whole.

A relative change in value may be measured as a percent increase (gain) or decrease (loss) from the base quantity. The first value always serves as the base quantity. Thus, a change from 40.00 to 50.00 is a 25% increase and a change from 50.00 to 40.00 is a 20% decrease. Successive percent changes cannot be added, e.g., a 20% gain followed by a 20% loss yields a net loss of 4%. A doubling is a 100% increase and a tripling is a 200% increase.

–%

The suffix symbol –% {percent sign} denotes a quantity as percent which means a relative portion represented as a ratio to one one-hundredth (1/100 = 0.010) of the whole.

Percentage of agreement The ratio of the number of times two judges agree, divided by the number of judgments. It is a measure that does not correct for chance agreement.

Percentage The noun percentage means a portion expressed as a percent of the whole.

Percentile The noun percentile means the percent of a frequency distribution that is less than or equal to a given value.

Percentile The point below which a specified percentage of the observations fall.

Perfect A number is perfect if sigma(n) = 2n, or, equivalently, if index(n) = 2. Another way of saying this is that sum of the proper divisors of n equals n.

Every multiple of a perfect number other than itself is abundant, and every proper divisor is deficient. A Mersenne prime multiplied by the preceding power of two is a perfect number; i.e., if 2^p – 1 is prime, then (2^p – 1)*2^(p – 1) is perfect. All even perfect numbers are of this form; it is not known if there any odd perfect numbers.

Perfect square trinomial The square of a binomial. with the form a2 ± 2ab + b2.

Perfective maintenance Changes made to a system to add new features or to improve performance.

Perimeter The perimeter of a polygon is the sum of the lengths of all its sides.

Example: What is the perimeter of a rectangle having side-lengths of 3.4 cm and 8.2 cm? Since a rectangle has 4 sides, and the opposite sides of a rectangle have the same length, a rectangle has 2 sides of length 3.4 cm, and 2 sides of length 8.2 cm. The sum of the lengths of all the sides of the rectangle is 3.4 + 3.4 + 8.2 + 8.2 = 23.2 cm.

Example: What is the perimeter of a square having side-length 74 cm? Since a square has 4 sides of equal length, the perimeter of the square is 74 + 74 + 74 + 74 = 4 × 74 = 296.

Example: What is the perimeter of a regular hexagon having side-length 2.5 m? A hexagon is a figure having 6 sides, and since this is a regular hexagon, each side has the same length, so the perimeter of the hexagon is 2.5 + 2.5 + 2.5 + 2.5 + 2.5 + 2.5 = 6 × 2.5 = 15m.

Example: What is the perimeter of a trapezoid having side-lengths 10 cm, 7 cm, 6 cm, and 7 cm? The perimeter is the sum 10 + 7 + 6 + 7 = 30cm.

Period The smallest value of q such that f(x) = f(x+q) where f(x) is a periodic function.

Periodic function A function y=f(x) is periodic if and only if there is a real number p such that for each x in the domain of f. f(x + p)=f(x).

Permutations The number of ways objects can be arranged taking ordering into account.

Perpendicular lines Two lines that meet at a right angle are perpendicular.

perpendicular(ITY) Lines in the same plane which intersect to form a right angle.

Perron frobenius theorem The Perron-Frobenius Theorem tells us a great deal about the largest eigenvalue of a positive or nonnegative matrix.

The Perron-Frobenius Theorem for a positive matrix: If the square matrix A is positive, then the spectral radius rho(A) > 0;

there is an eigenvalue lambda equal to rho(A);

there is a corresponding positive eigenvector x, normalized so that the sum of its entries is one, called the Perron vector, so that A * x = lambda * x = rho(A) * x;

lambda = rho(A) is an algebraically and geometrically simple eigenvalue of A;

all other eigenvalues of A are strictly smaller in magnitude than lambda;

$[A/\text{rho}(A)]^m$ –> L > 0 as m –> infinity, where L=x*y', with x the (right) eigenvector of lambda, and y a left eigenvector of lambda, with the properties that y > 0 and x'*y=1.

The Perron-Frobenius Theorem for a nonnegative matrix: If the nontrivial matrix A is nonnegative, then

there is an eigenvalue lambda equal to rho(A);

there is a corresponding nonnegative (and nonzero) eigenvector x, so that A * x = lambda * x = rho(A) * x;

The Perron-Frobenius Theorem for a nonnegative irreducible matrix: If the nontrivial matrix A is nonnegative and irreducible, then in addition:

the spectral radius rho(A) > 0;

there is an eigenvalue lambda equal to rho(A);

there is a corresponding positive eigenvector x, so that A * x = lambda * x = rho(A) * x;

lambda = rho(A) is an algebraically and geometrically simple eigenvalue of A;

The Perron-Frobenius Theorem for a nonnegative irreducible primitive matrix: If the nontrivial matrix A is nonnegative and irreducible and primitive, then in addition: $[A/\text{rho}(A)]^m$ –> L > 0 as m –> infinity, where L=x*y', with x the (right) eigenvector of lambda, and y a left eigenvector of lambda, with the properties that y > 0 and x'*y=1.

Personal view An approach taken by mathematicians and philosophers to calculate probability. Using their knowledge and reasoning skills, they think through the problem.

Persymmetric matrix A persymmetric matrix A is a square matrix whose values are reflected across its main anti-diagonal, that is, for all I and J:

$A_{i,j}$ = A(N+1–J,N+1–I).

Here is an example of a persymmetric matrix:

4 3 2 1

7 6 5 2

9 8 6 3

10 9 7 4

PERT chart A diagram that depicts project tasks and their interrelationships. PERT stands for Program Evaluation Review Technique.

Peta The metric prefix peta- (prefix symbol: P-) denotes a multiple of ten to the fifteenth power (10^{15} = 10 raised to power +15 = 1,000,000,000, 000,000) as defined by Le Système International d'Unités (SI).

Petabinary The adjective petabinary means two to the fiftieth power (2^{50} = 1,125,899,906,842, 624). See also binary prefix pebi- and compound binary prefix symbol

Pharmacoepidemiology Application of epidemiological reasoning, methods and knowledge to the study of the uses and effects (beneficial and adverse) of drugs in human populations. A relatively new field in epidemiology becoming more closely related to pharmacogenetics.

Phase shift The horizontal displacement of a function to the right or left of the vertical axis.

Phi (φ) The correlation coefficient when both of the variables are measured as dichotomies.

Phi coefficient A measure of association of two variables calculated from a contingency table as $(X^2 / N)^{1/2}$. Its value varies between 0 (no association) and 1 (strongest association) for 2x2 tables where it is an accurate statistics .In a way, it is a corrected Chi-squared value for the number of observations.

Physical file A named set of table rows stored in a contiguous section of secondary memory.

Physical table A named set of rows and columns that specifies the fields in each row of the table.

Pi (¼) The ratio of the circumference of a circle to its diameter. It is the whole number 3 and a decimal fraction that has endless decimal places. We usually round it off to four decimal places (3.1416) for our work.

Pico The metric prefix pico- (prefix symbol: p-) denotes a multiple of ten to the negative twelfth power (10^{-12} = 10 raised to power -12 = 0.000 000 000 001) as defined by Le Système International d'Unités (SI).

Pictograph A graphical representation that shows numerical information by using picture symbols.

Picture (or template) A pattern of codes that restricts the width and possible values for each position of a field.

Pie graph A diagram showing a system of connections or interrelations between two or more things by using a circle divided into segments that look like pieces of pie.

Pivoting The attempt to improve the accuracy of a linear algebra algorithm by choosing the most suitable column or row for use during a single step.

The most common instance of pivoting occurs in Gauss elimination. During the first step, we wish to eliminate all the entries in column one except for one entry. We will do this by adding carefully chosen multiples of row one to the other rows. For example, if the first two rows were

2 3 5 8

6 8 9 2

we could add –3 times row 1 to row 2, and thus zero out the 6 in the first column of row 2. This scheme would not work, however, if the first entry in the first row were zero, because no multiple of zero can be anything but zero.

Thus, we allow ourselves to interchange the first row with some other row which has a nonzero entry in the first column. But since we're going to interchange rows anyway, it turns out that best accuracy occurs if we choose to bring in the entry with the largest absolute value. Such a scheme, which considers each column in order, and uses the largest entry in the row as the pivot, is called partial pivoting. This form of pivoting is used in most linear algebra software.

It can be shown that Gauss elimination can be carried out without pivoting if each of the principal minors of the matrix is nonsingular. If the matrix is positive definite, then this condition is guaranteed, and it is common to factor the matrix without pivoting.

A complete pivoting or full pivoting scheme would search for the entry of largest magnitude anywhere in the uneliminated portion of the matrix, and use that entry as the next pivot. Such a scheme requires a lot more searching, and auxiliary storage of the row and column numbers of the pivots. It is little used, since it does not seem to bring a great improvement in accuracy.

QR factorization can also use pivoting. At step I, columns I through N are examined. The column with the greatest norm is interchanged with column I, and then the QR operations for that step are carried out. In this case as well, pivoting is done to try to ensure stability.

Place value The value of a digit as determined by its position in a number (e.g., in the number 11 the one is worth either 10 or 1, depending on the position).

Plane A flat, level or even surface.

Platykurtic A distribution that is relatively thick in the shoulders.

Plex The suffix -plex means a building divided into the prefix number of spaces.

The suffix -plex most commonly refers to housing and theatre complexes. Duplex is an alternate name for a twoplex (2–plex) and triplex is an alternate name for a threeplex (3–plex). A multiplex is a building with many spaces.

Origin: The suffix -plex is a back formation of the noun complex.

Plumb Exactly Vertical. A line that is perpendicular to ground level is plumb.

Plural The adjective plural means more than one (>1).

Plurality 1. The noun plurality means the state of being more than one (>1).

2. The noun plurality means the largest number of votes cast in an election for any candidate or question, especially if less than a majority.

Plus . The adjective and noun plus (prefix symbol: + {plus sign}) means positive or greater than zero (>0).

±

The symbol ± {plus or minus sign} is used to denote a quantity which should be both added and subtracted. The ± symbol can be used to denote a range of uncertainty, or to denote a pair of quantities, such as the two roots of a quadratic equation.

Point biserial correlation (r_{pb}) The correlation coefficient when one of the variables is measured as a dichotomy.

Point estimate The specific value taken as the estimate of a parameter.

Point matrix A 2 x 1 matrix whose ele- ments are the coordinates of a point.

Point of tangency The point where a circle and a tangent line touch.

Pointer A field of data that can be used to locate a related field or row of data.

Points A point is one of the basic terms in geometry. We may think of a point as a dot on a piece of paper. We identify this point with a number or letter. A point has no length or width, it just specifies an exact location.

Example: The following is a diagram of points A, B, C, and Q:

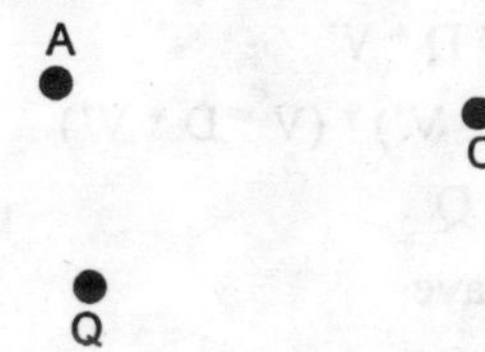

Point-slope form If (x1, y1) is a point on the line y = mx + b. then y – y1 = m(x – xl) is the point-slope form of the equation.

Poisson distribution The probability distribution of the number of (rare) occurrences of some random event in an interval of time or space. Poisson distribution is used to represent distribution of counts like number of defects in a piece of material, customer arrivals, insurance claims, incoming telephone calls, or alpha particles emitted. A transformation that often changes Poisson data approximately normal is the square root.

Poisson regression Analysis of the relationship between an observed count with a Poisson distribution (i.e., outcome variable) and a set of explanatory variables. In general it is appropriate to fit a Poisson model to the data if the sample size is > 100 and the mean for the occurrence of the event is <0.10xN.

Polar axis A ray extending from the pole in a polar coordinate system.

Polar coordinate system A coordinate system using distance and angle for position.

Polar coordinates An ordered pair consisting of a radius and an angle.

Polar decomposition The polar decomposition of any matrix A has the form

A = P * Q

where P is positive semidefinite and has the same rank as A, and Q is unitary.

The polar decomposition of a matrix can be determined from its singular value decomposition:

A = U * D * V'

= (U * V') * (V * D * V')

= P * Q

where we have

P = U * V'

Q = V * D * V'

Pole A pole is a singularity of an analytic function where it behaves very nearly like a rational function. If the function is bounded around a singularity the latter is called removable.

Political feasibility The process of evaluating how key stakeholders within the organization view the proposed system.

Polygamic The adjective polygamic means being engaged in multiple marriage.

Polygamist The noun and adjective polygamist means a person who is engaged in multiple marriage.

Polygamous The adjective polygamous means being engaged in multiple marriage.

Polygamy The noun polygamy means marriage to more than one (1) spouse at the same time.

Polygamy usually refers to multiple marriage that is socially accepted, at least among a social subgroup. Bigamy usually refers to illegal multiple marriage.

Polygon A polygon is a closed figure made by joining line segments, where each line segment intersects exactly two others.

Examples: The following are examples of polygons:

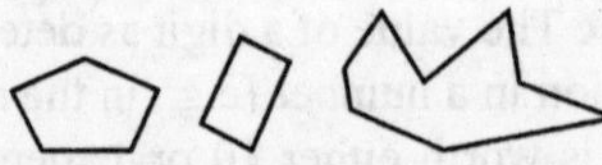

The figure below is not a polygon, since it is not a closed figure:

The figure below is not a polygon, since it is not made of line segments:

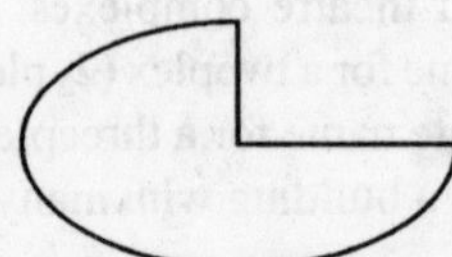

The figure below is not a polygon, since its sides do not intersect in exactly two places each:

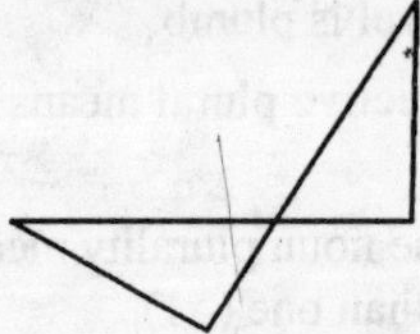

Polyhedra Any solid figure with an outer surface composed of polygon faces.

Polynomial A sum of multiples of integer powers of a variable. The highest power in the expression (n) is the degree of the polynomial. When n=1, for example, $f(x)=2x^1+3$, this is a linear expression. If n=2, it is quadratic (for example, $x^2 + 2x + 4$); if n=3, it is cubic, if n=4, it is quartic and if n=5, it is quintic.

Polynomial degree The highest power or sum of powers in any term of a given polynomial.

Polynomial function A function P whose values are defined by a polynomial.

Polynomial term An algebraic expression that is an addend in a polynomial expression.

Polynomial trend coefficient A set of coefficients used for testing for polynomial (e.g., linear, quadratic, ...) trend.

Pooled variance A weighted average of the separate sample variances. The degrees of freedom serve as the weights.

Population Statisticians define a population as the entire collection of items that is the focus of concern. The branch of Statistics called Descriptive Statistics, provides us with ways to describe the characteristics of a given population by measuring each of its items and then summarizing the set of measures in various ways. The branch of Statistics called Inferential Statistics consists of procedures to make educated inferences about the characteristics of a population by drawing a random sample and appropriately analyzing the information it provides.

A population can be of any size and while the items need not be uniform, the items must share at least one measurable feature. For example here is a population of 9 persons. While no two of the persons are identical they have many features in common. Each of the persons in this population has a weight, a height, a hat size and a shoe size, among many other potential features. The set of 9 measurements of any one of these features would, in statistical terms, be defined as a population.

The critical difference between a population and a sample, is that with a population our interest is to identify its characteristics whereas with a sample, our interest is to make inferences about the characteristics of the population from which the sample was drawn.

It is noteworthy that while the illustrated population is fairly small, it contains only 9 items, a different population might be extremely large, It might, for example, consist of all of the persons in a given city, country, planet, or even universe.

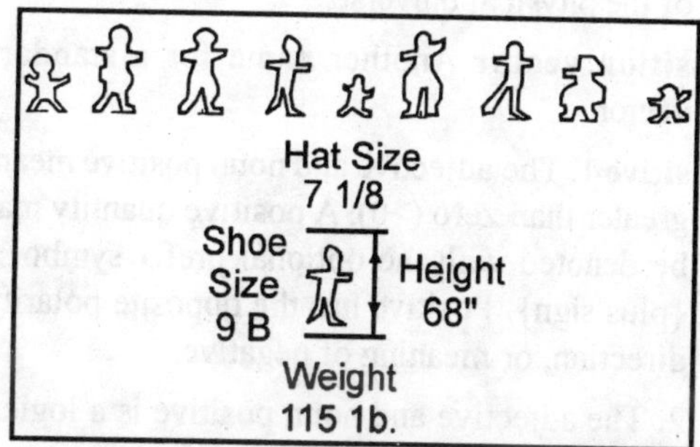

Population attributable risk The proportion of a disease in a specified population attributable to a specific factor (such as a genetic risk factor).

Population stratification (substructure) An example of 'confounding by ethnicity' in which the co-existence of different disease rates and allele frequencies within population sub-sections lead to an association between the two at a whole population level. Differing allele frequencies in ethnically different strata in a single population may lead to a spurious association or mask an association by artificially modifying allele frequencies in cases and controls when there is no real association (for this to happen, the subpopulations should differ not only in allele frequencies but also in baseline risk to the disease being studied). Case-control association studies can still be conducted by using genomic controls (Devlin, 1999; Pritchard, 1999) even when population stratification is present. The software STRUCTURE & STRAT or ADMIXMAP can be used to analyse case-control data with genomic control.

Population variance Variance of the population. It is usually estimated, rarely computed.

Position function This is a function that depends on time and tells you what your position is along a number line as time varies. For instance, if f(t) = t^2, in units of feet and seconds, then at time t=0, you are at the origin, at time t=1 second you are 1 foot to the right of the origin and at time t= 2 seconds, you are 4 feet to the right of the origin. Of course at time t= 52 hours, you have travelled farther than the speed of light would allow, breaking one of the most basic laws of the physical universe.

Position vector Another name for a standard vector.

Positive 1. The adjective and noun positive means greater than zero (>0). A positive quantity may be denoted with the optional prefix symbol + {plus sign}. Positive has the opposite polarity, direction, or meaning of negative.

2. The adjective and noun positive is a logical value. Negative and positive are sometimes used as a binary logic pair.

Positive angle Results from counterclockwise rotation.

Positive definite matrix A (complex) matrix A is positive definite if it is true that for every nonzero (complex) vector x, the product:

$x^* * A * x > 0$,

where, in particular, we are requiring that this product be a real number.

A real matrix A is restricted positive definite if it is true that for every nonzero real vector x:

$x' * A * x > 0$.

If a real matrix is positive definite, then it is restricted positive definite.

Every complex positive definite matrix is guaranteed to be Hermitian. Thus, the phrase positive definite Hermitian is, strictly speaking, redundant. However, a restricted positive definite matrix is not guaranteed to be symmetric. This can happen because we only use real vectors in the test product.

Here is a simple example of a restricted positive definite matrix which is not symmetric:

1 1

–1 1

because $x' * A * x = x_1^2+x_2^2$

Whenever real positive definite matrices occur in practical applications, however, they are assumed or required to be symmetric. If necessary, one can decompose a positive definite matrix in the restricted sense into its symmetric and antisymmetric parts. The symmetric part will actually be (fully) positive definite.

Simple facts about a positive definite matrix A:

A is nonsingular;

The inverse of A is positive definite;

every eigenvalue of A is positive;

Gauss elimination can be performed on A without pivoting;

if A is symmetric as well, it has a Cholesky factorization A = L * L', where L is lower triangular.

A matrix about which no such information is known is called indefinite. A matrix for which the product is always nonnegative is sometimes called positive indefinite or positive semidefinite. There are similar definitions of negative definite and negative indefinite.

Positive predictive value Probability of a true positive as in a person identified as diseased by a test is really diseased.

Positively skewed A distribution that trails off to the right.

Post hoc test A test following another one. The most common example is performing multiple comparisons between groups when the overall comparison between groups shows a significant difference. For example, when an ANOVA analysis yields a small P value, post hoc tests (such as Newman-Keuls, Duncan's or Dunnett's tests) are done to narrow down exactly which pairs differ most significantly (similarly, Dunn's test is done in a non-parametric ANOVA setting) (GraphPad Post ANOVA Test Calculator). In genetic association studies, multiple comparisons are justified only when performed as a post hoc test following a significant deviation in overall gene/marker frequencies.

Posterior probability Probability density is used to describe probability in a continuous probability distribution. For example, you can't say that the probability of a man being six feet tall is 20%, but you can say he has 20% of chances of being between five and six feet tall. Probability density is given by a probability density function. Contrast with probability mass.

Power For a statistician, the power of a test is the probability that the test will reject the hypothesis tested when a specific alternative hypothesis is true. To calculate the power of a given test it is necessary to specify alpha (the probability that the test will lead to the rejection of the hypothesis tested when that hypothesis is true) and to specify a specific alternative hypothesis.

Ppb In American use, the compound symbol ppb represents parts per billion designating a relative portion represented as a ratio to ten to the negative ninth power (10^{-9} = 10 raised to power –09 = 0.000 000 001) of the whole, usually by mass.

Ppm The compound symbol ppm represents parts per million designating a relative portion represented as a ratio to one one-millionth (1/1,000,000 = 0.000 001) of the whole, usually by mass.

Ppq In American use, the compound symbol ppq represents parts per quadrillion designating a relative portion represented as a ratio to ten to the negative fifteenth power (10^{-15} = 10 raised to power –15 = 0.000 000 000 000 001) of the whole, usually by mass.

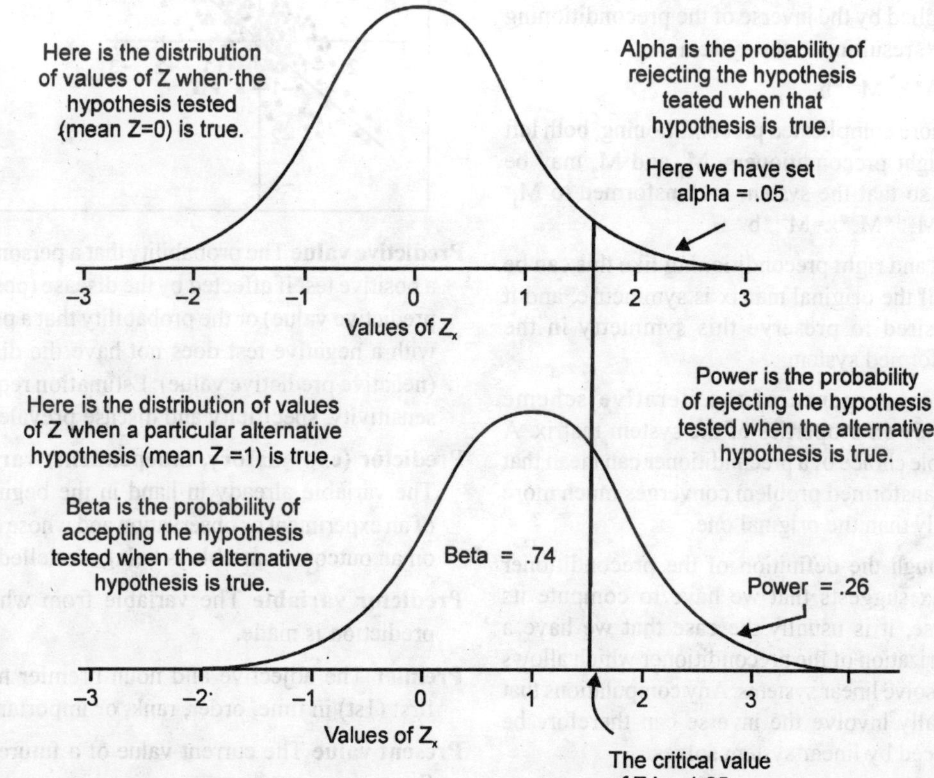

Power function Functions with equations of the form y = x^n, with it a positive integer.

Power rule In calculus, the derivative of x^n equals nx^{n-1}.

Ppt 1. The compound symbol ppt represents parts per thousand designating a relative portion represented as a ratio to one one-thousandth (1/1,000 = 0. 001) of the whole, usually by mass.

2. In American use, the compound symbol ppt represents parts per trillion designating a relative portion represented as a ratio to ten to the negative twelfth power (10^{-12} = 10 raised to power −12 = 0.000 000 000 001) of the whole, usually by mass.

Precision The smallest place value to which an approximate number or measurement is expressed (e.g., if pi is represented as 3.14, then its precision is .01).

Preconditioner A preconditioner is a matrix M most commonly used to improve the performance of an iterative method for linear equations.

The simplest preconditioning is done using a left preconditioner. The original linear system is left-multiplied by the inverse of the preconditioning matrix, resulting in the system

$M^{-1}*A*x=M^{-1}*b$

For more complicated preconditioning, both left and right preconditioners, M_1 and M_2 may be used, so that the system is transformed to $M_1^{-1}*A*M_2^{-1}*M_2*x=M^{-1}*b$

A left and right preconditioning like this can be used if the original matrix is symmetric, and it is desired to preserve this symmetry in the transformed system.

The convergence of the iterative scheme depends on properties of the system matrix. A suitable choice of a preconditioner can mean that the transformed problem converges much more rapidly than the original one.

Although the definition of the preconditioner matrix suggests that we have to compute its inverse, it is usually the case that we have a factorization of the preconditioner which allows us to solve linear systems. Any computations that formally involve the inverse can therefore be replaced by linear system solves.

Desirable properties of a preconditioner matrix include: M should approximate A in some way;

It should be significantly easier to solve linear systems involving M than A (otherwise the problem has just gotten harder);

M should not require much more storage space than A.

Examples of preconditioners include

The Jacobi preconditioner;

Incomplete Cholesky factorization

The incomplete LU factorization;

The incomplete Cholesky factorization;

Prediction In the concept of regression, the regression line indicates the predicted value of Y for each value of X.

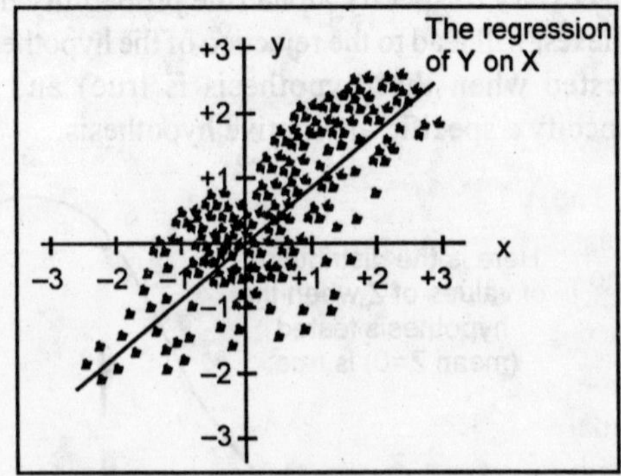

Predictive value The probability that a person with a positive test if affected by the disease (positive predictive value) or the probability that a person with a negative test does not have the disease (negative predictive value). Estimation requires sensitivity, specificity and disease prevalence.

Predictor (explanatory, independent) variable The variable already in hand in the beginning of an experiment or observation and whose effect on an outcome variable is being modelled.

Predictor variable The variable from which a prediction is made.

Premier The adjective and noun premier means first (1st) in time, order, rank, or importance.

Present value The current value of a future cash flow.

Preteen The noun and adjective preteen means less than thirteen (13) years of age. Preteen commonly means between nine (9) and thirteen (13) years of age.

Prevented fraction The amount of a health problem that actually has been prevented by a prevention strategy in real world.

Preventive maintenance Changes made to a system to avoid possible future problems.

Primary key An attribute whose value is unique across all occurrences of a relation.

Primary solutions Solutions defined over a limited domain.

Primary The adjective and noun primary means first (1st) in order or importance.

Primate The noun and adjective primate means a member of the first (1st) rank or order.

Prime 1. The adjective prime means of first (1st) rank.

2. The suffix symbol –2 {prime} denotes a linear measurement in feet. One foot is equal to 0.304 800 meter.

3. The suffix symbol –2 {prime} denotes an angular measurement in minutes. One minute is equal to one sixtieth (1/60) of a degree (suffix symbol: –° {degree symbol}) or one twenty-one-thousand-six-hundredth (1/21,600) of a circle.

Prime factorization A representation of a number as the product of its prime factors.

Prime number A prime number is an integer greater than one (1) that is a multiple of no integer other than itself and one (1), e.g., 2, 3, 5, 7, 11, 13, 17, 19, 23, 29, 31, 37, 41, 43, 47, 53, 59, 61, 67, 71, 73, 79, 83, 89, 97, etc.

Primitive DFD The lowest level of decomposition for a data flow diagram.

Primitive matrix A primitive matrix is a (square) matrix which is irreducible and which has only one eigenvalue of maximum modulus.

Note that we are not requiring that the eigenvalue be simple (that is, of multiplicity 1). Rather, we are requiring that there not be two distinct eigenvalues of the same modulus.

The Perron Frobenius Theorem applies to positive matrices. It is possible to extend most of the Perron-Frobenius theorem to a nonnegative matrix, if it is assumed to be a primitive matrix.

Princely The adjective and adverb princely means of the first (1st) or highest rank of citizenship.

Princess The noun princess means a woman of the first (1st) or highest rank of citizenship.

Principal cube root The one real-number cube root of a real number.

Principal nth root The unary root of a complex number.

Principal square root The positive square root, denoted by .

Principle square root function The function with the equation?.

Prior probability A population or statistical population is a set of entities about which statistical inferences are to be drawn, often based on random sampling. One can also talk about a population of measurements or values.

Prior state A prior state is a state which preceeds a transition. A state may be a prior state of some transitions and a subsequent state of others.

Prior state conjunction A transition may require more than one state to be on before it begins. The set of prior states which must be on is a prior state conjunction.

Prism A prism is a space figure with two congruent, parallel bases that are polygons.

Examples: The figure below is a pentagonal prism (the bases are pentagons). The grayed lines are edges hidden from view.

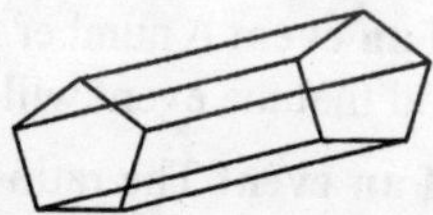

The figure below is a triangular prism (the bases are triangles). The grayed lines are edges hidden from view.

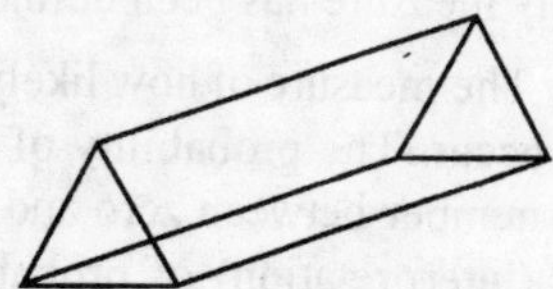

The figure below is a hexagonal prism (the bases are hexagons). The grayed lines are edges hidden from view.

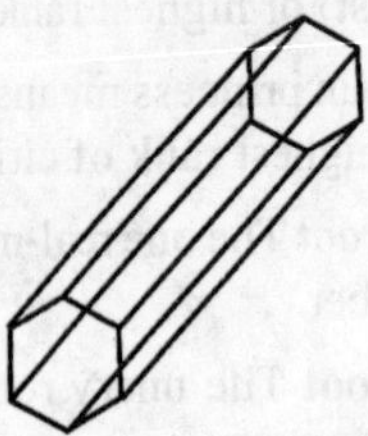

Prisoners values for c in the Julia Set or Mandelbrot set where at each iteration the resulting value becomes smaller and smaller, approaching zero.

Probability density function Gives the probability distribution for a continuous random variable.

Probability density function When a curve is used to model the variation in a population and the total area between the curve and the x-axis is 1, then the function that defines the curve is a probability density function.

Probability distribution A function that gives the probability of all elements in a given space.

Probability distribution function A function which gives for each number x, the probability that the value of a continuous random variable X is less than or equal to x. For discrete random variables, the probability distribution function is given as the probability associated with each possible discrete value of the variable.

Probability measure Gives the probability of events in a probability space.

Probability of an event A number that represents the likelihood that the event will occur.

Probability of an event The ratio of the number of outcomes in the event to the number of outcomes in the sample space.

Probability space A sample space over which a probability measure has been defined.

Probability The measure of how likely it is for an event to occur. The probability of an event is always a number between zero and 100%. The meaning (interpretation) of probability is the subject of theories of probability. However, any rule for assigning probabilities to events has to satisfy the axioms of probability.

Probability vector Any vector with non-negative entries whose sum is equal to 1.0.

Process modeling Graphically representing the processes that capture, manipulate, store, and distribute data between a system and its environment and among components within a system.

Process The work or actions performed on data so that they are transformed, stored, or distributed.

Processing logic The steps by which data are transformed or moved and a description of the events that trigger these steps.

Process-oriented approach An approach to developing information systems that focuses on how and when data are moved and changed.

Product-sum identities Useful in writing the product of trig functions as the sum and difference of trig functions.

Program A program is a sequence of instructions that accomplish a certain task. Often instructions are written in a machine language so that a computer may perform the task required.

Project A planned undertaking of related activities to reach an objective that has a beginning and an end.

Project closedown The final phase of the project management process, which focuses on bringing a project to an end.

Project execution The third phase of the project management process, in which the plans created in the prior phases (project initiation and planning) are put into action.

Project initiation The first phase of the project management process, in which activities are performed to assess the size, scope, and complexity of the project and to establish procedures to support later project activities.

Project management A controlled process of initiating, planning, executing, and closing down a project.

Project manager A systems analyst with a diverse set of skills—management, leadership, technical, conflict management, and customer relationship—who is responsible for initiating, planning, executing, and closing down a project.

Project planning The second phase of the project management process, which focuses on defining clear, discrete activities and the work needed to complete each activity within a single project.

Project workbook An on-line or hard-copy repository for all project correspondence, inputs, outputs, deliverables, procedures, and standards that is used for performing project audits, orientation of new team members, communication with management and customers, identifying future projects, and performing postproject reviews.

Projection matrix A matrix A is a projection matrix if it is idempotent and hermitian (or symmetric in the real case).

A projection matrix is sometimes referred to as an orthogonal projection matrix, but in this case, the matrix is not generally what is termed an orthogonal matrix; rather, what is being referred to is the fact that A may be used to determine the projection of a vector x onto the orthogonal complement of a given subspace. Thus, the vectors A*x and (I–A)*x comprise a decomposition of x into two orthogonal components, one lying in the space spanned by the columns of A, and one in the space orthogonal to that one.

Simple facts about a projection matrix A:

I – A is also a projection matrix;

A * A * x = A * x (idempotent);

A may be written as A = sum (1 <= i <= k) $v_i v_i^H$, where the v_i are an orthonormal set of vectors that span the range of A.

Given a set of k linearly independent vectors v_i, with F=[v_1, v_2,...,v_k] then the projection onto the space spanned by these vectors is

A = F * inverse (F^H * F) * F^H

The previous statement be rephrased as follows: Given an m by n rectangular matrix F, with m>=n, with maximal column rank (that is, the columns are linearly independent), then the matrix

A = F * inverse (F^H * F) * F^H

can be used to determine the projection of any vector onto the column space of F.

If, in fact, the columns of F happen to be of unit L2 norm and pairwise orthogonal, (that is, the columns are an orthonormal set of vectors), then F^H * F = I, so the formula for the projector simplifies to:

A = F * F^H

This particular formulation often occurs when a QR factorization has been used to extract an orthonormal basis for the column space.

Note, in particular, that it must be the case that A * F = F. When solving the overdetermined linear system,

F * x = b

it is natural to consider the related projected linear system

A * F * y = F * y = A * b

which can be solved exactly, because A*b lies in the column space of F. Moreover, this solution y is the best solution to the original problem F * x = b, in the least squares sense, that is, it achieves the residual of minimum L2 norm. The matrix A is sometimes called the hat matrix for matrix F. It is often used in statistics; in some cases, the diagonal entries of A are of considerable interest.

In the simple case where we are given one (nonzero) vector v, then F=v, and the projection onto the space spanned by v is

A = v * inverse (v' * v) * v' = v * v' / $||v||^2$

using the l2 norm; hence the matrix that projects a vector into the orthogonal complement of v is

I – A = I – v * v' / $||v||^2$,

or, if v is already of unit l2 norm, simply I–v*v'.

Projection Our shadow is our projection on the ground. x is the projection on the X axis and y is the projection on the Y axis, of the point (x, y).

Proper multiperfect A number is proper multiperfect if its index is an integer greater than two; i.e., if it is a multiperfect number greater than one which is not perfect. Often multiperfect is used to mean proper multiperfect.

Properties of operations Mathematical principals that are always true (e.g., commutative, associative, distributive, inverses).

Property A A matrix A has property A if the indices 1 to N can be divided into two sets S and T so that, for any $A_{i,j}$ which is not zero, it must be the case that:

I = J, or

I is in S and J is in T, or

I is in T and J is in S.

This is equivalent to saying that, by listing the S rows and columns first, the matrix can be rewritten in the block form:

| D1 F |

| G D2 |

where D1 and D2 are diagonal.

Proportion A relationship between two ratios in which the first ratio is always equal to the second.

Proportional improvement in prediction (PIP) $1-\sqrt{1-r^2}$. The degree to which the prediction of Y is improved by using X.

Proportional odds ratio When the response variable in an ordered/ordinal logistic regression model has more than two ordered response categories, odds ratio obtained for each category is called a proportional odds ratio.

Proportional reduction in error (PRE) The degree to which the residual error is reduced after taking X into account, relative to the error without X.

Proportionality A condition in a factorial analysis of variance where a certain proportionality exists among sample sizes.

Protected t A technique in which we run t tests between pairs of means only if the analysis of variance was significant. Also known as Fisher's LSD test.

Prototyping Building a scaled-down version of the desired information system.

Protractor An instrument for laying down and measuring angles on paper, used in drawing and plotting.

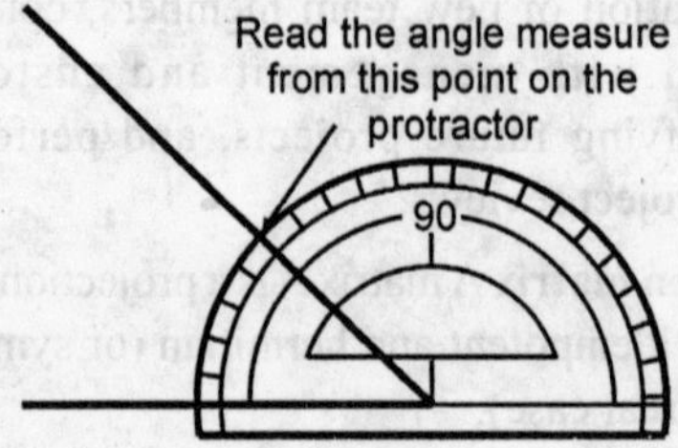

Proving the identity Showing the validity of one identity by using previously known facts.

Pure imaginary number Any number of the form ai. where a is real.

Pure quadratic equation A quadratic equation with no x-term.

Purpose The overall goal or function of a system.

Pyramid A pyramid is a space figure with a square base and 4 triangle-shaped sides.

Example: The picture below is a pyramid. The grayed lines are edges hidden from view.

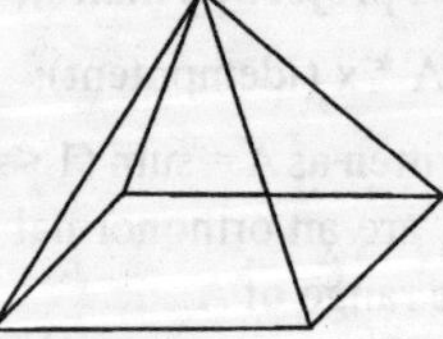

Pythagoras' constant Pythagoras' Constant is the ratio of the length of a diagonal of a square to the length of a side of the square. Pythagoras' Constant is an irrational number equal to the square root of two or approximately 1.414 213 562 373.

Pythagorean identities Fundamental identities that relate the sine and cosine functions and the Pythagorean Theorem.

Pythagorean theorem Used to find side lengths of right triangles, the Pythagorean Theorem states that the square of the hypotenuse is equal to the squares of the two sides, or $A^2 + B^2 = C^2$, where C is the hypotenuse.

Q

Q

QR Method for eigenvalues The QR method is used to find the eigenvalues of a square matrix.

For efficiency, the method begins by transforming the matrix A using Householder matrices until we have determined an upper Hessenberg matrix A1 which is orthogonally similar to A.

The QR factorization of A1 is then computed, A1 = Q1*R1. Then the factors Q1 and R1 are reversed, to compute a new matrix A2 = R1*Q1. A2 itself can be QR factored into A2 = Q2*R2, and then reversal of factors results in a matrix A3, and so on.

Each matrix A1, A2, and so on is orthogonally similar to A, and so shares the same eigenvalues. The sequence of matrices A, A1, A2, A3, ... will generally converge to a matrix B of a very simple type: B will consist entirely of diagonal entries and two by two blocks. The diagonal entries are the real eigenvalues, and the eigenvalues of the 2 by 2 blocks are the complex eigenvalues.

In the particular case where the eigenvalues are all real, (and hence B is diagonal), and if the orthogonal transformations have been accumulated along the way, the result can be written:

A = Q * B * Q'

which is an orthogonal similarity transformation of A into a diagonal matrix, or

A * Q = Q * B

which means that the columns of Q are the eigenvectors.

If the matrix B is not diagonal, then the pair of columns corresponding to any 2 by 2 block can be used to construct the pair of corresponding complex eigenvectors. Alternatively, once the eigenvalues have been determined, the inverse power method can be used to compute any particular eigenvector.

Quadra The numeric prefix quadra- means four (4). This prefix is derived from the Latin cardinal number quattuor (IIII).

Quadragenarian The noun and adjective quadragenarian means between forty (40) and fifty (50) years of age.

Quadragesimal The adjective quadragesimal means based on the number forty (40).

Quadraginta The numeric prefix quadraginta- means forty (40). This prefix is derived from the Latin cardinal number quadraginta (XXXX).

Quadragintennary The noun and adjective quadragintennary means a group of forty (40) or a period of forty (40) years.

Quadragintennial 1. The adjective and noun quadragintennial means recurring every forty (40) years.

2. The adjective and noun quadragintennial means lasting forty (40) years.

3. The noun and adjective quadragintennial means a forty (40) year anniversary.

Quadragintennium The noun and adjective quadragintennium means a period of forty (40) years.

Quadragintillion The noun and adjective quadragintillion means multiplied by 10^{123} in the American number naming system, but means multiplied by 10^{240} in the Chuquet number naming system. .

Quadrancentennial 1. The adjective and noun quadrancentennial means recurring every twenty-five (25) years.

2. The adjective and noun quadrancentennial means lasting twenty-five (25) years.

3. The noun and adjective quadrancentennial means a twenty-five (25) year anniversary.

Quadrangle The noun quadrangle means a plane figure with four (4) vertices.

Quadrangular The adjective quadrangular means having four (4) vertices.

Quadranmillenary 1. The noun and adjective quadranmillenary means a group of two hundred fifty (250).

2. The noun and adjective quadranmillenary means a period of two hundred fifty (250) years.

3. The noun and adjective quadranmillenary means a two hundred fifty (250) year anniversary.

4. The adjective and noun quadranmillenary means two hundred fiftieth (250th) in rank or order.

Quadranmillennial 1. The adjective and noun quadranmillennial means recurring every two hundred fifty (250) years.

2. The adjective and noun quadranmillennial means lasting two hundred fifty (250) years.

3. The noun and adjective quadranmillennial means a two hundred fifty (250) year anniversary.

Quadranmillennium The noun and adjective quadranmillennium means a period of two hundred fifty (250) years.

Quadrant The four parts of a grid divided by the axes. Each of these quadrants have a number designation:

First quadrant - contains all the points with positive x and positive y coordinates.

Second quadrant - contains all the points with negative x and positive y coordinates.

Third quadrant - contains all the points with negative x and negative y coordinates.

Fourth quadrant - contains all the points with positive x and negative y coordinates.

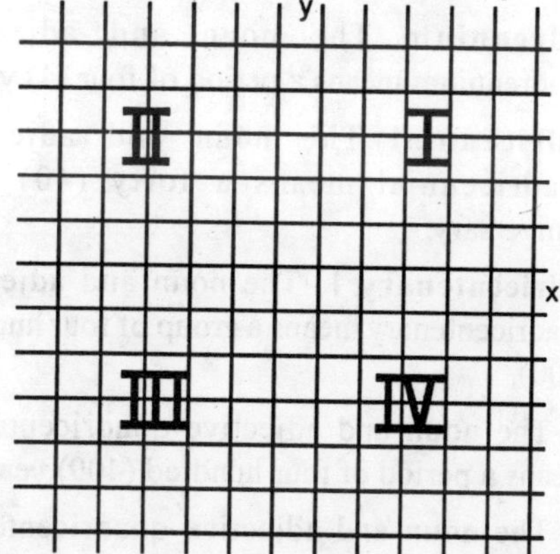

Quadrantal angle An angle in standard position with its terminal side on a coordinate axis.

Quadratic formula That amazing formula for finding all values of x that satisfy the equation $ax^2+bx+c=0$. Works even if you can't factor the left-hand side. There are two solutions, which may be equal.

$$x = \frac{-b+\sqrt{b^2-4ac}}{2a}$$

or

$$x = \frac{-b-\sqrt{b^2-4ac}}{2a}$$

Also slices and dices.

Quadratic function A function of the second degree [i.e., a function of the form f(x) = ax

2 + bx + c]; in a

Rectangular coordinate system, the graph of a quadratic function is a parabola.

Quadrennary The noun and adjective quadrennary means a group of four (4) or a period of four (4) years.

Quadrennial 1. The adjective and noun quadrennial means recurring every four (4) years.

2. The adjective and noun quadrennial means lasting four (4) years.

3. The noun and adjective quadrennial means a four (4) year anniversary.

Quadrennium The noun and adjective quadrennium means a period of four (4) years.

Quadricennial The noun and adjective quadricennial means a forty (40) year anniversary.

Quadricentenary 1. The noun and adjective quadricentenary means a group of four hundred (400).

2. The noun and adjective quadricentenary means a period of four hundred (400) years.

3. The noun and adjective quadricentenary means a four hundred (400) year anniversary.

4. The adjective and noun quadricentenary means four hundredth (400th) in rank or order.

Quadricentennial 1. The adjective and noun quadricentennial means recurring every four hundred (400) years.

2. The adjective and noun quadricentennial means lasting four hundred (400) years.

3. The noun and adjective quadricentennial means a four hundred (400) year anniversary.

Quadricentennium The noun and adjective quadricentennium means a period of four hundred (400) years.

Quadricycle The noun, adjective, and verb quadricycle means having four (4) wheels.

Quadriennial The adjective quadriennial means recurring every four (4) years.

Quadrikibinal The adjective quadrikibinal means based on the number four thousand ninety-six ($2^{12} = 4096$).

Quadrilateral A four-sided polygon. The sum of the angles of a quadrilateral is 360 degrees.

Examples:

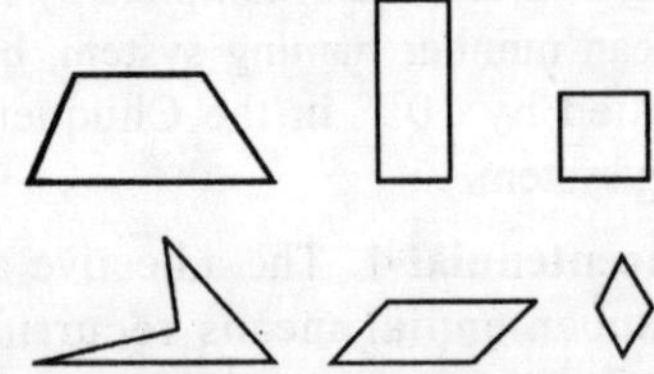

Quadrilateral A polygon that has four sides.

Quadrilateral The adjective quadrilateral means having four (4) sides.

Quadrilliard The noun and adjective quadrilliard means multiplied by ten to the twenty-seventh power (10^{27} = 10 raised to power +27 = 1,000,000,000,000,000,000,000,000,000) in the modified Chuquet number naming system.

Quadrillion The noun and adjective quadrillion means multiplied by ten to the fifteenth power (10^{15} = 10 raised to power +15 = 1,000,000,000,000,000) in the American number naming system, but means multiplied by ten to the twenty-fourth power (10^{24} = 10 raised to power +24 = 1,000,000,000,000,000,000,000,000) in the Chuquet number naming system. .

Quadrimillenary 1. The noun and adjective quadrimillenary means a group of four thousand (4000).

2. The noun and adjective quadrimillenary means a period of four thousand (4000) years.

3. The noun and adjective quadrimillenary means a four thousand (4000) year anniversary.

4. The adjective and noun quadrimillenary means four thousandth (4000th) in rank or order.

Quadrimillennial 1. The adjective and noun quadrimillennial means recurring every four thousand (4000) years.

2. The adjective and noun quadrimillennial means lasting four thousand (4000) years.

3. The noun and adjective quadrimillennial means a four thousand (4000) year anniversary.

Quadrimillennium The noun and adjective quadrimillennium means a period of four thousand (4000) years.

Quadriplegia The noun quadruplegia means a paralysis of all four (4) limbs.

Quadriplegic The adjective and noun quadruplegic means having a paralysis of all four (4) limbs.

Quadruple The adjective and noun quadruple means multiplied by four (4).

Quadruplet The noun and adjective quadruplet means one of four (4) siblings born together or one of four (4) very similar things.

Quadruplicate The verb, noun, and adjective quadruplicate means to make three (3) copies of one original, for a total of four (4) objects.

Qualitative data Non-numerical data, often in the form of categorical data.

Qualitative Qualitative (categorical) variables define different categories or classes of an attribute. Examples are gender, blood groups or disease states. A qualitative (categorical) variable may be nominal or ordinal. When there are only two categories, it is termed binary (like sex, dead or alive).

Quantiles A generic name for statistics such as deciles, percentiles, and quartiles.

Quantitative data Data obtained by measuring objects or events.

Quantitative Quantitative variables are variables for which a numeric value representing an amount is measured. They may be discrete (for example, taking values of integers) or continuous (such as weight, height, blood pressure). If a quantitative variable is categorised, it becomes an ordinal variable.

Quantity 1. The noun and adjective quantity means either an abstract number or a hypothetically measurable physical attribute.

2. The noun and adjective quantity means a significantly large quantity.

Quart The noun and adjective quart means one quarter (1/4) of a gallon.

Quartal The adjective quartal means based on the number four (4).

Quarter 1. The noun and adjective quarter (1/4) means divided by four (4). A quarter coin has a value of one fourth of a dollar ($0.25).

2. The verb quarter means to divide by four (1/4).

Quarter century The compound noun quarter century means twenty-five (25) years.

Quarterly The adjective quarterly means recurring once (1) in each quarter (1/4) of the year or every three (3) months.

Quartile The noun quartile means a frequency distribution divided into four (4) equal parts.

Quartiles The points which break the distribution into fourths.

Quarto The noun quarto (4to) means a book with leaves printed on one fourth (1/4) of the printer sheet.

Quasquicentenary 1. The noun and adjective quasquicentenary means a group of one hundred twenty-five (125).

2. The noun and adjective quasquicentenary means a period of one hundred twenty-five (125) years.

3. The noun and adjective quasquicentenary means a one hundred twenty-five (125) year anniversary.

4. The adjective and noun quasquicentenary means one hundred twenty-fifth (125th) in rank or order.

Quasquicentennial 1. The adjective and noun quasquicentennial means recurring every one hundred twenty-five (125) years.

2. The adjective and noun quasquicentennial means lasting one hundred twenty-five (125) years.

3. The noun and adjective quasquicentennial means a one hundred twenty-five (125) year anniversary.

Quasquicentennium The noun and adjective quasquicentennium means a period of one hundred twenty-five (125) years.

Quaternary 1. The adjective and noun quaternary means fourth (4th) in order or importance.

2. The adjective quaternary means having four (4) parts.

Quattuordecennary The noun and adjective quattuordecennary means a group of fourteen (14) or a period of fourteen (14) years.

Quattuordecennial 1. The adjective and noun quattuordecennial means recurring every fourteen (14) years.

2. The adjective and noun quattuordecennial means lasting fourteen (14) years.

3. The noun and adjective quattuordecennial means a fourteen (14) year anniversary.

Quattuordecennium The noun and adjective quattuordecennium means a period of fourteen (14) years.

Quattuordecentenary 1. The noun and adjective quattuordecentenary means a group of one thousand four hundred (1400).

2. The noun and adjective quattuordecentenary means a period of one thousand four hundred (1400) years.

3. The noun and adjective quattuordecentenary means a one thousand four hundred (1400) year anniversary.

4. The adjective and noun quattuordecentenary means one thousand four hundredth (1400th) in rank or order.

Quattuordecentennial 1. The adjective and noun quattuordecentennial means recurring every one thousand four hundred (1400) years.

2. The adjective and noun quattuordecentennial means lasting one thousand four hundred (1400) years.

3. The noun and adjective quattuordecentennial means a one thousand four hundred (1400) year anniversary.

Quattuordecentennium The noun and adjective quattuordecentennium means a period of one thousand four hundred (1400) years.

Quattuordeci The numeric prefix quattuordeci- means fourteen (14). This prefix is derived from the Latin cardinal number quattuordecim (XIIII).

Quattuordecillion The noun and adjective quattuordecillion means multiplied by 10^{45} in the American number naming system, but means multiplied by 10^{84} in the Chuquet number naming system. .

Quattuornonaginta The numeric prefix quattuornonaginta- means ninety-four (94). This prefix is derived from the Latin cardinal number quattuor et nonaginta (LXXXXIIII).

Quattuornonagintennary The noun and adjective quattuornonagintennary means a group of ninety-four (94) or a period of ninety-four (94) years.

Quattuornonagintennial 1. The adjective and noun quattuornonagintennial means recurring every ninety-four (94) years.

2. The adjective and noun quattuornonagintennial means lasting ninety-four (94) years.

3. The noun and adjective quattuornonagintennial means a ninety-four (94) year anniversary.

Quattuornonagintennium The noun and adjective quattuornonagintennium means a period of ninety-four (94) years.

Quattuornonagintillion The noun and adjective quattuornonagintillion means multiplied by 10^{285} in the American number naming system, but means multiplied by 10^{564} in the Chuquet number naming system..

Quattuoroctoginta The numeric prefix quattuoroctoginta- means eighty-four (84). This prefix is derived from the Latin cardinal number quattuor et octoginta (LXXXIIII).

Quattuoroctogintennary The noun and adjective quattuoroctogintennary means a group of eighty-four (84) or a period of eighty-four (84) years.

Quattuoroctogintennial 1. The adjective and noun quattuoroctogintennial means recurring every eighty-four (84) years.

2. The adjective and noun quattuoroctogintennial means lasting eighty-four (84) years.

3. The noun and adjective quattuoroctogintennial means an eighty-four (84) year anniversary.

Quattuoroctogintennium The noun and adjective quattuoroctogintennium means a period of eighty-four (84) years.

Quattuoroctogintillion The noun and adjective quattuoroctogintillion means multiplied by 10^{255} in the American number naming system, but means multiplied by 10^{504} in the Chuquet number naming system.

Quattuorquadraginta The numeric prefix quattuorquadraginta- means forty-four (44). This prefix is derived from the Latin cardinal number quattuor et quadraginta (XXXXIIII).

Quattuorquadragintennary The noun and adjective quattuorquadragintennary means a group of forty-four (44) or a period of forty-four (44) years.

Quattuorquadragintennial 1. The adjective and noun quattuorquadragintennial means recurring every forty-four (44) years.

2. The adjective and noun quattuorquadragintennial means lasting forty-four (44) years.

3. The noun and adjective quattuorquadragintennial means a forty-four (44) year anniversary.

Quattuorquadragintennium The noun and adjective quattuorquadragintennium means a period of forty-four (44) years.

Quattuorquadragintillion The noun and adjective quattuorquadragintillion means multiplied by 10^{135} in the American number naming system, but means multiplied by 10^{264} in the Chuquet number naming system. .

Quattuorquinquaginta The numeric prefix quattuorquinquaginta- means fifty-four (54). This prefix is derived from the Latin cardinal number quattuor et quinquaginta (LIIII).

Quattuorquinquagintennary The noun and adjective quattuorquinquagintennary means a group of fifty-four (54) or a period of fifty-four (54) years.

Quattuorquinquagintennial 1. The adjective and noun quattuorquinquagintennial means recurring every fifty-four (54) years.

2. The adjective and noun quattuorquinquagintennial means lasting fifty-four (54) years.

3. The noun and adjective quattuorquinquagintennial means a fifty-four (54) year anniversary.

Quattuorquinquagintennium The noun and adjective quattuorquinquagintennium means a period of fifty-four (54) years.

Quattuorquinquagintillion The noun and adjective quattuorquinquagintillion means multiplied by 10^{165} in the American number naming system, but means multiplied by 10^{324} in the Chuquet number naming system. .

Quattuorseptuaginta The numeric prefix quattuorseptuaginta- means seventy-four (74). This prefix is derived from the Latin cardinal number quattuor et septuaginta (LXXIIII).

Quattuorseptuagintennary The noun and adjective quattuorseptuagintennary means a group of seventy-four (74) or a period of seventy-four (74) years.

Quattuorseptuagintennial 1. The adjective and noun quattuorseptuagintennial means recurring every seventy-four (74) years.

2. The adjective and noun quattuorseptuagintennial means lasting seventy-four (74) years.

3. The noun and adjective quattuorseptuagintennial means a seventy-four (74) year anniversary.

Quattuorseptuagintennium The noun and adjective quattuorseptuagintennium means a period of seventy-four (74) years.

Quattuorseptuagintillion The noun and adjective quattuorseptuagintillion means multiplied by 10^{225} in the American number naming system, but means multiplied by 10^{444} in the Chuquet number naming system. .

Quattuorsexaginta The numeric prefix quattuorsexaginta- means sixty-four (64). This prefix is derived from the Latin cardinal number quattuor et sexaginta (LXIIII).

Quattuorsexagintakibinal The adjective quattuorsexagintakibinal means based on the number sixty-five thousand five hundred thirty-six (2^{16} = 65,536).

Quattuorsexagintennary The noun and adjective quattuorsexagintennary means a group of sixty-four (64) or a period of sixty-four (64) years.

Quattuorsexagintennial 1. The adjective and noun quattuorsexagintennial means recurring every sixty-four (64) years.

2. The adjective and noun quattuorsexagintennial means lasting sixty-four (64) years.

3. The noun and adjective quattuorsexagintennial means a sixty-four (64) year anniversary.

Quattuorsexagintennium The noun and adjective quattuorsexagintennium means a period of sixty-four (64) years.

Quattuorsexagintillion The noun and adjective quattuorsexagintillion means multiplied by 10^{195} in the American number naming system, but means multiplied by 10^{384} in the Chuquet number naming system. .

Quattuortriginta The numeric prefix quattuortriginta- means thirty-four (34). This prefix is derived from the Latin cardinal number quattuor et triginta (XXXIIII).

Quattuortrigintacentenary 1. The noun and adjective quattuortrigintacentenary means a group of three thousand four hundred (3400).

2. The noun and adjective quattuortrigintacentenary means a period of three thousand four hundred (3400) years.

3. The noun and adjective quattuortrigintacentenary means a three thousand four hundred (3400) year anniversary.

4. The adjective and noun quattuortrigintacentenary means three thousand four hundredth (3400th) in rank or order.

Quattuortrigintacentennial 1. The adjective and noun quattuortrigintacentennial means recurring every three thousand four hundred (3400) years.

2. The adjective and noun quattuortrigintacentennial means lasting three thousand four hundred (3400) years.

3. The noun and adjective quattuortrigintacentennial means a three thousand four hundred (3400) year anniversary.

Quattuortrigintacentennium The noun and adjective quattuortrigintacentennium means a period of three thousand four hundred (3400) years.

Quattuortrigintennary The noun and adjective quattuortrigintennary means a group of thirty-four (34) or a period of thirty-four (34) years.

Quattuortrigintennial 1. The adjective and noun quattuortrigintennial means recurring every thirty-four (34) years.

2. The adjective and noun quattuortrigintennial means lasting thirty-four (34) years.

3. The noun and adjective quattuortrigintennial means a thirty-four (34) year anniversary.

Quattuortrigintennium The noun and adjective quattuortrigintennium means a period of thirty-four (34) years.

Quattuortrigintillion The noun and adjective quattuortrigintillion means multiplied by 10^{105} in the American number naming system, but means multiplied by 10^{204} in the Chuquet number naming system. .

Quattuorvigesimal The adjective quattuorvigesimal means based on the number twenty-four (24).

Quattuorvigintenary The noun and adjective quattuorvigintenary means a group of twenty-four (24) or a period of twenty-four (24) years.

Quattuorvigintennial 1. The adjective and noun quattuorvigintennial means recurring every twenty-four (24) years.

2. The adjective and noun quattuorvigintennial means lasting twenty-four (24) years.

3. The noun and adjective quattuorvigintennial means a twenty-four (24) year anniversary.

Quattuorvigintennium The noun and adjective quattuorvigintennium means a period of twenty-four (24) years.

Quattuorviginti The numeric prefix quattuorviginti- means twenty-four (24). This prefix is derived from the Latin cardinal number quattuor et viginti (XXIIII).

Quattuorviginticentenary 1. The noun and adjective quattuorviginticentenary means a group of two thousand four hundred (2400).

2. The noun and adjective quattuorviginticentenary means a period of two thousand four hundred (2400) years.

3. The noun and adjective quattuorviginticentenary means a two thousand four hundred (2400) year anniversary.

4. The adjective and noun quattuorviginticentenary means two thousand four hundredth (2400th) in rank or order.

Quattuorviginticentennial 1. The adjective and noun quattuorviginticentennial means recurring every two thousand four hundred (2400) years.

2. The adjective and noun quattuorviginticentennial means lasting two thousand four hundred (2400) years.

3. The noun and adjective quattuorviginticentennial means a two thousand four hundred (2400) year anniversary.

Quattuorviginticentennium The noun and adjective quattuorviginticentennium means a period of two thousand four hundred (2400) years.

Quattuorvigintillion The noun and adjective quattuorvigintillion means multiplied by 10^{75} in the American number naming system, but means multiplied by 10^{144} in the Chuquet number naming system. .

Quinary 1. The adjective and noun quinary means fifth (5th) in order or importance.

2. The adjective quinary means having five (5) parts.

Quincentenary 1. The noun and adjective quincentenary means a group of five hundred (500).

2. The noun and adjective quincentenary means a period of five hundred (500) years.

3. The noun and adjective quincentenary means a five hundred (500) year anniversary.

4. The adjective and noun quincentenary means five hundredth (500th) in rank or order.

Quincentennial 1. The adjective and noun quincentennial means recurring every five hundred (500) years.

2. The adjective and noun quincentennial means lasting five hundred (500) years.

3. The noun and adjective quincentennial means a five hundred (500) year anniversary.

Quincentennium The noun and adjective quincentennium means a period of five hundred (500) years.

Quincunx The noun quincunx means five (5) objects arranged in a square or rectangle with one (1) at each corner and one (1) in the center.

Quincux The noun and adjective quincux means having five (5) of something.

Quindecennary The noun and adjective quindecennary means a group of fifteen (15) or a period of fifteen (15) years.

Quindecennial 1. The adjective and noun quindecennial means recurring every fifteen (15) years.

2. The adjective and noun quindecennial means lasting fifteen (15) years.

3. The noun and adjective quindecennial means a fifteen (15) year anniversary.

Quindecennium The noun and adjective quindecennium means a period of fifteen (15) years.

Quindeci The numeric prefix quindeci- means fifteen (15). This prefix is derived from the Latin cardinal number quindecim (XV).

Quindecillion The noun and adjective quindecillion means multiplied by 10^{48} in the American number naming system, but means multiplied by 10^{90} in the Chuquet number naming system.

Quinmillenary 1. The noun and adjective quinmillenary means a group of five thousand (5000).

2. The noun and adjective quinmillenary means a period of five thousand (5000) years.

3. The noun and adjective quinmillenary means a five thousand (5000) year anniversary.

4. The adjective and noun quinmillenary means five thousandth (5000th) in rank or order.

Quinmillennial 1. The adjective and noun quinmillennial means recurring every five thousand (5000) years.

2. The adjective and noun quinmillennial means lasting five thousand (5000) years.

3. The noun and adjective quinmillennial means a five thousand (5000) year anniversary.

Quinmillennium The noun and adjective quinmillennium means a period of five thousand (5000) years.

Quinnonaginta The numeric prefix quinnonaginta- means ninety-five (95). This prefix is derived from the Latin cardinal number quinque et nonaginta (LXXXXV).

Quinnonagintennary The noun and adjective quinnonagintennary means a group of ninety-five (95) or a period of ninety-five (95) years.

Quinnonagintennial 1. The adjective and noun quinnonagintennial means recurring every ninety-five (95) years.

2. The adjective and noun quinnonagintennial means lasting ninety-five (95) years.

3. The noun and adjective quinnonagintennial means a ninety-five (95) year anniversary.

Quinnonagintennium The noun and adjective quinnonagintennium means a period of ninety-five (95) years.

Quinnonagintillion The noun and adjective quinnonagintillion means multiplied by 10^{288} in the American number naming system, but means multiplied by 10^{570} in the Chuquet number naming system.

Quinoctoginta The numeric prefix quinoctoginta- means eighty-five (85). This prefix is derived from the Latin cardinal number quinque et octoginta (LXXXV).

Quinoctogintennary The noun and adjective quinoctogintennary means a group of eighty-five (85) or a period of eighty-five (85) years..

Quinoctogintennial 1. The adjective and noun quinoctogintennial means recurring every eighty-five (85) years.

2. The adjective and noun quinoctogintennial means lasting eighty-five (85) years.

3. The noun and adjective quinoctogintennial means an eighty-five (85) year anniversary.

Quinoctogintennium The noun and adjective quinoctogintennium means a period of eighty-five (85) years.

Quinoctogintillion The noun and adjective quinoctogintillion means multiplied by 10^{258} in the American number naming system, but means multiplied by 10^{510} in the Chuquet number naming system.

Quinquadraginta The numeric prefix quinquadraginta- means forty-five (45). This prefix is derived from the Latin cardinal number quinque et quadraginta (XXXXV).

Quinquadragintennary The noun and adjective quinquadragintennary means a group of forty-five (45) or a period of forty-five (45) years.

Quinquadragintennial 1. The adjective and noun quinquadragintennial means recurring every forty-five (45) years.

2. The adjective and noun quinquadragintennial means lasting forty-five (45) years.

3. The noun and adjective quinquadragintennial means a forty-five (45) year anniversary.

Quinquadragintennium The noun and adjective quinquadragintennium means a period of forty-five (45) years.

Quinquadragintillion The noun and adjective quinquadragintillion means multiplied by 10^{138} in the American number naming system, but means multiplied by 10^{270} in the Chuquet number naming system.

Quinquagenarian The noun and adjective quinquagenarian means between fifty (50) and sixty (60) years of age.

Quinquagenary The adjective quinquagenary means having fifty (50) parts.

Quinquagesimal The adjective quinquagesimal means based on the number fifty (50).

Quinquaginta The numeric prefix quinquaginta- means fifty (50). This prefix is derived from the Latin cardinal number quinquaginta (L).

Quinquagintennary The noun and adjective quinquagintennary means a group of fifty (50) or a period of fifty (50) years.

Quinquagintennial 1. The adjective and noun quinquagintennial means recurring every fifty (50) years.

2. The adjective and noun quinquagintennial means lasting fifty (50) years.

3. The noun and adjective quinquagintennial means a fifty (50) year anniversary.

Quinquagintennium The noun and adjective quinquagintennium means a period of fifty (50) years.

Quinquagintillion The noun and adjective quinquagintillion means multiplied by 10^{153} in the American number naming system, but means multiplied by 10^{300} in the Chuquet number naming system.

Quinquennary The noun and adjective quinquennary means a group of five (5) or a period of five (5) years.

Quinquennial 1. The adjective and noun quinquennial means recurring every five (5) years.

2. The adjective and noun quinquennial means lasting five (5) years.

3. The noun and adjective quinquennial means a five (5) year anniversary.

Quinquennium The noun and adjective quinquennium means a period of five (5) years.

Quinquinquaginta The numeric prefix quinquinquaginta- means fifty-five (55). This prefix is derived from the Latin cardinal number quinque et quinquaginta (LV).

Quinquinquagintennary The noun and adjective quinquinquagintennary means a group of fifty-five (55) or a period of fifty-five (55) years.

Quinquinquagintennial 1. The adjective and noun quinquinquagintennial means recurring every fifty-five (55) years.

2. The adjective and noun quinquinquagintennial means lasting fifty-five (55) years.

3. The noun and adjective quinquinquagintennial means a fifty-five (55) year anniversary.

Quinquinquagintennium The noun and adjective quinquinquagintennium means a period of fifty-five (55) years.

Quinquinquagintillion The noun and adjective quinquinquagintillion means multiplied by 10^{168} in the American number naming system, but means multiplied by 10^{330} in the Chuquet number naming system.

Quinseptuaginta The numeric prefix quinseptuaginta- means seventy-five (75). This prefix is derived from the Latin cardinal number quinque et septuaginta (LXXV).

Quinseptuagintennary The noun and adjective quinseptuagintennary means a group of seventy-five (75) or a period of seventy-five (75) years.

Quinseptuagintennial 1. The adjective and noun quinseptuagintennial means recurring every seventy-five (75) years.

2. The adjective and noun quinseptuagintennial means lasting seventy-five (75) years.

3. The noun and adjective quinseptuagintennial means a seventy-five (75) year anniversary.

Quinseptuagintennium The noun and adjective quinseptuagintennium means a period of seventy-five (75) years.

Quinseptuagintillion The noun and adjective quinseptuagintillion means multiplied by 10^{228} in the American number naming system, but means multiplied by 10^{450} in the Chuquet number naming system.

Quinsexaginta The numeric prefix quinsexaginta- means sixty-five (65). This prefix is derived from the Latin cardinal number quinque et sexaginta (LXV).

Quinsexagintennary The noun and adjective quinsexagintennary means a group of sixty-five (65) or a period of sixty-five (65) years.

Quinsexagintennial 1. The adjective and noun quinsexagintennial means recurring every sixty-five (65) years.

2. The adjective and noun quinsexagintennial means lasting sixty-five (65) years.

3. The noun and adjective quinsexagintennial means a sixty-five (65) year anniversary.

Quinsexagintennium The noun and adjective quinsexagintennium means a period of sixty-five (65) years.

Quinsexagintillion The noun and adjective quinsexagintillion means multiplied by 10^{198} in the American number naming system, but means multiplied by 10^{390} in the Chuquet number naming system. .

Quint The numeric prefix quint- means five (5). This prefix is derived from the Latin cardinal number quinque (V).

Quintal The adjective quintal means based on the number five (5).

Quintessence The noun quintessence means the fifth (5th) and highest essence of an object or being.

Origin: Aristotle taught that all things are composed of four material elements and one immaterial essence. The four material elements are earth, water, air, and fire. The fifth essence is the form of an object or the spirit of a being. The fifth essence creates the structure and logic of the universe.

Quintessential The adjective quintessential means embodying the fifth (5th) and highest essence of an object or class.

Quintile The noun quintile means a frequency distribution divided into five (5) equal parts.

Quintillion The noun and adjective quintillion means multiplied by 10^{18} in the American number naming system, but means multiplied by 10^{30} in the Chuquet number naming system. .

Quintriginta The numeric prefix quintriginta- means thirty-five (35). This prefix is derived from the Latin cardinal number quinque et triginta (XXXV).

Quintrigintacentenary 1. The noun and adjective quintrigintacentenary means a group of three thousand five hundred (3500).

2. The noun and adjective quintrigintacentenary means a period of three thousand five hundred (3500) years.

3. The noun and adjective quintrigintacentenary means a three thousand five hundred (3500) year anniversary.

4. The adjective and noun quintrigintacentenary means three thousand five hundredth (3500th) in rank or order.

Quintrigintacentennial 1. The adjective and noun quintrigintacentennial means recurring every three thousand five hundred (3500) years.

2. The adjective and noun quintrigintacentennial means lasting three thousand five hundred (3500) years.

3. The noun and adjective quintrigintacentennial means a three thousand five hundred (3500) year anniversary.

Quintrigintacentennium The noun and adjective quintrigintacentennium means a period of three thousand five hundred (3500) years.

Quintrigintennary The noun and adjective quintrigintennary means a group of thirty-five (35) or a period of thirty-five (35) years.

Quintrigintennial 1. The adjective and noun quintrigintennial means recurring every thirty-five (35) years.

2. The adjective and noun quintrigintennial means lasting thirty-five (35) years.

3. The noun and adjective quintrigintennial means a thirty-five (35) year anniversary.

Quintrigintennium The noun and adjective quintrigintennium means a period of thirty-five (35) years.

Quintrigintillion The noun and adjective quintrigintillion means multiplied by 10^{108} in the American number naming system, but means multiplied by 10^{210} in the Chuquet number naming system.

Quintuple The verb, adjective, adverb, and noun quintuple means multiply by five (5).

Quintuplet 1. The noun quintuplet means five (5).

2. The noun and adjective quintuplet means one of five (5) siblings born together or one of five (5) very similar things.

Quintuplicate The verb, noun, and adjective quintuplicate means to make four (4) copies of one original, for a total of five (5) objects.

Quinvigintennary The noun and adjective quinvigintennary means a group of twenty-five (25) or a period of twenty-five (25) years.

Quinvigintennial 1. The adjective and noun quinvigintennial means recurring every twenty-five (25) years.

2. The adjective and noun quinvigintennial means lasting twenty-five (25) years.

3. The noun and adjective quinvigintennial means a twenty-five (25) year anniversary.

Quinvigintennium The noun and adjective quinvigintennium means a period of twenty-five (25) years.

Quinviginti The numeric prefix quinviginti- means twenty-five (25). This prefix is derived from the Latin cardinal number quinque et viginti (XXV).

Quinvigínticentenary 1. The noun and adjective quinviginticentenary means a group of two thousand five hundred (2500).

2. The noun and adjective quinviginticentenary means a period of two thousand five hundred (2500) years.

3. The noun and adjective quinviginticentenary means a two thousand five hundred (2500) year anniversary.

4. The adjective and noun quinviginticentenary means two thousand five hundredth (2500th) in rank or order.

Quinviginticentennial 1. The adjective and noun quinviginticentennial means recurring every two thousand five hundred (2500) years.

2. The adjective and noun quinviginticentennial means lasting two thousand five hundred (2500) years.

3. The noun and adjective quinviginticentennial means a two thousand five hundred (2500) year anniversary.

Quinviginticentennium The noun and adjective quinviginticentennium means a period of two thousand five hundred (2500) years.

Quinvigintillion The noun and adjective quinvigintillion means multiplied by 10^{78} in the American number naming system, but means multiplied by 10^{150} in the Chuquet number naming system.

Quorum The noun quorum means the minimum number of members required to be present to conduct business. An organization or committee may establish a quorum, which by default is a majority of members.

Quotient identities Fundamental identities that involve the quotient of basic trig functions.

Quotient When performing division, the number of times one value can be multiplied to reach the other value represents the quotient. For example, when dividing 7 by 3, 3 can be multiplied twice, making 6, and the remainder is 1, so the quotient is 2 .

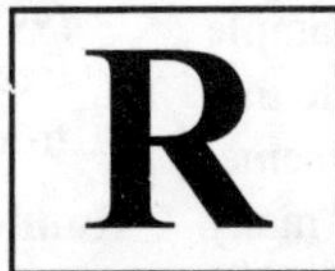

Radians A measurement of an angle or an arc length based on the length of the radius. One radian is an arc equal in length to the radius. One radian = 180° / π = 57.295° .

Radical Another name for the roots of numbers, such as the square root of 5 or "5.

Radii More than one radius.

Radius A straight line from the center of a circle to a point on the circle. Also the measurement of that line.

Radius of an arc (or elbow/bend) The distance from the vertex of an arc to the center line of the elbow or bend.

Radius vector Another name for a standard vector, a vector in standard position.

Random marginal totals The situation in which the marginal totals in a contingency table are not known before the data are collected and are subject to sampling error.

Random model Anova An analysis of variance model in which the levels of the independent variable are treated as a random variable.

Random number generators A device used to produce a selection of numbers in a fair manner, in no particular order and with no favor being given to any numbers. Examples include dice, spinners, coins, and computer programs designed to randomly pick numbers.

Random sample In statistical terms a random sample is a set of items that have been drawn from a population in such a way that each time an item was selected, every item in the population had an equal opportunity to appear in the sample. In practical terms, it is not so easy to draw a random sample. First, the only factor operating when a given item is selected, must be chance.

If, for example, numbered pieces of cardboard are drawn from a hat, it is important that they be thoroughly mixed, that they be identical in every respect except for the number printed on them and that the person selecting them be well blindfolded.

Second in order to meet the equal opportunity requirement, it is important that the sampling be done with replacement. That is, each time an item is selected, the relevant measure is taken and recorded. Then the item must be replaced in the population and be thoroughly mixed with the other items before the next item is drawn. If the items are not replaced in the population, each time an item is withdrawn, the probability of being selected, for each of the remaining items, will have been increased.

For example, with the illustrated population, the initial probability that a given item will be selected is 1/9. If, however, an item is drawn and not returned before drawing a second item, the probability that a given item will be drawn will have been increased to 1/8. Of course, this kind of change in probability becomes trivial if our population is very large, but it is important to recognize the principle illustrated here, to fully understand the concept of a random sample.

It is also important to recognize that when sampling with replacement, it is possible for the same item to appear more than once in a sample and it is possible to draw a random sample that is larger than the population from which it came. Notice also, that it is possible to draw as many random samples as we like from a give population. The key idea here is that we either sample with replacement or we draw our samples from a population that is so large that the withdrawal of successive items changes probability by an amount that is too small to be of concern.

Random variable A random variable is a numerical value which is determined by the outcomes or events of an experiment. A random independent variable is one whose levels vary from one replication to another, and are not determined by the experimenter.

Random vector (or multivariate random variable) A vector whose components are random variables on the same probability space.

Randomised (complete) block design An experimental design in which the treatments in each block are assigned to the experimental units in random order. Blocks are all of the same size and each treatment appears in the same number of times within each block (usually once). A different level of the factor is assigned to each member of the block randomly. The data can be analysed using the paired t-test (when there are two units per block) or by randomised block ANOVA (in blocks of any size). The results are substantially more precise than a completely randomised design of comparable size. In studies with a block design, more assumptions are required for the model: no interactions between treatments and blocks, and constant variance from block to block.

Randomized blocks design A design in which subjects are matched against one another and put into blocks of subjects of the same size as the number of treatments. Members of each block are then randomly assigned to treatments.

Range The range is one of several indices of variability that statisticians use to characterize the dispersion among the measures in a given population.

The range is the distance between the highest and lowest score.

Numerically, the range equals the highest score minus the lowest score.

Range of the function f The set of all the numbers f(x) for x in the domain of f.

Range restrictions Refers to cases in which the range over which X or Y vary is artificially limited.

Rank one update A rank one update to a matrix is a simple but useful perturbatation whose effects can be calculated.

If we suppose that we have an m by n matrix A, then a rank one update is accomplished by adding the outer product of an m vector u and an n vector v:

B = A + u v'

If the matrix A is actually square, then the rank of A only changes by +1, -1 or 0 when a rank one update is applied. If A is actually invertible, then the Sherman Morrison Formula shows how to cheaply compute the inverse of the updated matrix.

If the matrix A is the identity matrix, then we have

det (I + u v') = 1 + v'u.

Ranked data Data for which the observations have been replaced by their numerical ranks from lowest to highest.

Rank-randomization tests A class of nonparametric tests based on the theoretical distribution of randomly assigned ranks.

Rapid Application Development (RAD) Systems development methodology created to radically decrease the time needed to design and implement information systems. RAD relies on extensive user involvement, Joint Application Design sessions, prototyping, integrated CASE tools, and code generators.

Rate A ratio comparing two different units (e.g., miles per hour).

Rate of change The rate of change is the speed at which a function is changing. If the function is measuring your position, then your speed(as measured by your speedometer) is your rate of change. Another name for the rate of change of a function IS the derivative of that function.

Ratio A comparison of one value to another value by division.

Ratio A comparison of two whole numbers by division.

Ratio A rational number of the form a/b where a is called the numerator and b is called the denominator.

Ratio scale A scale with a true zero point — ratios are meaningful.

Ratio variable A quantitative variable that has a zero point as its origin (like 0 cm = 0 inch) so that ratios between values are meaningfully defined. Unlike the interval variables, which do not have a true zero point, the ratio of any two values in the scales is independent of the unit of measurement. For example, 2/12 cm has the same ratio as the corresponding values in inch (but the same cannot be said for 2/12 Celsius and 2/12 Fahrenheit which are interval variables).

Rational fraction A rational fraction is a rational number expressed as the ratio or fraction of one integer divided by another integer. Examples include 1/2, 3/6, 500/1000, –5/64, 0/19, 4/3, –475/1, etc.

Rational number A rational number is a number of the form a/b where a and b are integers. For instance, a few famous ones include 1/2 and 3/4. Each rational number has a decimal representation that is either terminating (consisting of finitely many decimal places) or repeating. Interestingly enough, there are a lot more irrational numbers than rational numbers.

Rationalizing the denominator The process of changing an irrational denominator to a rational one.

Ratios of the sides The division of one side of a right triangle by another side. The ratios of the sides are directly related to the number of degrees in the reference angle.

Rayleigh quotient The Rayleigh quotient of a matrix A and vector x is

R(A,x) = (x'*A*x) / (x'*x).

If the matrix A is symmetric, then the Rayleigh quotient is an excellent estimate of the nearest eigenvalue; in particular,

$\text{Lambda}_{min} <= R(A,x) <= \text{Lambda}_{max}$

On the other hand, if the matrix is not symmetric, then we cannot guarantee that the eigenvalues will be real; moreover, we cannot guarantee that the eigenvectors will be orthogonal. Thus for a general matrix, the information obtained from the Rayleigh quotient must be interpreted more carefully.

A frequent use of the Rayleigh quotient is in the inverse power method. Starting with approximations for the eigenvalue and eigenvector, one step of the inverse power method gives an improved estimate of the eigenvector. Then the Rayleigh quotient improves the eigenvalue estimate, and will be used to shift the next step of the inverse power method. This pair of steps can be rapidly convergent.

If the matrix A is actually positive definite symmetric, then the quantity (y'*A*x) has all the properties of an inner product. In that case, the Rayleigh quotient may be generalized to involve any pair of vectors:

R2(y,A,x) = (y'*A * x) / (y'*x).

If the matrix is not symmetric, but we still want a sensible way to estimate the eigenvalues, then the Unsymmetric Rayleigh Quotient could be defined as:

URQ(A,x) = sqrt (((A * x)'*(A * x)) / (x'*x)).

This amounts to the square root of the Rayleigh quotient for the symmetric matrix A'*A, but can also be regarded as the ratio of the L2 norms of A*x and x.

Rays A ray is one of the basic terms in geometry. We may think of a ray as a straight line that begins at a certain point and extends forever in one direction. The point where the ray begins is known as its endpoint. We write the name of a ray with endpoint A and passing through a point B as ray AB or as $\overrightarrow{AB}$. Note how the arrow heads denotes the direction the ray extends in: there is no arrow head over the endpoint.

Example: The following is a diagram of two rays: ray HG and ray AB.

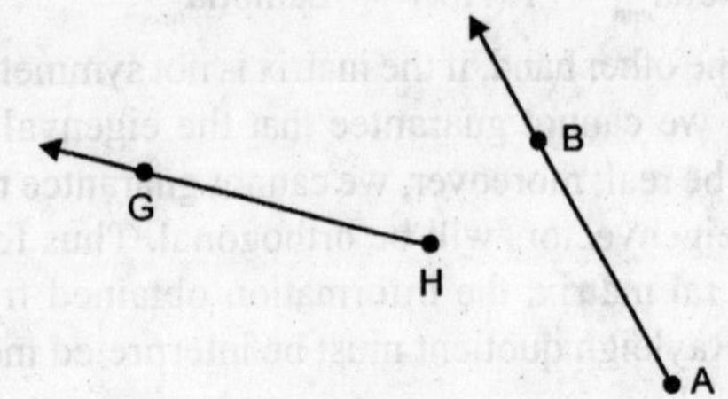

Real axis An axis in the complex plane.

Real lower limit The points halfway between the bottom of one interval and the top of the interval below it.

Real number A real number is a number that can be represented as a decimal number of unlimited precision. The set of real numbers can be represented by a number line of infinite length. We use the real numbers to count and measure our physical world.

Real time constraint A real-time constraint is a maximum time in which something should be completed. Real-time constraints are shown in diagrams inside braces ({}) and may be applied to states, triggers, actions, entire transitions, or state nets (using path markers).

Real upper limit The points halfway between the top of one interval and the bottom of the next.

Reciprocal identities Fundamental identities that involve the reciprocals of basic trig functions.

Reciprocals Pairs of real numbers that have the product 1. Also known as mulliplicarive inverses.

Rectangle A four-sided polygon having all right angles. The sum of the angles of a rectangle is 360 degrees.

Examples:

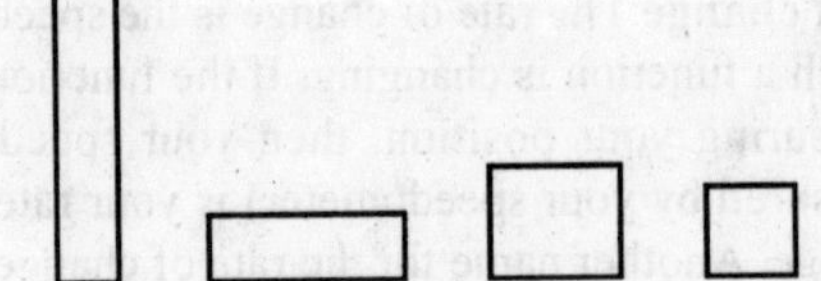

Rectangular distribution A distribution in which all outcomes are equally likely. Also known as a uniform distribution.

Rectangular matrix A rectangular matrix is a matrix which is not square, that is, a matrix whose row order and column order are different.

While many operations and algorithms of linear algebra only apply to a square matrix, a rectangular matrix does have an LU factorization, and a singular value decomposition.

A rectangular matrix can occur when solving an under-determined or over-determined linear system.

Recurring costs A cost resulting from the ongoing evolution and use of a system.

Recursion Given some starting information and a rule for how to use it to get new information, the rule is then repeated using the new information.

Recursive foreign key A foreign key in a relation that references the primary key values of that same relation.

Recursive relationship A function rule which uses the value of the preceding term in the definition.

Reduction formulas for cosine Useful in rewriting cosines of angles greater than 90° as functions of acute angles.

Reduction formulas for sine Useful in rewriting sines of angles greater than 90° as functions of acute angles.

Reduction formulas for tangent Useful in rewriting tangents greater than 90° as functions of acute angles.

Reference angle An acute angle whose trigonometric ratios are the same (except for sign) as the given angle. A

Referential integrity An integrity constraint specifying that the value (or existence) of an attribute in one relation depends on the value (or existence) of an attribute in the same or another relation.

Reflect In a tessellation, reflect means to repeat an image by flipping it across a line so it appears as it would in a mirror.

Reflection (flip) A transformation which produces the mirror image of a figure (i.e., flipping a figure across a line).

Reflection If we flip (or mirror) along some line, we say the figure is a reflection along that line.

Examples: Reflections along a vertical line:

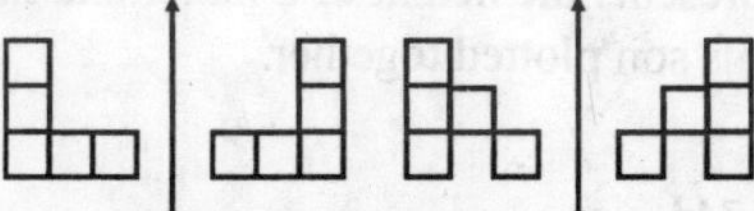

Reflections along a horizontal line:

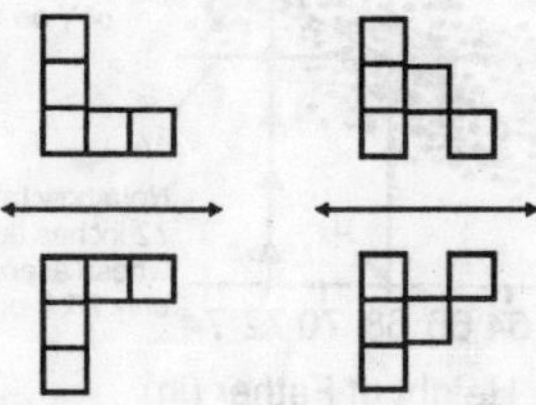

Reflections along a diagonal line:

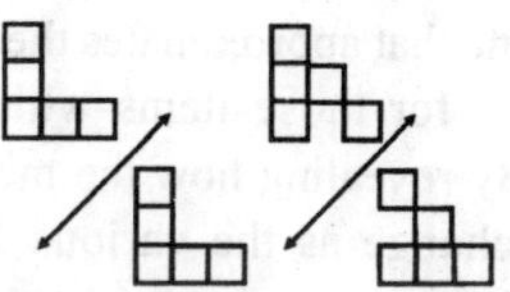

Reflection matrix A reflection matrix A has the property it carries out the reflection (or negation) of the portion of every vector that is perpendicular to some hyperplane, while leaving the parallel portion of the vector unchanged.

A reflection matrix A is involutory, that is,

$A * A = I$,

which strongly restricts the eigenvalues.

Examples of reflection matrices include the identity matrix, a diagonal matrix whose diagonal entries are +1 or –1, any matrix which rotates two coordinate axes by 180 degrees, and the Householder matrices.

Simple facts about a reflection matrix A:

$A = A^{-1}$;

the matrix I - A is an idempotent matrix;

All eigenvalues of A have magnitude 1.

Regression Given a pair of related measures (X and Y) on each of a set of items, the term regression is used to characterize the manner in which one of the measures (for example the Y measures) change as the other measure (in this case, the X measure) changes.

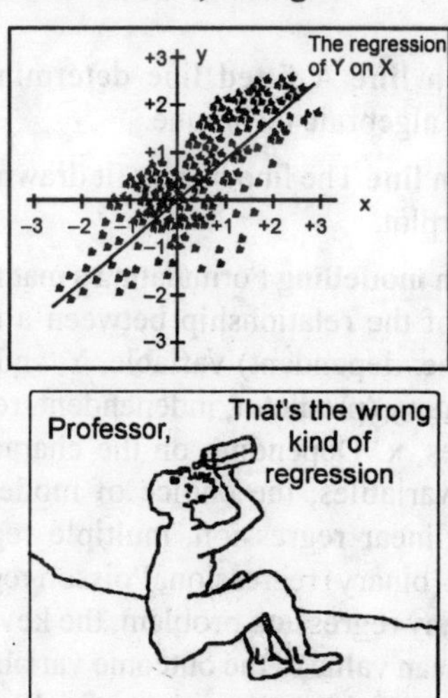

For any set of related measures, it is possible to specify a line that approximates the mean of the Y measures for those items with a given X measure. By revealing how the mean of the Y measures change as the various X measures change, this line is understood to describe the regression of Y on X.

It is noteworthy that for the same set of related measures there is always a second regression line that describes the regression of X on Y.

The regression line is the predicted value of Y for each value of X.

Regression coefficients The general name given to the slope and the intercept ‹most often refers just to the slope).

Regression diagnostics Tests to identify the main problem areas in regression analysis: normality, common variance and independence of the error terms; outliers, influential data points, collinearity, independent variables being subject to error, and inadequate specification of the functional form of the model. The purpose of the diagnostic techniques is to identify weaknesses in the regression model or the data. Remedial measures, correction of errors in the data, elimination of true outliers, collection of better data, or improvement of the model, will allow greater confidence in the final product.

Regression equation The equation that predicts Y from X.

Regression line A fitted line determined by a specific algebraic technique.

Regression line The line of best fit drawn through a scatterplot.

Regression modelling Formulating a mathematical model of the relationship between a response (outcome, dependent) variable, Y, and a set of explanatory (predictor, independent, regressor) variables, x. Depending on the characteristics of the variables, the choice of model can be simple linear regression, multiple regression, logistic (binary) regression, Poisson regression, etc. In any regression problem, the key quantity is the mean value of the outcome variable, given the value of the independent variable(s). This quantity is called the conditional mean and expressed as E (Yáx) where Y is the response (outcome), x is the explanatory (predictor) variable. The question is whether the variable(s) in question tells us more about the outcome variable than a model that does not include that variable. In other words, whether the coefficient of the variable(s) is zero and the outcome is equal to a constant (which is the mean for Y) or not. The aim of model building is to arrive at a meaningful (say, biologically relevant) and parsimonious model that explains the data. The model may be linear if the parameters are linear, or nonparametric if the parameters are not linear. No matter how strong is the statistical relationship between x and Y, no cause-and-effect pattern is necessarily implied by the regression models.

Regression surface The generalization of the regression line, or the regression plane, to multidimensional space.

Regression The prediction of one variable from knowledge of one or more other variables.

Regression towards the mean Regression toward the mean is a property of any scatterplot where the linear relationship is less than perfect. Under such circumstances the regression of Y on X will have a slope that is less than 45 degrees.

As a result, the mean of the values of Y for a given value of X will be closer to the mean of all values of Y than that value of X is to the mean of all the values of X.

In the scatterplot shown below, each dot represents the height of a father and the height if his son plotted together.

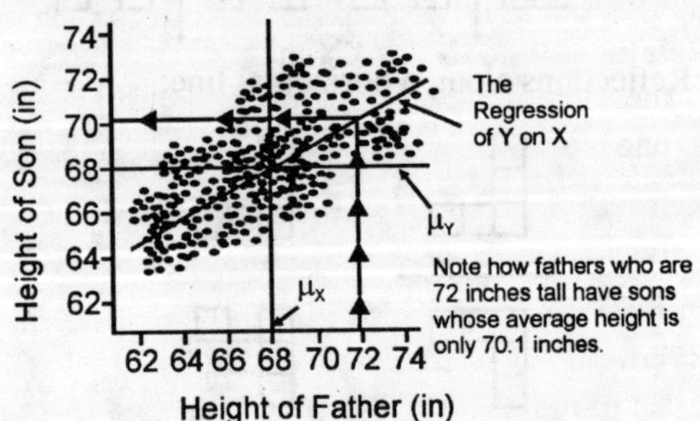

Regular fractals see fractal .

Regular polygon A regular polygon is a polygon whose sides are all the same length, and whose angles are all the same. The sum of the angles of a polygon with n sides, where n is 3 or more, is 180° × (n – 2) degrees.

Examples: The following are examples of regular polygons:

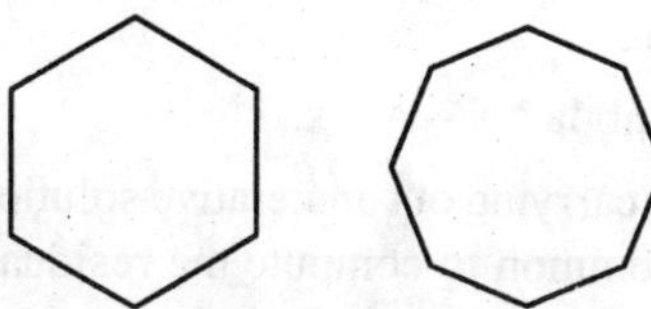

Examples: The following are not examples of regular polygons:

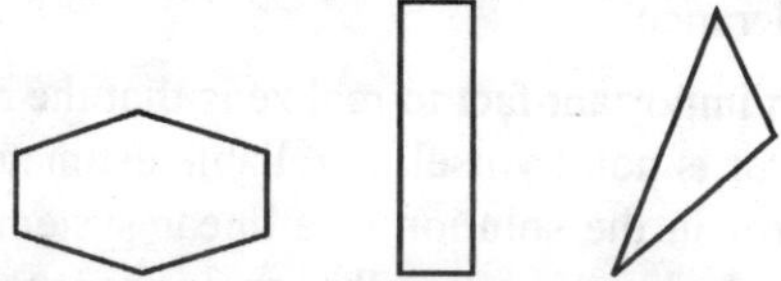

Rejection level The probability with which we are willing to reject H_0 when it is in fact true.

Rejection region The set of outcomes of an experiment that will lead to rejection of H_0 .

Related measures With any set of items (for example the population of 9 persons we have discussed elsewhere) we could, if we chose to, measure more than one feature, We could, for example, measure each persons weight as well as their height. The population of paired numbers would be described as related measures provided that we always knew which weight was related to which height. That is, we always associated the weight of a given person with that same person's height.

Related samples An experimental design in which the same subject is observed under more than one treatment.

Relation A named, two-dimensional table of data. Each relation consists of a set of named columns and an arbitrary number of unnamed rows.

Relational database model Data represented as a set of related tables or relations.

Relational object class An object class which contains relationship sets. Shown on diagrams by drawing an object class rectangle around a relationship set and all participating object classes.

Relationship An association between the instances of one or more entity types that is of interest to the organization.

Relationship between alpha, beta, and power The figure shown below provides a graphical illustration of the relationship between alpha, beta, and Power.

Relationship set A relationship set is a set of similar relationships.

Relative frequency Relative frequency is the number of items of a certain type divided by the number of all the numbers being considered.

Relative frequency view Definition of probability in terms of past performance.

Relative risk (RR) Also known as risk ratio. The RR shows how many times more or less the individuals with the risk factor are likely to get the disease relative to those who do not have the risk factor. RR gives the strength of association in prospective cohort studies. It cannot be estimated in retrospective case-control studies, and its use to describe the cross-product ratio (as frequently done in HLA association studies) is inappropriate.

Remainders After dividing one number by another, if any amount is left that does not divide evenly, that amount is called the remainder. For example, when 8 is divided by 3, three goes in to eight twice (making 6), and the remainder is 2. When dividing 9 by 3, there is no remainder, because 3 goes in to 9 exactly 3 times, with nothing left over.

Remove interaction Remove deletes an object from an object class, but not necessarily from the analysis model.

Repeated measures design In this design, the same experimental unit is subjected to the different treatments under consideration at different points in time. Each unit, therefore, serves as a block. If for example, two different treatments and placebo treatment are applied to the same patient sequentially, this is a repeated measures design.

Repeated-measures designs An experimental design in which each subject receives all levels of at least one independent variable.

Repeating decimal A decimal in which a digit or a group of digits repeats forever, e.g. 0.66666 ... or 0.846846846 ...

Repeating group A set of two or more multivalued attributes that are logically related.

Repetend The portion of a repeating decimal that repeats.

Replicate To repeat an experiment.

Report A business document that contains only predefined data; it is a passive document used only for reading or viewing. A form typically contains data from many unrelated records or transactions.

Repository A centralized database that contains all diagrams, forms and report definitions, data structure, data definitions, process flows and logic, and definitions of other organizational and system components; it provides a set of mechanisms and structures to achieve seamless data-to-tool and data-to-data integration.

Request for proposal (RFP) A document provided to vendors to ask them to propose hardware and system software that will meet the requirements of a new system.

Research hypothesis The hypothesis that the experiment was designed to investigate.

Residual (error) sum of squares (RSS) The measure of within treatment groups sum of squares (variability) in ANOVA. It is the deviation around the fitted regression line. The sum of squared differences between each observed Y value (Y_i) and the fitted Y value (Y_i –hat) equals to the residual (error) sum of squares.

Residual error The residual error is a measure of the error that occurs when a given approximate solution vector x is substituted into the equation of the problem being solved.

For a system of linear equations, the residual error is the vector r defined as

r = b – A * x

For the eigenvalue problem, the residual error is a vector which is a function of the approximate eigenvector x and the approximate eigenvalue lambda:

r = lambda * x – A * x.

When carrying out an iterative solution process, it is common to compute the residual error for each new approximate solution, and to terminate the iteration successfully if the vector norm of the residual error decreases below some tolerance.

An important fact to realize is that the residual error is not, by itself, a reliable estimate of the error in the solution of a linear system. If the residual error has small norm, we can really only hope that the solution error (between our computed x and the true solution x*) is small. We need to know the norm of the inverse of the matrix, in which case the following restriction holds:

$$\|x - x^*\| <= \|A^{-1}\| * \|A(x - x^*)\|$$

$$= \|A^{-1}\| * \|A * x - b + b - A * x^*\|$$

$$= \|A^{-1}\| * \|r\|,$$

so that when the norm of the residual error, ‖r‖ is small, we have a precise upper bound on the error in the solution. Since such matrix norms are generally not known or computable, what we really have is a promise of continuity in the errors: as we drive the residual down, we are forcing down the upper limit on the approximation error.

Residual plot A graph that plots residuals against fitted values. It is used to check equal variance assumption of the error terms in linear regression. Residual analysis for logistic regression is more difficult than for general linear regression models because the responses Y_i can take only the values 0 and 1.

Consequently, the residuals will not be normally distributed. Plots of residuals against fitted values or explanatory variables will be uninformative. Residual plots are generally unhelpful for generalised linear models (where index plots and half-normal plots are preferred).

Residual variance The square of the standard error of estimate.

Residuals Residuals reflect the overall badness-of-fit of the model. They are the differences between the observed values of the outcome variable and the corresponding fitted values predicted by the regression line (the vertical distance between the observed values and the fitted line). In a regression analysis, a large residual for a data point indicates that the data point concerned is not close to its fitted value. If there are too many large (standardised) residuals either the model fitted is not adequate or there is overdispersion of the data. Ideally, the residuals should have constant variance along the line. This can be checked by a normal probability plot of the residuals. In the plot of residuals against the explanatory variable (or the fitted values), there should not be any pattern if the assumption of constant variation is met, i.e., residuals do not tend to get larger as the variable values get larger or smaller.

Resistance The degree to which an estimator is not influenced by the presence of outliers..

Resources Any person, group of people, piece of equipment, or material used in accomplishing an activity.

Restrictions The set of possible values for the variables for which an expression is real, or for which it is defined.

Resultant The sum of two vector.

Resultant vector The result obtained after vector manipulation.

Revo The nonstandard prefix revo- denotes a multiple of ten to the negative thirty-third power (10^{-33} = 10 raised to power –33 = 0.000 000 000 000 000 000 000 000 000 000 001).

Rho Correlation coefficient on the population. Also occasionally used for Spearman's rank-order correlation.

Rhombus A four-sided polygon having all four sides of equal length. The sum of the angles of a rhombus is 360 degrees.

Examples:

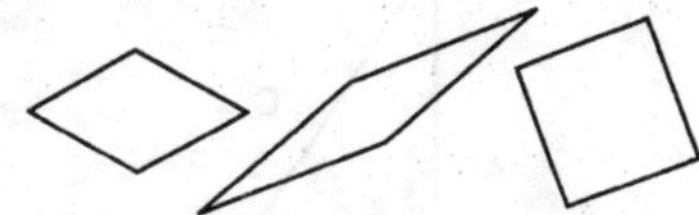

RHS This section contains the elements of the right-hand side. The section begins with a card with RHS in columns 1–3. Since the right-hand side can be regarded as another column of the matrix, the data cards specifying the nonzero entries are in exactly the same format as the COLUMNS data cards, except that field 2 (columns 5–12) has a label for the right-hand side. More than one right-hand side may thus be specified in this section.

Right angles A right angle is an angle measuring 90 degrees. Two lines or line segments that meet at a right angle are said to be perpendicular. Note that any two right angles are supplementary angles (a right angle is its own angle supplement).

Example: The following angles are both right angles.

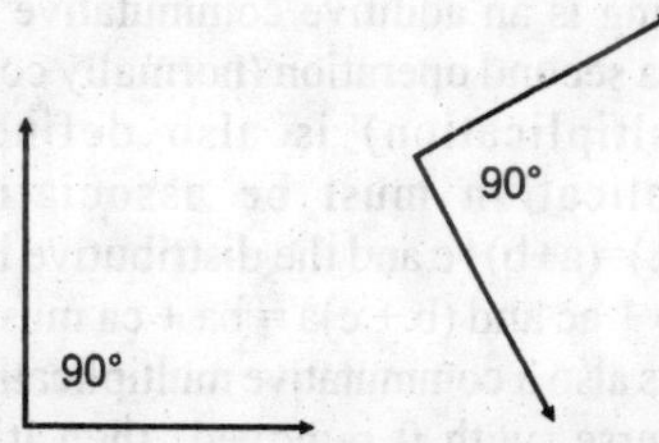

Right triangle A triangle having a right angle. One of the angles of the triangle measures 90 degrees. The side opposite the right angle is called the hypotenuse. The two sides that form the right angle are called the legs. A right triangle has the special property that the sum of the squares of the lengths of the legs equals the square of the length of the hypotenuse. This is known as the

Pythagorean Theorem.

Examples:

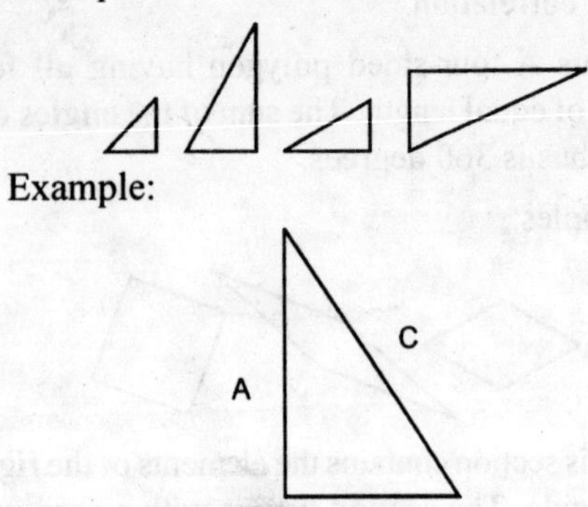

Example:

For the right triangle above, the lengths of the legs are A and B, and the hypotenuse has length C. Using the Pythagorean Theorem, we know that $A^2 + B^2 = C^2$.

Example:

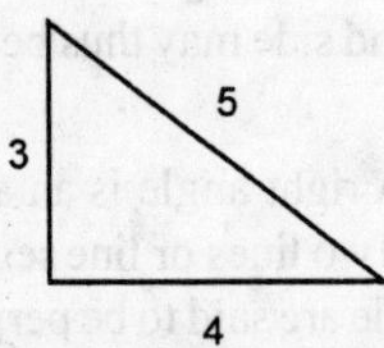

In the right triangle above, the hypotenuse has length 5, and we see that $3^2 + 4^2 = 5^2$ according to the Pythagorean Theorem.

Right triangle A triangle with one angle equal to 90° and two angles whose sum is 90° .

Ring A ring is an additive commutative group in which a second operation (normally considered as multiplication) is also defined. The multiplication must be associative, i.e. a+(b+c)=(a+b)+c and the distributive law a(b + c) = ab + ac and (b + c)a = ba + ca must hold. If a ring is also a commutative multiplicative group (of course, with 0 removed) then it's called a field.

Risk ratio (Relative risk) The risk ratio is the percentage difference in classification between two groups obtained as the ratio of two risks or proportions. For example, the proportion of people recovering after stroke with one treatment equals 0.10, while the proportion after a different treatment equals 0.16. The risk ratio equals 0.625 (0.10/0.16); 37.5% ((1–0.625)*100 or (0.16–0.10)/0.16) fewer patients treated by the first method recover. The risk ratio takes on values between zero ('0') and infinity. One ('1') is the neutral value and means that there is no difference between the groups compared.

Robust A test is robust if it is not seriously disturbed by the violation of underlying assumptions.

Role A role is a shorthand way of expressing a specialization.

Rolling offset A simple offset that is rolled to one side.

Root of unity An N-th root of unity is any complex number W such that $W^N = 1$.

For a given N, there are N such roots, which can be summarized as:

THETA = 2 * pi / N

W_J = cos ((J – 1) * THETA) + i * sin ((J – 1) * THETA)

Roots of unity are especially useful in discussing Fourier matrices and Circulant matrices.

Root The Nth root of a number,is that which when multiplied by itself N times produces the desired number.The square (2nd) root of 4 is 2 (2 x 2 = 4).The cube (3rd) root of 27 is 9 (3 x 3 x 3 =9).

Rotate To rotate an object in a tessellation means to repeat the object by spinning it on a point a certain angle.

Rotation When a figure is turned, we call it a rotation of the figure. We can measure this rotation in terms of degrees; a 360 degree turn rotates a figure around once back to its original position.

Example: For the following pairs of figures, the figure on the right is a rotation of the figure on the left.

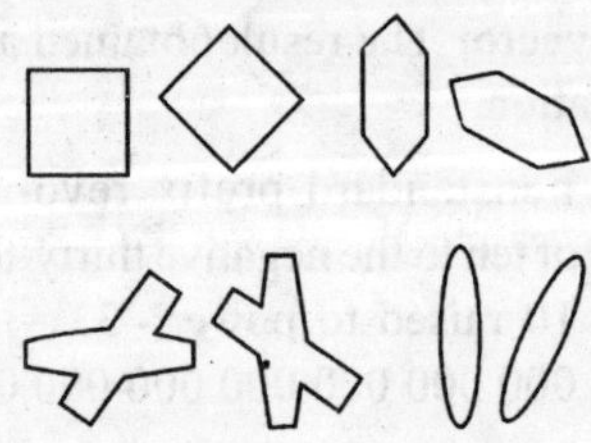

Rounding off a number Simplifying a number by slightly raising or lowering its value. Accuracy in work is restricted by the tools used in that work. Calculations can not be any closer than what the tools allow, so numbers are rounded off to a point useful for that tool. Needed accuracy will vary among trades or jobs.

Row echelon form A special matrix structure which is usually arrived at by a form of Gauss elimination.

Any matrix, including singular and rectangular matrices, can be transformed into this form, using a series of elementary row operations. Once the form is computed, it is easy to compute the determinant, inverse, the solution of linear systems (even for underdetermined or overdetermined systems), the rank, and solutions to linear programming problems.

Moreover, the process can be considered a matrix factorization, of the form

A = B * E

where B is nonsingular, and E is in row echelon form.

A matrix (whether square or rectangular) is in row echelon form if:

Each nonzero row of the matrix has a 1 as its first nonzero entry.

The leading 1 in a given row occurs in a column to the right of the leading 1 in the previous row.

Rows that are completely zero occur last.

A matrix is in row reduced echelon form if it is in row echelon form, and it is also true that:

Each column containing a leading 1 has no other nonzero entries.

Row echelon form is primarily of use for teaching, and analysis of small problems, using exact arithmetic. It is of little interest numerically, because very slight errors in numeric representation or arithmetic can result in completely erroneous results.

Row matrix A matrix with one row.

Row rank The row rank of a matrix is the number of linearly independent rows it contains.

The matrix

1 2 3 4

5 6 7 8

9 10 11 12

has row rank 2, because row 3 is equal to twice row 2 minus row 1.

For any matrix, the row rank is the same as the column rank. This common value is also equal to the rank of the matrix, which is defined for both square and rectangular matrices.

For a square matrix of order n, the rank is a number between 0 and n. A square matrix whose rank is n is said to be nonsingular. For a rectangular matrix of order m by n, the rank must be a number between 0 and the minimum of m and n. Rectangular matrices which attain this maximal value are not called nonsingular. Instead, they are said to have full row rank, full column rank, or maximal rank.

Row space The row space of an M by N matrix A is the set of all possible linear combinations of rows of A.

If the N vector v is a linear combination of rows of A, then there is a M vector c of coefficients with the property that

v = c * A

In other words, the row space is the set of all possible results of premultiplying A by an arbitrary vector.

Row totals The total number of observations occurring in a row of a contingency table.

Rule of probabilities multiplication for simultaneous independent events When finding the probability of two independent events (two things happening where the outcomes are not affected by each other), multiply the probabilities of each event happening to get the probability of both events happening. For example, to find the probability of getting heads and then tails when flipping a coin twice, multiply the probability of getting heads once by the probability of getting tails once..

Rules That part of a decision table that specifies which actions are to be followed for a given set of conditions.

Russell's paradox Popularly : The Cretan who said that all Cretan's are liars - is a liar - was he lying or telling the truth?

In a mathematical context : the set of all sets that are not members of themselves. This concept embodies a contradiction.

Ryan procedure (REGWQ) A multiple comparison procedure that holds the familywise error rate while having greater power than Tukey's test.

S

Sacred ratio The Sacred Ratio (symbol: φ {Greek lower case letter phi}) is the irrational number that is one greater than its multiplicative inverse, i.e., $\varphi = 1 + 1/\varphi = (1+5^{1/2})/2$ = approximately 1.618 033 988 750. The ancient Egyptians discovered the Sacred Ratio which they used extensively in architecture. The ancient Greeks treasured φ as their Golden Ratio. The ancient Egyptians discovered φ, which they honored as their Sacred Ratio. The ancient Greeks treasured φ as their Golden Ratio. The ratio φ is used extensively in art and architecture.

Sample Set of actual observations. Subset of the population.

Sample mean ($\bar{x}$) The mean of a random sample is an unbiased estimate of the mean of the population from which it was drawn. Another way to say this is to assert that regardless of the size of the population and regardless of the size of the random sample, it can be shown (through The Central Limit Theorem) that if we repeatedly took random samples of the same size from the same population, the sample means would cluster around the exact value of the population mean. As illustrated here, our random sample contains 4 items and it was drawn from a population that contains 9 items. Most statisticians use (n) to represent the number of items in a sample, whereas they use the symbol (N) to represent the number of items in a population. For this sample from this population n=4, N=9.

Sample s square (Sample variance) (S^2) The statistic s square (S^2) is a measure on a random sample that is used to estimate the variance of the population from which the sample is drawn. Numerically, it is the sum of the squared deviations around the mean of a random sample divided by the sample size minus one.

Regardless of the size of the population, and regardless of the size of the random sample, it can be algebriacally shown that if we repeatedly took random samples of the same size from the same population and calculated the variance estimate on each sample, these values would cluster around the exact value of the population variance. In short, the statistic s squared is an unbiased estimate of the variance of the population from which a sample is drawn.

The illustration shows the formula for the calculation of s square and it provides an example of its calculation. It is a useful exercise to carry out these calculations by oneself.

Sample size determination Mathematical process of deciding how many subjects should be studied (at the planning phase of a study). Among the factors to consider are the incidence or prevalence of the condition, the magnitude of difference expected between cases and controls, the power that is desired and the allowable magnitude of type I error (pre-determined significance probability). Sample size calculator, sample size calculation.

Sample space The set or collection of all possible outcomes of a probability experiment.

Sample statistics Statistics calculated from a sample and used primarily to describe the sample.

Sample variance (s^2) Sum of the squared deviations about the mean divided by N - 1.

Sample with replacement Sampling in which the item drawn on trial N is replaced before the drawing on trial N + 1.

Sampling A process of selecting observations to obtain knowledge about a population. There are many methods to choose on which sample to do the observations.

Sampling distribution of differences between means The distribution of the differences between means over repeated sampling from the same population(s).

Sampling distribution of the mean The distribution of sample means over repeated sampling from one population.

Sampling distribution The Sampling Distribution of a statistic is the set of values that we would obtain if we drew an infinite number of random samples from a given population and calculated the statistic on each sample. In doing so, all samples must be of the same size (n). While it is not possible for anyone to actually draw an infinite number of samples, the concept of a sampling distribution can be understood by taking the time to carefully consider the following theoretical exercise.

Imagine that our population consists of only three numbers: the number 2, the number 3 and the number 4. Our plan is to draw an infinite number of random samples of size n = 2 and form a sampling distribution of the sample means. The accompanying illustration shows this population and it's first two columns show each of the possible random samples (of size n = 2), that might be drawn from this population. If the first item is a 2, the second item can be either a 2 again, or it can be a 3 or a 4. Remember, we are drawing a random sample and our population is small, hence we are sampling with replacement. If our first item happened to be a 3, the second item might be a 2, a 3 again, or a 4 and if our first item happened to be a 4 the second item could be either a 2 or a 3 or a 4 again.

As seen here, there are only 9 possible combinations of two numbers in a given sample and each of the combinations is equally likely. The same is not true, however, for the means of the samples.

The third column in the illustration shows the means of each of the possible samples and the histogram shows the relative frequency of each of these means. In doing so the histogram provides a detailed representation of the sampling distribution of means (of size n = 2) that would be obtained if we were, in fact, able to draw an infinite number of random samples from the indicated population and graphically represent their frequency distribution in a histogram.

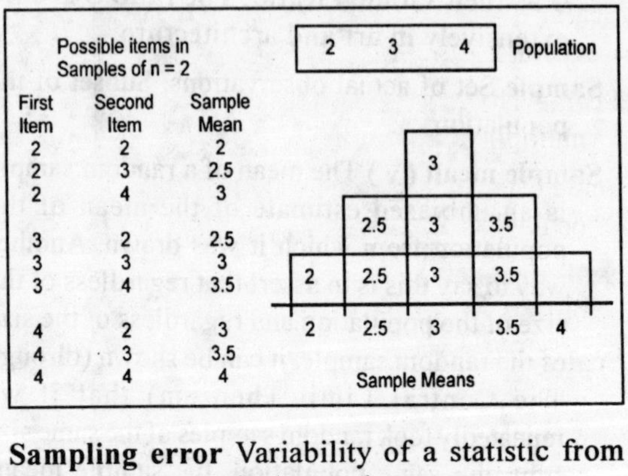

Possible items in Samples of n = 2

First Item	Second Item	Sample Mean
2	2	2
2	3	2.5
2	4	3
3	2	2.5
3	3	3
3	4	3.5
4	2	3
4	3	3.5
4	4	4

2 3 4 Population

3
2.5 3 3.5
2 2.5 3 3.5
2 2.5 3 3.5 4
Sample Means

Sampling error Variability of a statistic from sample to sample due to chance.

Sampling fraction The fraction of the number of levels actually used in an experiment to the potential number of levels that could have been used. In a fixed model the sampling fraction is 1.0, and in a random model it approaches 0.0.

SAS (Statistical Analysis System) A comprehensive computer software system for data processing and analysis. It can be used for almost any type of statistical analysis. Produced by SAS Institute.

Saturated model A model that contains as many parameters as there are data points. This model contains all main effects and all possible interactions between factors. For categorical data, this model contains the same number of parameters as cells and results in a perfect fit for a data set. The (residual) deviance is a measure of the extent to which a particular model differs from the saturated model.

Scalable The ability to upgrade seamlessly the capabilities of the system through either hardware upgrades, software upgrades, or both.

Scalar algebra The plain old kind of algebra you learned in high school—as opposed to matrix algebra.

Scalar multiplication For any scalar c, number, and any m x n matrix A with elements aij, the product of c and A is the in m x n matrix whose elements are caij.

Scalar multiplication of algebraic vectors A processes of multiplying vector components.

Scalar quantity The value of a dot product of two vectors.

Scalar The real number that is a factor in scalar multiplication.

Scale Choice of increments and range of numbers on an axis.

Scale drawing A scaled representation of physical objects or drawings.

Scales of measurement The type of data is always one of the following four scales of measurement: nominal, ordinal, interval, or ratio. Each of these can be discrete or continuous.

Scatter diagram A figure in which the individual data points are plotted in two-dimensional space.

Scatter plot A graphical representation consisting of ordered pairs possibly showing a relationship between two variable quantities.

Scattergram A figure in which the individual data points are plotted in two-dimensional space.

Scatterplot A figure in which the individual data points are plotted in two-dimensional space.

Schedule feasibility The process of assessing the degree to which the potential time frame and completion dates for all major activities within a project meet organizational deadlines and constraints for affecting change.

Scheffé test A relatively conservative multiple comparison procedure.

Schoenfeld residual test One of the diagnostic tests to check the proportionality assumption (covariates are time independent) in proportional hazard modelling. A variation is the use of scaled Schoenfeld residuals.

Scientific notation Representation of a number as the product of a number between 1 and 10 and a power of 10; used especially for very small or very large numbers.

Scribe The person who makes detailed notes of the happenings at a Joint Application Design session.

Secant One of the functions of an angle. It is found by dividing the length of the hypotenuse of a right triangle by the length of the adjacent side of a reference angle.

Second 1. The adjective, adverb, and noun second (2nd or 2[d]) is the standard name for the ordinal number that follows first (1st). Second corresponds with the cardinal number two (2).

2. The noun second means divided by sixty twice ((1/60)·(1/60) = 1/3,600). In time measurement, a second is one thirty-six-hundredth (1/3,600) of an hour or one eighty-six-thousand-four-hundredth (1/86,400) of a day. In angular measurement, a second (suffix symbol: –3 {double prime}) is one thirty-six-hundredth (1/3,600) of a degree (suffix symbol: –° {degree symbol}) or one one-million-two-hundred-ninety-six-thousandth (1/1,296,000) of a circle.

Second normal form (2NF) A relation is in second normal form if every nonprimary key attribute is functionally dependent on the whole primary key.

Second order determinant A determinant with two rows and two columns.

Second order interaction The interaction of three variables.

Secondary key One or a combination of fields for which more than one row may have the same combination of values.

Secondary The adjective and noun secondary means second (2nd or 2d) in order or importance.

Sector A piece of an object. In the spinner, any of the numbered segments is a sector.

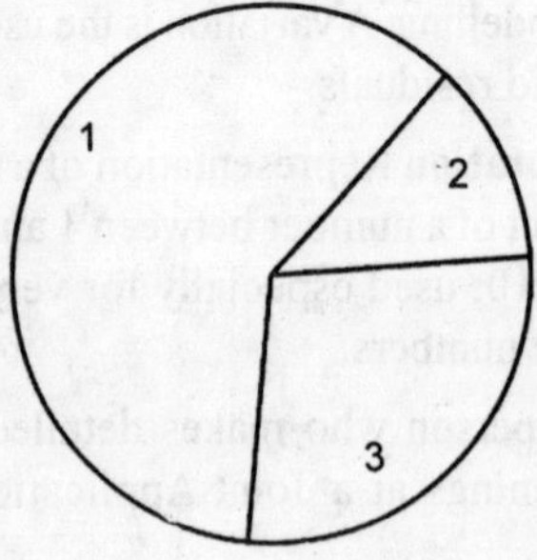

Sector A portion of a circle enclosed by a central angle and its subtended arc.

Seed In the successive adjustment method, a seed is a product of powers of the first k primes, for some small value of k.

Self-similarity Two or more objects having the same characteristics. In fractals, the shapes of lines at different iterations look like smaller versions of the earlier shapes .

Self-support A prime power has self-support if it provides support for itself.

Semi partial correlation ($r_{0(1.2)}$) The correlation between the dependent and the independent variables with the effect of another variable or variables removed from just the independent variables. Also known as the part correlation.

Semi The noun and adjective semi means a tractor and trailer combination freight truck (the tractor and trailer are the two (2) halves (1/2) of the truck.).

Semiannual The adjective semiannual means recurring every one half (1/2) year or every six (6) months.

Semicentennial 1. The adjective and noun semicentennial means recurring every fifty (50) years.

2. The adjective and noun semicentennial means lasting fifty (50) years.

3. The noun and adjective semicentennial means a fifty (50) year anniversary.

Semidiurnal The adjective semidiurnal means recurring twice (2) per day.

Semimonthly The adjective semimonthly means recurring twice (2) per month.

Semiperimeter One-half the perimeter of a triangle.

Semitrailer The noun and adjective semitrailer means a freight trailer designed to be towed by a tractor in a combination freight truck (the tractor and semitrailer are the two (2) halves (1/2) of the truck.)

Semiweekly The adjective semiweekly means recurring twice (2) per week.

Senary 1. The adjective and noun senary means sixth (6th) in order or importance.

2. The adjective senary means having six (6) parts.

Sensitivity Sensitivity is the proportion of true positives that are correctly identified by a diagnostic test.

Septe The numeric prefix septe- means seven (7). This prefix is derived from the Latin cardinal number septem (VII).

September The month name September literally means the seventh (7th) month. The Roman year began on the first (1st) day of March.

Septemcentenary 1. The noun and adjective septemcentenary means a group of seven hundred (700).

2. The noun and adjective septemcentenary means a period of seven hundred (700) years.

3. The noun and adjective septemcentenary means a seven hundred (700) year anniversary.

4. The adjective and noun septemcentenary means seven hundredth (700th) in rank or order.

Septemcentennial 1. The adjective and noun septemcentennial means recurring every seven hundred (700) years.

2. The adjective and noun septemcentennial means lasting seven hundred (700) years.

3. The noun and adjective septemcentennial means a seven hundred (700) year anniversary.

Septemcentennium The noun and adjective septemcentennium means a period of seven hundred (700) years.

Septemcentivigesimal The adjective septemcentivigesimal means based on the number seven hundred twenty (720).

Septemmillenary 1. The noun and adjective septemmillenary means a group of seven thousand (7000).

2. The noun and adjective septemmillenary means a period of seven thousand (7000) years.

3. The noun and adjective septemmillenary means a seven thousand (7000) year anniversary.

4. The adjective and noun septemmillenary means seven thousandth (7000th) in rank or order.

Septemmillennial 1. The adjective and noun septemmillennial means recurring every seven thousand (7000) years.

2. The adjective and noun septemmillennial means lasting seven thousand (7000) years.

3. The noun and adjective septemmillennial means a seven thousand (7000) year anniversary.

Septemmillennium The noun and adjective septemmillennium means a period of seven thousand (7000) years.

Septenary 1. The adjective and noun septenary means seventh (7th) in order or importance.

2. The adjective septenary means having seven (7) parts.

Septendecennary The noun and adjective septendecennary means a group of seventeen (17) or a period of seventeen (17) years.

Septendecennial 1. The adjective and noun septendecennial means recurring every seventeen (17) years.

2. The adjective and noun septendecennial means lasting seventeen (17) years.

3. The noun and adjective septendecennial means a seventeen (17) year anniversary.

Septendecennium The noun and adjective septendecennium means a period of seventeen (17) years.

Septendecentenary 1. The noun and adjective septendecentenary means a group of one thousand seven hundred (1700).

2. The noun and adjective septendecentenary means a period of one thousand seven hundred (1700) years.

3. The noun and adjective septendecentenary means a one thousand seven hundred (1700) year anniversary.

4. The adjective and noun septendecentenary means one thousand seven hundredth (1700th) in rank or order.

Septendecentennial 1. The adjective and noun septendecentennial means recurring every one thousand seven hundred (1700) years.

2. The adjective and noun septendecentennial means lasting one thousand seven hundred (1700) years.

3. The noun and adjective septendecentennial means a one thousand seven hundred (1700) year anniversary.

Septendecentennium The noun and adjective septendecentennium means a period of one thousand seven hundred (1700) years.

Septendeci The numeric prefix septendeci- means seventeen (17). This prefix is derived from the Latin cardinal number septendecim (XVII).

Septendecillion The noun and adjective septendecillion means multiplied by 10^{54} in the

American number naming system, but means multiplied by 10^{102} in the Chuquet number naming system. .

Septendecimal The adjective septendecimal means based on the number seventeen (17).

Septennary The noun and adjective septennary means a group of seven (7) or a period of seven (7) years.

Septennial 1. The adjective and noun septennial means recurring every seven (7) years.

2. The adjective and noun septennial means lasting seven (7) years.

3. The noun and adjective septennial means a seven (7) year anniversary.

Septennonaginta The numeric prefix septennonaginta- means ninety-seven (97). This prefix is derived from the Latin cardinal number septem et nonaginta (LXXXXVII).

Septennonagintennary The noun and adjective septennonagintennary means a group of ninety-seven (97) or a period of ninety-seven (97) years.

Septennonagintennial 1. The adjective and noun septennonagintennial means recurring every ninety-seven (97) years.

2. The adjective and noun septennonagintennial means lasting ninety-seven (97) years.

3. The noun and adjective septennonagintennial means a ninety-seven (97) year anniversary.

Septennonagintennium The noun and adjective septennonagintennium means a period of ninety-seven (97) years.

Septennonagintillion The noun and adjective septennonagintillion means multiplied by 10^{294} in the American number naming system, but means multiplied by 10^{582} in the Chuquet number naming system. .

Septenquadraginta The numeric prefix septenquadraginta- means forty-seven (47). This prefix is derived from the Latin cardinal number septem et quadraginta (XXXXVII).

Septenquadragintennary The noun and adjective septenquadragintennary means a group of forty-seven (47) or a period of forty-seven (47) years.

Septenquadragintennial 1. The adjective and noun septenquadragintennial means recurring every forty-seven (47) years.

2. The adjective and noun septenquadragintennial means lasting forty-seven (47) years.

3. The noun and adjective septenquadragintennial means a forty-seven (47) year anniversary.

Septenquadragintennium The noun and adjective septenquadragintennium means a period of forty-seven (47) years.

Septenquadragintillion The noun and adjective septenquadragintillion means multiplied by 10^{144} in the American number naming system, but means multiplied by 10^{282} in the Chuquet number naming system. .

Septenquinquaginta The numeric prefix septenquinquaginta- means fifty-seven (57). This prefix is derived from the Latin cardinal number septem et quinquaginta (LVII).

Septenquinquagintennary The noun and adjective septenquinquagintennary means a group of fifty-seven (57) or a period of fifty-seven (57) years.

Septenquinquagintennial 1. The adjective and noun septenquinquagintennial means recurring every fifty-seven (57) years.

2. The adjective and noun septenquinquagintennial means lasting fifty-seven (57) years.

3. The noun and adjective septenquinquagintennial means a fifty-seven (57) year anniversary.

Septenquinquagintennium The noun and adjective septenquinquagintennium means a period of fifty-seven (57) years.

Septenquinquagintillion The noun and adjective septenquinquagintillion means multiplied by 10^{174} in the American number naming system, but means multiplied by 10^{342} in the Chuquet number naming system. .

Septenseptuaginta The numeric prefix septenseptuaginta- means seventy-seven (77).

This prefix is derived from the Latin cardinal number septem et septuaginta (LXXVII).

Septenseptuagintennary The noun and adjective septenseptuagintennary means a group of seventy-seven (77) or a period of seventy-seven (77) years.

Septenseptuagintennial 1. The adjective and noun septenseptuagintennial means recurring every seventy-seven (77) years.

2. The adjective and noun septenseptuagintennial means lasting seventy seven (77) years.

3. The noun and adjective septenseptuagintennial means a seventy-seven (77) year anniversary.

Septenseptuagintennium The noun and adjective septenseptuagintennium means a period of seventy-seven (77) years.

Septenseptuagintillion The noun and adjective septenseptuagintillion means multiplied by 10^{234} in the American number naming system, but means multiplied by 10^{462} in the Chuquet number naming system.

Septensexaginta The numeric prefix septensexaginta- means sixty-seven (67). This prefix is derived from the Latin cardinal number septem et sexaginta (LXVII).

Septensexagintennary The noun and adjective septensexagintennary means a group of sixty-seven (67) or a period of sixty-seven (67) years.

Septensexagintennial 1. The adjective and noun septensexagintennial means recurring every sixty-seven (67) years.

2. The adjective and noun septensexagintennial means lasting sixty-seven (67) years.

3. The noun and adjective septensexagintennial means a sixty-seven (67) year anniversary.

Septensexagintennium The noun and adjective septensexagintennium means a period of sixty-seven (67) years.

Septensexagintillion The noun and adjective septensexagintillion means multiplied by 10^{204} in the American number naming system, but means multiplied by 10^{402} in the Chuquet number naming system.

Septentriginta The numeric prefix septentriginta- means thirty-seven (37). This prefix is derived from the Latin cardinal number septem et triginta (XXXVII).

Septentrigintacentenary 1. The noun and adjective septentrigintacentenary means a group of three thousand seven hundred (3700).

2. The noun and adjective septentrigintacentenary means a period of three thousand seven hundred (3700) years.

3. The noun and adjective septentrigintacentenary means a three thousand seven hundred (3700) year anniversary.

4. The adjective and noun septentrigintacentenary means three thousand seven hundredth (3700th) in rank or order.

Septentrigintacentennial 1. The adjective and noun septentrigintacentennial means recurring every three thousand seven hundred (3700) years.

2. The adjective and noun septentrigintacentennial means lasting three thousand seven hundred (3700) years.

3. The noun and adjective septentrigintacentennial means a three thousand seven hundred (3700) year anniversary.

Septentrigintacentennium The noun and adjective septentrigintacentennium means a period of three thousand seven hundred (3700) years.

Septentrigintennary The noun and adjective septentrigintennary means a group of thirty-seven (37) or a period of thirty-seven (37) years.

Septentrigintennial 1. The adjective and noun septentrigintennial means recurring every thirty-seven (37) years.

2. The adjective and noun septentrigintennial means lasting thirty-seven (37) years.

3. The noun and adjective septentrigintennial means a thirty-seven (37) year anniversary.

Septentrigintennium The noun and adjective septentrigintennium means a period of thirty-seven (37) years.

Septentrigintillion The noun and adjective septentrigintillion means multiplied by 10^{114} in the American number naming system, but means multiplied by 10^{222} in the Chuquet number naming system. .

Septenvigintennary The noun and adjective septenvigintennary means a group of twenty-seven (27) or a period of twenty-seven (27) years.

Septenvigintennial 1. The adjective and noun septenvigintennial means recurring every twenty-seven (27) years.

2. The adjective and noun septenvigintennial means lasting twenty-seven (27) years.

3. The noun and adjective septenvigintennial means a twenty-seven (27) year anniversary.

Septenvigintennium The noun and adjective septenvigintennium means a period of twenty-seven (27) years.

Septenviginti The numeric prefix septenviginti- means twenty-seven (27). This prefix is derived from the Latin cardinal number septem et viginti (XXVII).

Septenviginticentenary 1. The noun and adjective septenviginticentenary means a group of two thousand seven hundred (2700).

2. The noun and adjective septenviginticentenary means a period of two thousand seven hundred (2700) years.

3. The noun and adjective septenviginticentenary means a two thousand seven hundred (2700) year anniversary.

4. The adjective and noun septenviginticentenary means two thousand seven hundredth (2700th) in rank or order.

Septenviginticentennial 1. The adjective and noun septenviginticentennial means recurring every two thousand seven hundred (2700) years.

2. The adjective and noun septenviginticentennial means lasting two thousand seven hundred (2700) years.

3. The noun and adjective septenviginticentennial means a two thousand seven hundred (2700) year anniversary.

Septenviginticentennium The noun and adjective septenviginticentennium means a period of two thousand seven hundred (2700) years.

Septenvigintillion The noun and adjective septenvigintillion means multiplied by 10^{84} in the American number naming system, but means multiplied by 10^{162} in the Chuquet number naming system. .

Septillion The noun and adjective septillion means multiplied by 10^{24} in the American number naming system, but means multiplied by 10^{42} in the Chuquet number naming system. .

Septoctoginta The numeric prefix septoctoginta- means eighty-seven (87). This prefix is derived from the Latin cardinal number septem et octoginta (LXXXVII).

Septoctogintennary The noun and adjective septoctogintennary means a group of eighty-seven (87) or a period of eighty-seven (87) years.

Septoctogintennial 1. The adjective and noun septoctogintennial means recurring every eighty-seven (87) years.

2. The adjective and noun septoctogintennial means lasting eighty-seven (87) years.

3. The noun and adjective septoctogintennial means an eighty-seven (87) year anniversary.

Septoctogintennium The noun and adjective septoctogintennium means a period of eighty-seven (87) years.

Septoctogintillion The noun and adjective septoctogintillion means multiplied by 10^{264} in the American number naming system, but means multiplied by 10^{522} in the Chuquet number naming system. .

Septuagenarian The noun and adjective septuagenarian means between seventy (70) and eighty (80) years of age.

Septuagenary The adjective septuagenary means having seventy (70) parts.

Septuagesimal The adjective septuagesimal means based on the number seventy (70).

Septuaginta The numeric prefix septuaginta- means seventy (70). This prefix is derived from the Latin cardinal number septuaginta (LXX).

Septuagintennary The noun and adjective septuagintennary means a group of seventy (70) or a period of seventy (70) years.

Septuagintennial 1. The adjective and noun septuagintennial means recurring every seventy (70) years.

2. The adjective and noun septuagintennial means lasting seventy (70) years.

3. The noun and adjective septuagintennial means a seventy (70) year anniversary.

Septuagintennium The noun and adjective septuagintennium means a period of seventy (70) years.

Septuagintillion The noun and adjective septuagintillion means multiplied by 10^{213} in the American number naming system, but means multiplied by 10^{420} in the Chuquet number naming system. .

Septuplet The noun and adjective septuplet means one of seven (7) siblings born together or one of seven (7) very similar things.

Sequence A function defined on the positive integers or a subset of consecutive positive integers starting with 1.

Sequence diagram Depicts the interactions among objects during a certain period of time.

Sequence effect The situation in which the presentation of one level of the independent variable has an effect on response to another level of that variable.

Sequential file organization The rows in the file are stored in sequence according to a primary key value.

Sesqui The Latin-based numeric prefix sesqui- means one and one half (1.5).

Sesquicentenary 1. The noun and adjective sesquicentenary means a group of one hundred fifty (150).

2. The noun and adjective sesquicentenary means a period of one hundred fifty (150) years.

3. The noun and adjective sesquicentenary means a one hundred fifty (150) year anniversary.

4. The adjective and noun sesquicentenary means one hundred fiftieth (150th) in rank or order.

Sesquicentennial 1. The adjective and noun sesquicentennial means recurring every one hundred fifty (150) years.

2. The adjective and noun sesquicentennial means lasting one hundred fifty (150) years.

3. The noun and adjective sesquicentennial means a one hundred fifty (150) year anniversary.

Sesquicentennium The noun and adjective sesquicentennium means a period of one hundred fifty (150) years.

Sesquimillenary 1. The noun and adjective sesquimillenary means a group of one thousand five hundred (1500).

2. The noun and adjective sesquimillenary means a period of one thousand five hundred (1500) years.

3. The noun and adjective sesquimillenary means a one thousand five hundred (1500) year anniversary.

4. The adjective and noun sesquimillenary means one thousand five hundredth (1500th) in rank or order.

Sesquimillennial 1. The adjective and noun sesquimillennial means recurring every one thousand five hundred (1500) years.

2. The adjective and noun sesquimillennial means lasting one thousand five hundred (1500) years.

3. The noun and adjective sesquimillennial means a one thousand five hundred (1500) year anniversary.

Sesquimillennium The noun and adjective sesquimillennium means a period of one thousand five hundred (1500) years.

Sesquipedalian The adjective sesquipedalian literally means having or being one and one half (1.5) feet. More commonly, sesquipedalian means very long (as of a word like anti-dis-es-tab-lish-men-tar-ian-ism that may look a foot and a half long.)

Set builder notation A mathematical sentence that builds the set by selecting the elements that belong to the set.

Set The notion of a set is the most fundamental in Mathematics. Careless use of sets leads to contradictions the most famous of which was discovered by Bertrand Russell. Without going into details, sets are collections of elements that share some property P - the characteristic property of a set. The following notation is quite common: {x: P(x)} is read the set of all x that have the property P. In view of the possibility of paradoxes, it must be remarked that not every possible property P defines a set.

Seven The noun, pronoun, and adjective seven (7) is the standard name for the cardinal number that follows six (6). Seven corresponds with the ordinal number seventh (7th).

Seventeen The noun, pronoun, and adjective seventeen (17) is the standard name for the cardinal number that follows sixteen (16). Seventeen corresponds with the ordinal number seventeenth (17th).

Seventh 1. The adjective, adverb, and noun seventh (7th) is the standard name for the ordinal number that follows sixth (6th). Seventh corresponds with the cardinal number seven (7).

2. The adjective and noun seventh (1/7) means divided by seven (7).

Seventies 1. The plural noun and adjective seventies (70s) means an integer multiple of seventy (70).

2. The plural noun and adjective seventies (70's) means the range of numbers at least seventy (70) but less than eighty (80). When refering to dates, seventies means a span of ten years in which the tens (10s) digit of the year number is seven (7).

Seventieth The adjective, adverb, and noun seventieth (70th) is the standard name for the ordinal number that follows sixty-ninth (69th). Seventieth corresponds with the cardinal number seventy (70).

Seventy The noun, pronoun, and adjective seventy (70) is the standard name for the cardinal number that follows sixty-nine (69). Seventy corresponds with the ordinal number seventieth (70th).

Seventy-eight The noun and adjective seventy-eight (78) is the standard name for the cardinal number that follows seventy-seven (77). Seventy-eight corresponds with the ordinal number seventy-eighth (78th).

Seventy-eighth The compound adjective, adverb, and noun seventy-eighth (78th) is the standard name for the ordinal number that follows seventy-seventh (77th). Seventy-eighth corresponds with the cardinal number seventy-eight (78).

Seventy-first The compound adjective, adverb, and noun seventy-first (71st) is the standard name for the ordinal number that follows seventieth (70th). Seventy-first corresponds with the cardinal number seventy-one (71).

Seventy-five The noun and adjective seventy-five (75) is the standard name for the cardinal number that follows seventy-four (74). Seventy-five corresponds with the ordinal number seventy-fifth (75th).

Seventy-four The noun and adjective seventy-four (74) is the standard name for the cardinal number that follows seventy-three (73). Seventy-four corresponds with the ordinal number seventy-fourth (74th).

Seventy-nine The noun and adjective seventy-nine (79) is the standard name for the cardinal number that follows seventy-eight (78). Seventy-nine corresponds with the ordinal number seventy-ninth (79th).

Seventy-one The noun and adjective seventy-one (71) is the standard name for the cardinal number that follows seventy (70). Seventy-one corresponds with the ordinal number seventy-first (71st).

Seventy-second 1. The compound adjective, adverb, and noun seventy-second (72nd or 72d) is the standard name for the ordinal number that follows seventy-first (71st). Seventy-second corresponds with the cardinal number seventy-two (72).

2. The compound adjective and noun seventy-second (1/72) means divided by seventy-two (72).

Seventy-seven The noun and adjective seventy-seven (77) is the standard name for the cardinal number that follows seventy-six (76). Seventy-seven corresponds with the ordinal number seventy-seventh (77th).

Seventy-six The noun and adjective seventy-six (76) is the standard name for the cardinal number that follows seventy-five (75). Seventy-six corresponds with the ordinal number seventy-sixth (76th).

Seventy-three The noun and adjective seventy-three (73) is the standard name for the cardinal number that follows seventy-two (72). Seventy-three corresponds with the ordinal number seventy-third (73rd or 73d).

Seventy-two The noun and adjective seventy-two (72) is the standard name for the cardinal number that follows seventy-one (71). Seventy-two corresponds with the ordinal number seventy-second (72nd or 72d).

Several The adjective and plural pronoun several means a number of individuals or discrete objects generally more than two (2) but not more than twelve (12).

Sex The numeric prefix sex- means six (6). This prefix is derived from the Latin cardinal number sex (VI).

Sexagenarian The noun and adjective sexagenarian means between sixty (60) and seventy (70) years of age.

Sexagesimal The adjective sexagesimal means based on the number sixty (60).

Sexagesimo-quarto The noun sexagesimo-quarto (64mo) means a book with leaves printed on one sixty-fourth (1/64) of the printer sheet.

Sexaginta The numeric prefix sexaginta- means sixty (60). This prefix is derived from the Latin cardinal number sexaginta (LX).

Sexagintennary The noun and adjective sexagintennary means a group of sixty (60) or a period of sixty (60) years.

Sexagintennial 1. The adjective and noun sexagintennial means recurring every sixty (60) years.

2. The adjective and noun sexagintennial means lasting sixty (60) years.

3. The noun and adjective sexagintennial means a sixty (60) year anniversary.

Sexagintennium The noun and adjective sexagintennium means a period of sixty (60) years.

Sexagintillion The noun and adjective sexagintillion means multiplied by 10^{183} in the American number naming system, but means multiplied by 10^{360} in the Chuquet number naming system. .

Sexcentenary 1. The noun and adjective sexcentenary means a group of six hundred (600)

2. The noun and adjective sexcentenary means a period of six hundred (600) years.

3. The noun and adjective sexcentenary means a six hundred (600) year anniversary.

4. The adjective and noun sexcentenary means six hundredth (600th) in rank or order.

Sexcentennial 1. The adjective and noun sexcentennial means recurring every six hundred (600) years.

2. The adjective and noun sexcentennial means lasting six hundred (600) years.

3. The noun and adjective sexcentennial means a six hundred (600) year anniversary.

Sexcentennium The noun and adjective sexcentennium means a period of six hundred (600) years.

Sexdecennary The noun and adjective sexdecennary means a group of sixteen (16) or a period of sixteen (16) years.

Sexdecennial 1. The adjective and noun sexdecennial means recurring every sixteen (16) years.

2. The adjective and noun sexdecennial means lasting sixteen (16) years.

3. The noun and adjective sexdecennial means a sixteen (16) year anniversary.

Sexdecennium The noun and adjective sexdecennium means a period of sixteen (16) years.

Sexdecentenary 1. The noun and adjective sexdecentenary means a group of one thousand six hundred (1600).

2. The noun and adjective sexdecentenary means a period of one thousand six hundred (1600) years.

3. The noun and adjective sexdecentenary means a one thousand six hundred (1600) year anniversary.

4. The adjective and noun sexdecentenary means one thousand six hundredth (1600th) in rank or order.

Sexdecentennial 1. The adjective and noun sexdecentennial means recurring every one thousand six hundred (1600) years.

2. The adjective and noun sexdecentennial means lasting one thousand six hundred (1600) years.

3. The noun and adjective sexdecentennial means a one thousand six hundred (1600) year anniversary.

Sexdecentennium The noun and adjective sexdecentennium means a period of one thousand six hundred (1600) years.

Sexdeci The numeric prefix sexdeci- means sixteen (16). This prefix is derived from the Latin cardinal number sedecim (XVI).

Sexdecillion The noun and adjective sexdecillion means multiplied by 10^{51} in the American number naming system, but means multiplied by 10^{96} in the Chuquet number naming system..

Sexdecimal The adjective sexdecimal means based on the number sixteen (16).

Sexennary The noun and adjective sexennary means a group of six (6) or a period of six (6) years.

Sexennial 1. The adjective and noun sexennial means recurring every six (6) years.

2. The adjective and noun sexennial means lasting six (6) years.

3. The noun and adjective sexennial means a six (6) year anniversary.

Sexennium The noun and adjective sexennium means a period of six (6) years.

Sexmillenary 1. The noun and adjective sexmillenary means a group of six thousand (6000).

2. The noun and adjective sexmillenary means a period of six thousand (6000) years.

3. The noun and adjective sexmillenary means a six thousand (6000) year anniversary.

4. The adjective and noun sexmillenary means six thousandth (6000th) in rank or order.

Sexmillennial 1. The adjective and noun sexmillennial means recurring every six thousand (6000) years.

2. The adjective and noun sexmillennial means lasting six thousand (6000) years.

3. The noun and adjective sexmillennial means a six thousand (6000) year anniversary.

Sexmillennium The noun and adjective sexmillennium means a period of six thousand (6000) years.

Sexnonaginta The numeric prefix sexnonaginta- means ninety-six (96). This prefix is derived from the Latin cardinal number sex et nonaginta (LXXXXVI).

Sexnonagintennary The noun and adjective sexnonagintennary means a group of ninety-six (96) or a period of ninety-six (96) years.

Sexnonagintennial 1. The adjective and noun sexnonagintennial means recurring every ninety-six (96) years.

2. The adjective and noun sexnonagintennial means lasting ninety-six (96) years.

3. The noun and adjective sexnonagintennial means a ninety-six (96) year anniversary.

Sexnonagintennium The noun and adjective sexnonagintennium means a period of ninety-six (96) years.

Sexnonagintillion The noun and adjective sexnonagintillion means multiplied by 10^{291} in the American number naming system, but means multiplied by 10^{576} in the Chuquet number naming system. .

Sexoctoginta The numeric prefix sexoctoginta- means eighty-six (86). This prefix is derived from the Latin cardinal number sex et octoginta (LXXXVI).

Sexoctogintennary The noun and adjective sexoctogintennary means a group of eighty-six (86) or a period of eighty-six (86) years.

Sexoctogintennial 1. The adjective and noun sexoctogintennial means recurring every eighty-six (86) years.

2. The adjective and noun sexoctogintennial means lasting eighty-six (86) years.

3. The noun and adjective sexoctogintennial means an eighty-six (86) year anniversary.

Sexoctogintennium The noun and adjective sexoctogintennium means a period of eighty-six (86) years.

Sexoctogintillion The noun and adjective sexoctogintillion means multiplied by 10^{261} in the American number naming system, but means multiplied by 10^{516} in the Chuquet number naming system. .

Sexquadraginta The numeric prefix sexquadraginta- means forty-six (46). This prefix is derived from the Latin cardinal number sex et quadraginta (XXXXVI).

Sexquadragintennary The noun and adjective sexquadragintennary means a group of forty-six (46) or a period of forty-six (46) years.

Sexquadragintennial 1. The adjective and noun sexquadragintennial means recurring every forty-six (46) years.

2. The adjective and noun sexquadragintennial means lasting forty-six (46) years.

3. The noun and adjective sexquadragintennial means a forty-six (46) year anniversary.

Sexquadragintennium The noun and adjective sexquadragintennium means a period of forty-six (46) years.

Sexquadragintillion The noun and adjective sexquadragintillion means multiplied by 10^{141} in the American number naming system, but means multiplied by 10^{276} in the Chuquet number naming system. .

Sexquinquaginta The numeric prefix sexquinquaginta- means fifty-six (56). This prefix is derived from the Latin cardinal number sex et quinquaginta (LVI).

Sexquinquagintennary The noun and adjective sexquinquagintennary means a group of fifty-six (56) or a period of fifty-six (56) years.

Sexquinquagintennial 1. The adjective and noun sexquinquagintennial means recurring every fifty-six (56) years.

2. The adjective and noun sexquinquagintennial means lasting fifty-six (56) years.

3. The noun and adjective sexquinquagintennial means a fifty-six (56) year anniversary.

Sexquinquagintennium The noun and adjective sexquinquagintennium means a period of fifty-six (56) years.

Sexquinquagintillion The noun and adjective sexquinquagintillion means multiplied by 10^{171} in the American number naming system, but means multiplied by 10^{336} in the Chuquet number naming system. .

Sexseptuaginta The numeric prefix sexseptuaginta- means seventy-six (76). This prefix is derived from the Latin cardinal number sex et septuaginta (LXXVI).

Sexseptuagintennary The noun and adjective sexseptuagintennary means a group of seventy-six (76) or a period of seventy-six (76) years.

Sexseptuagintennial 1. The adjective and noun sexseptuagintennial means recurring every seventy-six (76) years.

2. The adjective and noun sexseptuagintennial means lasting seventy-six (76) years.

3. The noun and adjective sexseptuagintennial means a seventy-six (76) year anniversary.

Sexseptuagintennium The noun and adjective sexseptuagintennium means a period of seventy-six (76) years.

Sexseptuagintillion The noun and adjective sexseptuagintillion means multiplied by 10^{231} in the American number naming system, but means multiplied by 10^{456} in the Chuquet number naming system. .

Sexsexaginta The numeric prefix sexsexaginta- means sixty-six (66). This prefix is derived from the Latin cardinal number sex et sexaginta (LXVI).

Sexsexagintennary The noun and adjective sexsexagintennary means a group of sixty-six (66) or a period of sixty-six (66) years.

Sexsexagintennial 1. The adjective and noun sexsexagintennial means recurring every sixty-six (66) years.

2. The adjective and noun sexsexagintennial means lasting sixty-six (66) years.

3. The noun and adjective sexsexagintennial means a sixty-six (66) year anniversary.

Sexsexagintennium The noun and adjective sexsexagintennium means a period of sixty-six (66) years.

Sexsexagintillion The noun and adjective sexsexagintillion means multiplied by 10^{201} in the American number naming system, but means multiplied by 10^{396} in the Chuquet number naming system. .

Sextal The adjective sextal means based on the number six (6).

Sextillion The noun and adjective sextillion means multiplied by 10^{21} in the American number naming system, but means multiplied by 10^{36} in the Chuquet number naming system. .

Sextodecimo The noun sextodecimo (16mo) means a book with leaves printed on one sixteenth (1/16) of the printer sheet.

Sextriginta The numeric prefix sextriginta- means thirty-six (36). This prefix is derived from the Latin cardinal number sex et triginta (XXXVI).

Sextrigintacentenary 1. The noun and adjective sextrigintacentenary means a group of three thousand six hundred (3600).

2. The noun and adjective sextrigintacentenary means a period of three thousand six hundred (3600) years.

3. The noun and adjective sextrigintacentenary means a three thousand six hundred (3600) year anniversary.

4. The adjective and noun sextrigintacentenary means three thousand six hundredth (3600th) in rank or order.

Sextrigintacentennial 1. The adjective and noun sextrigintacentennial means recurring every three thousand six hundred (3600) years.

2. The adjective and noun sextrigintacentennial means lasting three thousand six hundred (3600) years.

3. The noun and adjective sextrigintacentennial means a three thousand six hundred (3600) year anniversary.

Sextrigintacentennium The noun and adjective sextrigintacentennium means a period of three thousand six hundred (3600) years.

Sextrigintennary The noun and adjective sextrigintennary means a group of thirty-six (36) or a period of thirty-six (36) years.

Sextrigintennial 1. The adjective and noun sextrigintennial means recurring every thirty-six (36) years.

2. The adjective and noun sextrigintennial means lasting thirty-six (36) years.

3. The noun and adjective sextrigintennial means a thirty-six (36) year anniversary.

Sextrigintennium The noun and adjective sextrigintennium means a period of thirty-six (36) years.

Sextrigintillion The noun and adjective sextrigintillion means multiplied by 10^{111} in the American number naming system, but means multiplied by 10^{216} in the Chuquet number naming system. .

Sextuplet The noun and adjective sextuplet means one of six (6) siblings born together or one of six (6) very similar things.

Sextuplicate The verb, noun, and adjective sextuplicate means to make five (5) copies of one original, for a total of six (6) objects.

Sexvigintennary The noun and adjective sexvigintennary means a group of twenty-six (26) or a period of twenty-six (26) years.

Sexvigintennial 1. The adjective and noun sexvigintennial means recurring every twenty-six (26) years.

2. The adjective and noun sexvigintennial means lasting twenty-six (26) years.

3. The noun and adjective sexvigintennial means a twenty-six (26) year anniversary.

Sexvigintennium The noun and adjective sexvigintennium means a period of twenty-six (26) years.

Sexviginti The numeric prefix sexviginti- means twenty-six (26). This prefix is derived from the Latin cardinal number sex et viginti (XXVI).

Sexviginticentenary 1. The noun and adjective sexviginticentenary means a group of two thousand six hundred (2600).

2. The noun and adjective sexviginticentenary means a period of two thousand six hundred (2600) years.

3. The noun and adjective sexviginticentenary means a two thousand six hundred (2600) year anniversary.

4. The adjective and noun sexviginticentenary means two thousand six hundredth (2600th) in rank or order.

Sexviginticentennial 1. The adjective and noun sexviginticentennial means recurring every two thousand six hundred (2600) years.

2. The adjective and noun sexviginticentennial means lasting two thousand six hundred (2600) years.

3. The noun and adjective sexviginticentennial means a two thousand six hundred (2600) year anniversary.

Sexviginticentennium The noun and adjective sexviginticentennium means a period of two thousand six hundred (2600) years.

Sexvigintillion The noun and adjective sexvigintillion means multiplied by 10^{81} in the American number naming system, but means multiplied by 10^{156} in the Chuquet number naming system. .

Shorter Way is to define the tree as a connected graph with no circuits. The absence of circuits means that there is always exactly one way to get from one vertex of the tree to any other.

As a basic data structure, tree is designed to easily store information about graph trees.

Sigma notation Expresses the sum of a series.

Sigma sigma is the sum-of-divisors function; sigma(n) is the sum of all the positive integers dividing n. A generalization is sigma_k(n), the sum of all the kth powers of the divisors of n. sigma_k, and thus sigma = sigma_1, is a multiplicative function.

Sign test A test based on the probabilities of different outcomes for any number of pluses and minuses, i.e., observations below or above a prespecified value. The sign test can be used to investigate the significance of the difference between a population median and a specified value for it, or between the observed sex/transmission ratio and the 50:50 expected value. It can also be used for paired data. This time, the differences between the pairs will be either negative or positive, and the smaller of the two total negatives or positives plus the total number of pairs will form the test statistics. For example, when the total number is 20, if the number for the less frequent sign is 5 or smaller, $P < 0.05$ (two-tailed). A sign test in disguise is McNemar's test, which is used for paired data for dichotomous response.

Significance level The probability with which we are willing to reject H_0 when it is in fact correct.

Significant digits The number of digits to consider when using measuring numbers. There are three rules in determining the number of digits considered significant in a number.

1. All non-zeros are significant.

2. Any zeros between two non-zeros are significant

3. Only trailing zeros behind the decimal are considered significant

Similar figures Figures that have the same shape are called similar figures. They may be different sizes or turned somewhat.

Example: The following pairs of figures are similar.

These pairs of figures are not similar.

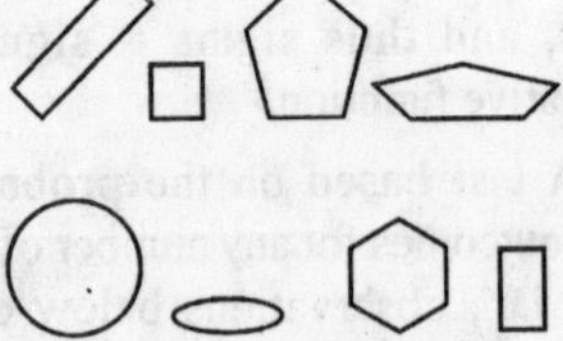

Similar triangles Two triangles whose angle measurements are the same.

Similarity Two triangles are similar if they have equal angles in which case their corresponding sides, say, a_1, b_1, c_1 and a_2, b_2, c_2, are proportional with a coefficient r: $a_1/a_2 = b_1/b_2 = c_1/c_2 = r$. Since side length could be looked at as the distances between the vertices, going from one triangle to another modifies these distances by a factor of r. Distances between other corresponding points in the triangles also change by the factor of r. This can be generalized. A similarity is such a transformation of the plane under which distance between any pair of points changes by the same factor. Similarity transformations play a central role in the Fractal Geometry and even more so in the theory of Iteration Function Systems (IFS). Consecutive Peano curves each consist of four images of itself each half the size of the whole. Similarity transformations are easily defined in spaces of higher dimensions.

Simple effect The effect of one independent variable at one level of another independent variable. (Also known as simple main effect.)

Simple eigenvalue A simple eigenvalue is an eigenvalue, or root of the characteristic equation, which has the property that it is not a repeated root. Another way of stating this is that a simple eigenvalue has an algebraic multiplicity of 1. Since every eigenvalue has at least one eigenvector, and the geometric multiplicity is never greater than the algebraic multiplicity, we have that a simple eigenvector is associated with a 1-dimensional eigenspace.

A semisimple eigenvalue is an eigenvalue for which the algebraic and geometric multiplicities are equal. A semisimple eigenvector is associated with a K-dimensional eigenspace, and there are K linearly independent eigenvectors. Of course, any simple eigenvalue is also a semisimple eigenvalue.

A matrix is diagonalizable if and only if every eigenvalue is semisimple.

The remaining possibility, called a nonsimple eigenvalue, is one for which the algebraic multiplicity is greater than one, and the geometric multiplicity is smaller than the algebraic multiplicity. In this case, the eigenvalue is associated with a K-dimensional eigenspace, but there are less than K linearly independent eigenvectors.

Simple harmonic motion A component of uniform circular motion.

Simple interaction The interaction of two variables at one level of a third variable.

Simple linear regression model The linear regression model for a normally distributed outcome (response) variable and a single predictor (explanatory) variable. The straight line models the mean value of the response variable for each value of the explanatory variable. The major assumption is constant variation of residuals along the fitted line which points out that the model is equally good across all x values. The null hypothesis stating that the explanatory variable has no effect on the response (in other words, the slope of the fitted line is zero) can be tested statistically. The two main aims of regression analysis are to predict the response and to understand the relationships between variables. As in all linear models, the error term (shown as W_i is additive and independent, and they are assumed to have a normal distribution. As an exception, the simple linear regression is a special case for generalised linear models.

Simple message A message that transfers control from the sender to the recipient without describing the details of the communication.

Simple offset A procedure by which a center line is moved up, down, or over to reach a new path going in the same direction. A simple offset uses two same angle turns.

Simulation A representation of a situation or problem with a similar but simpler model or a more easily manipulated model in order to determine experimental results.

Sine One of the functions of an angle. It is found by dividing the length of the opposite side of a reference angle by the length of the hypotenuse of the right triangle.

Singular matrix A matrix that does not have a unique inverse.

Singular The adjective singular means relating to one (1).

Singular value decomposition The singular value decomposition of a rectangular M by N matrix A is a factorization of the form:

A = U * S * V'

where, in the EISPACK version:

U is an M by N matrix with columns that are pairwise orthogonal;

S is an N by N diagonal matrix, containing the nonnegative singular values of A;

V is an N by N orthogonal matrix.

or, in the LINPACK version:

U is an M by M orthogonal matrix;

S is an M by N rectangular matrix whose diagonal contains the nonnegative singular values of A;

V is an N by N orthogonal matrix.

The LINPACK form has the advantage that U and V are both orthogonal, and S retains the shape of A. Moreover, this format allows us to consider U and V to be composed of left and right singular vectors, in analogy to the factorization of a square matrix via two orthogonal matrices of left and right eigenvectors.

The solution of A * x = b for non-square A can be found by seeking that x which minimizes the ||A*x–b||. That x is equal to

$V * S^{-1} * U'$,

where, since S may have zeroes on its diagonal, S^{-1} is constructed by replacing each nonzero diagonal element by its inverse.

For any column I no greater than the minimum of M and N, let u_i be the I–th column of U, and v_i be the I–th column of V, and s_{ii} be the I–th diagonal element of S. Then it is a fact that

$A * v_i = s_{ii} * u_i$

and

$A' * u_i = s_{ii} * v_i$

which allows us to conclude that

$A * A' * u_i = s_{ii} * s_{ii} * u_i$

and

$A' * A * v_i = s_{ii} * s_{ii} * v_i$

In other words, U, V and S contain information about the eigenvalues and eigenvectors of A * A'.

Conversely, if we know the eigenvalues and eigenvectors of A * A', then we know the squares of the singular values of A, and the left singular vectors of A (the U matrix).

The singular value decomposition can be used to construct the the pseudoinverse of the rectangular or singular matrix A.

Routines for singular value decomposition are included in EISPACK, LAPACK, and LINPACK.

Singularity The noun singularity means the state of being one (1).

Six feet under The compound adjective six (6) feet under means dead and buried. Six (6) feet (approximately 1.8 meters) is a common burial depth.

Six The noun, pronoun, and adjective six (6) is the standard name for the cardinal number that follows five (5). Six corresponds with the ordinal number sixth (6th).

Sixteen The noun, pronoun, and adjective sixteen (16) is the standard name for the cardinal number that follows fifteen (15). Sixteen equals two to the fourth power (2^4). Sixteen corresponds with the ordinal number sixteenth (16th).

Sixteenmo The noun sixteenmo (16mo) means a book with leaves printed on one sixteenth (1/16) of the printer sheet.

Sixteenth 1. The adjective, adverb, and noun sixteenth (16th) is the standard name for the ordinal number that follows fifteenth (15th). Sixteenth corresponds with the cardinal number sixteen (16).

2. The adjective and noun sixteenth (1/16) means divided by sixteen (16).

Sixth 1. The adjective, adverb, and noun sixth (6th) is the standard name for the ordinal number that follows fifth (5th). Sixth corresponds with the cardinal number six (6).

2. The adjective and noun sixth (1/6) means divided by six (6).

Sixties 1. The plural noun and adjective sixties (60s) means an integer multiple of sixty (60).

2. The plural noun and adjective sixties (60's) means the range of numbers at least sixty (60) but less than seventy (70). When refering to dates, sixties means a span of ten years in which the tens (10s) digit of the year number is six (6).

Sixtieth The adjective, adverb, and noun sixtieth (60th) is the standard name for the ordinal number that follows fifty-ninth (59th). Sixtieth corresponds with the cardinal number sixty (60).

2. The adjective and noun sixtieth (1/60) means divided by sixty (60).

Sixty The noun, pronoun, and adjective sixty (60) is the standard name for the cardinal number that follows fifty-nine (59). Sixty corresponds with the ordinal number sixtieth (60th).

Sixty-eight The noun and adjective sixty-eight (68) is the standard name for the cardinal number that follows sixty-seven (67). Sixty-eight corresponds with the ordinal number sixty-eighth (68th).

Sixty-five The noun and adjective sixty-five (65) is the standard name for the cardinal number that follows sixty-four (64). Sixty-five corresponds with the ordinal number sixty-fifth (65th).

Sixty-fourth 1. The compound adjective, adverb, and noun sixty-fourth (64th) is the standard name for the ordinal number that follows sixty-third (63rd or 63^{d}). Sixty-fourth corresponds with the cardinal number sixty-four (64).

2. The compound adjective and noun sixty-fourth (1/64) means divided by sixty-four (64).

Sixty-nine The noun and adjective sixty-nine (69) is the standard name for the cardinal number that follows sixty-eight (68). Sixty-nine corresponds with the ordinal number sixty-ninth (69th).

Sixty-one The noun and adjective sixty-one (61) is the standard name for the cardinal number that follows sixty (60). Sixty-one corresponds with the ordinal number sixty-first (61st).

Sixty-second The compound adjective, adverb, and noun sixty-second (62nd or 62^{d}) is the standard name for the ordinal number that follows sixty-first (61st). Sixty-second corresponds with the cardinal number sixty-two (62).

Sixty-seven The noun and adjective sixty-seven (67) is the standard name for the cardinal number that follows sixty-six (66). Sixty-seven corresponds with the ordinal number sixty-seventh (67th).

Sixty-six The noun and adjective sixty-six (66) is the standard name for the cardinal number that follows sixty-five (65). Sixty-six corresponds with the ordinal number sixty-sixth (66th).

Sixty-something The compound noun and adjective sixty-something means at least sixty (60) but less than seventy (70). Sixty-something most commonly means between sixty (60) and seventy (70) years of age.

Sixty-three The noun and adjective sixty-three (63) is the standard name for the cardinal number that follows sixty-two (62). Sixty-three corresponds with the ordinal number sixty-third (63rd or 63d).

Sixty-two The noun and adjective sixty-two (62) is the standard name for the cardinal number that follows sixty-one (61). Sixty-two corresponds with the ordinal number sixty-second (62nd or 62d).

Skew hermitian matrix A complex matrix A is skew Hermitian if it is equal to the negative of its transpose complex conjugate:

A = - (conjugate (A))'.

A skew Hermitian matrix must have a purely imaginary diagonal.

Skewness The degree of (lack of) asymmetry about a central value of a distribution. A distribution with many small values and few large values is positively (right) skewed (long tail in the distribution curve or stemplot is to the right); the opposite (left tail) is negatively (left) skewed. The measures of location median, midinterquartile range (midQ) and midrange decrease in this order for a left-skewed distribution.

Skillion The slang noun and adjective skillion means an extremely large number. Skillion is a parody of the Chuquet number naming system.

Slack time The amount of time that an activity can be delayed without delaying the project.

Slides A transformation in which the figures moves in a linear direction.

Slope For a straight line in the plane, the slope is the tangent of the angle it forms with the positive X axis. For a curve (e.g., a graph of a function), the slope is, by definition, the slope of the tangent line. Therefore, if the slope is constant a line is straight. Slope being a number should not be confused with a related notion of gradient, which is a vector.

Slope intercept form The form Y = mx + b of a linear equation in two variables. The slope is m and the y-intercept is b.

Slope of the linear function The slope of the line y = mx + b is the rate at which y is changing per unit of change in x. The units of measurement of the slope are units of y per unit of x .

Solitary The adjective solitary means one (1) isolated from any others.

Solo The noun, adjective, and adverb solo means one (1) member alone.

Solutions The values of a variable that make an equation or inequality a true statement.

Solving the triangle A process for finding the values of sides and angles of a triangle given the values of the remaining sides and angles.

Some The noun, pronoun, and adjective some means not none ('"0). When applied to individuals or discrete objects, some also usually implies more than one (>1) but not all (<100%).

Source/Sink The origin and/or destination of data; also called external entities.

Space figure A space figure or three-dimensional figure is a figure that has depth in addition to width and height. Everyday objects such as a tennis ball, a box, a bicycle, and a redwood tree are all examples of space figures. Some common simple space figures include cubes, spheres, cylinders, prisms, cones, and pyramids. A space figure having all flat faces is called a polyhedron. A cube and a pyramid are both polyhedrons; a sphere, cylinder, and cone are not.

Space There are too many different spaces in Mathematics to enumerate all. Generally, space is a set of points with additional features.

Span of a set of vectors The span of a set of vectors { v(i)| 1 <= I <= N } is the set of vectors that can be constructed from the set via linear combinations. The span is sometimes called the spanning space.

The span of a set of vectors is a linear space; it includes the 0 vector, each of the original vectors, and sums and multiples of them.

As an example of a vector span, consider the eigenvectors associated with a particular eigenvalue of a matrix. If the eigenvalue has algebraic multiplicity of 1, then we usually think there is only one associated eigenvector; if the eigenvalue has a higher multiplicity, there may be 2 or more linearly independent eigenvectors. In any case, the eigenspace associated with the eigenvalue is the span of these eigenvectors. Any element of this span is itself also an eigenvector for the eigenvalue.

Given a set of vectors which define a spanning space, it is often desired to know whether a subset of those vectors will define the same spanning space; that is, whether we can toss out some of the vectors because they don't add anything to the span. If we toss out as many vectors as possible, we end up with a basis for the span; the number of vectors remaining tells us the dimension of that space.

Sparse matrix storage A sparse storage scheme is a matrix storage method of storing the nonzero entries of a sparse matrix in a convenient and efficient way.

There are many different kinds of sparse matrix storage scheme. Their features depend on the properties of the underlying problem. In this discussion, a few simple non-specific schemes will be discussed.

If a direct linear equation solver will be used, storage must also be made available for fill-in values that will occur during the factorization process. An iterative solver, on the other hand, has no fill-in values, and never needs to change the entries of the matrix.

A simple storage scheme would require the user to supply NZ, the number of nonzero entries, and three arrays of size NZ:

IROW(I) and ICOL(I) specify the row and column indices for entry I;

A(I) specifies the value of entry I. Thus the first nonzero entry of the matrix occurs in row IROW(1), column ICOL(1), with the value A(1).

If this scheme seems excessive, the elements of the matrix could be listed in order of rows, the ROW array could be replaced by a ROWEND array of length N. ROWEND(I) is the index in A of the last nonzero element of row I of the original matrix. This small saving comes at the price of complicating access to the matrix.

Sparseness A contingency table is sparse when many cells have small values. When N is the total sample size, and r and c are the number of rows and columns, N / rc is an index of sparseness. Smaller values refer to more sparse tables. Sparse tables often contain zero values (empty cell).

Sparskit A tool kit for sparse matrix computations.

SPARSKIT can manipulate sparse matrices in a variety of formats, and can convert from one to another. For example, a matrix can be converted from the generalized diagonal format used by ELLPACK and ITPACK to the format used by the Harwell-Boeing Sparse Matrix Collection (HBSMC) or even into LINPACK banded format.

Utilities available include converting data structures, printing simple statistics on a matrix, plotting a matrix profile, performing basic linear algebra operations (similar to the BLAS for dense matrix), and so on.

Spearman's rank correlation A non-parametric correlation coefficient (rho) that is calculated by computing the Pearson's correlation coefficient (r) for the association between the ranks given to the values of the variables involved. It is used for ordinal data and interval/

ratio data. It is also possible to compare two correlation coefficients.

Specialization A specialization is an object class which is a subset of another object class (or classes). Specialization models the is a relationship set since members of the specialization class (or classes) are always members of the generalization class. This means that members of the specialization class have all of the same properties of the generalization class including relationships with other objects as well as behaviour. Therefore, specialization plays an important role in the ORM, OBM, and the OIM.

Specificity Specificity is the proportion of true negatives that are correctly identified by the test..

Spectral radius The spectral radius of a matrix is the magnitude of the largest eigenvalue of a matrix.

The spectral radius is often easy to compute, and it is a useful measure of the size or strength of a matrix. However, the spectral radius is not a matrix norm.

The spectral radius violates several rules for matrix norms. In particular, it is possible to have any of the following situations:

rho(A)=0 but A is not the zero matrix;

rho(A+B) > rho(A) + rho(B);

rho(A*B) > rho(A) * rho(B);

On the other hand, the spectral radius is a lower bound for the value of any vector-bound matrix norm on A, because there must be an eigenvalue lambda and a vector of unit norm x with the property that

A * x = lambda * x

so the norm of A*x divided by the norm of x is lambda. Therefore, the matrix norm of A must be at least | lambda |.

The Euclidean norm of a real symmetric matrix is equal to its spectral radius.

Spectrum The spectrum of a square matrix is the set of eigen values.

Speed This is the absolute value of velocity.

Sphere A sphere is a space figure having all of its points the same distance from its center. The distance from the center to the surface of the sphere is called its radius. Any cross-section of a sphere is a circle.

If r is the radius of a sphere, the volume V of the sphere is given by the formula V = 4/3 × pi ×r^3.

The surface area S of the sphere is given by the formula S = 4 × pi ×r^2.

Example: The space figure pictured below is a sphere.

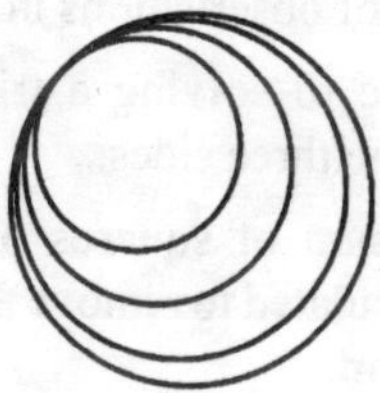

Example: To the nearest tenth, what is the volume and surface area of a sphere having a radius of 4cm?

Using an estimate of 3.14 for pi,

the volume would be $4/3 \times 3.14 \times 4^3 = 4/3 \times 3.14 \times 4 \times 4 \times 4 = 268$ cubic centimeters.

Using an estimate of 3.14 for pi, the surface area would be $4 \times 3.14 \times 4^2 = 4 \times 3.14 \times 4 \times 4 = 201$ square centimeters.

Sphericity A condition very like compound symmetry that is required for repeated-measures designs.

Square A four-sided polygon having equal-length sides meeting at right angles. The sum of the angles of a square is 360 degrees.

Examples:

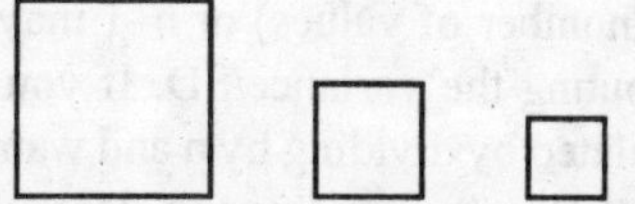

Square matrix of order n A matrix with n rows and n columns.

Square root A factor of a number which when squared produces that number. A square root multiplied by itself equals the number under the square root symbol.

SSA Reference to solving a triangle given the lengths of two sides and the measure of a non-included angle.

SS_{cells} The sum of squares assessing differences among cell means.

SS_{error} The sum of the squared residuals. In Anova the sum of the sums of squares within each group.

SS_{group} The sum of the squared deviation of the group means from the grand mean divided the the number of observations in each cell.

SSS Reference to solving a triangle given the lengths of the three sides.

$SS_{subjects}$ The sum of squares of subject totals. Usually calculated to remove those effects from the error term.

SS_{total} The sum of squares of all of the scores, regardless of group membership.

SS_Y The sum of the squared deviations about the Y mean.

Stability Solution to a problem is stable if a small modification in the conditions of the problem does not change the solution too much. How much is much depends of course on the problem.

Standard deviation Like variance, the standard deviation (SD) is a measure of spread (scatter) of a set of data. Unlike variance, which is expressed in squared units of measurement, the SD is expressed in the same units as the measurements of the original data. It is calculated from the deviations between each data value and the sample mean. It is the square root of the variance. For different purposes, n (the total number of values) or n–1 may be used in computing the variance/SD. If you have a SD calculated by dividing by n and want to convert it to a SD corresponding to a denominator of n–1, multiply the result by the square root of n/(n–1). If a distribution's SD is greater than its mean, the mean is inadequate as a representative measure of central tendency. For normally distributed data values, approximately 68% of the distribution falls within ? 1 SD of the mean, 95% of the distribution falls within ? 2 SDs of the mean, and 99.7% of the distribution falls within ? 3 SDs of the mean (empirical rule). SD should not be confused with the standard error of the mean (SEM), which is the SD of the sampling distribution of a statistics and quantifies how accurately the mean is known .

Standard error of differences between means The standard deviation of the sampling distribution of the differences between means.

Standard error of estimate The average of the squared deviations about the regression line.

Standard error The Standard Error, or Standard Error of the Mean, is an estimate of the standard deviation of the sampling distribution of means, based on the data from one or more random samples.

Numerically, it is equal to the square root of the quantity obtained when s squared is divided by the size of the sample.

$$\text{S.E.M.} = \sqrt{\frac{S_x^2}{n}} = \sqrt{S_{\bar{x}}^2}$$

Standard normal distribution A normal distribution with a mean equal to 0 and variance equal to 1. Denoted N (0, 1).

Standard position (angle) An angle with its initial side on the positive x–axis and vertex at the origin.

Standard position (vector) A vector that has been translated so that its initial point is at the origin.

Standard residual The standardised residual value (observed minus predicted divided by the square root of the residual mean square).

Standard scores Scores with a predetermined mean and standard deviation.

Standard vector A vector in standard position.

Standardized regression coefficient The regression coefficient that would result from data that have been standardized.

Stata A powerful statistical package particularly useful for epidemiologic and longitudinal data management and analysis. It is mainly a command driven program produced by Stata Corporation.

State A state is the status, phase, situation, or activity of an object at any given point in time.

State net A state net is the representation of the states of an object and the flow of control within an object or between objects.

State transition Changes in the attributes of an object or in the links an object has with other objects.

Statement of work (SOW) A document prepared for the customer during project initiation and planning that describes what the project will deliver and outlines generally and at a high level all work required to complete the project.

Static equilibrium The sum of all the force vectors add up to zero.

Statistic A statistic is a measure on the items in a random sample. Since the only reason to ever draw a random sample is to infer something about the population from which it came, it should be clear that when we calculate a given statistic we only do so in order to estimate a corresponding parameter of the population from which the sample was drawn. An example of a statistic is the mean (i.e. average) of the measures in the sample.

As indicated by the central limit theorem, the mean of a random sample can be used to estimate the mean of the population from which the sample was taken.

$\bar{x} = \frac{\sum x}{n}$ is a statistic

so is $S_x^2 = \frac{\sum (x-\bar{x})^2}{n-1}$

Note: n is the number of items in the random sample

Statistical dispersion (also called statistical variability) A measure of how diverse some data is. It can be expressed by the variance or the standard deviation. .

Statistical inference Inference about a population from a random sample drawn from it or, more generally, about a random process from its observed behavior during a finite period of time.

Statistical investigation A procedure for obtaining data and drawing conclusions or making decisions on the basis of available data.

Statistical parameter A parameter that indexes a family of probability distributions.

Statistical power The probability that a test will produce a significant difference at a given significance level is called the power of the test. This is equal to the probability of rejecting the null hypothesis when it is untrue, i.e., making the correct decision. It is 1 minus the probability of a type II error. The true differences between the populations compared, the sample size and the significance level chosen affect the power of a statistical test. Ideally, power should be at least 0.80 to detect a reasonable departure from the null hypothesis.

Statistical significance A finding (for example the observed difference between the means of two random samples) is described as statistically significant, when it can be demonstrated that the probability of obtaining such a difference by chance only, is relatively low. In Psychology, and in many other domains, it is customary to describe one's finding as statistically significant, when the obtained result is among those that (theoretically) would occur no more than 5 out of 100 times when the only factors operating are the chance variations that occur whenever random samples are drawn.

Statistical test A statistical test is a procedure for deciding whether an assertion (e.g. an Hypothesis) about a quantitative feature of a population is true or false. We test an hypothesis of this sort by drawing a random sample from the population in question and calculating an appropriate statistic on its items. If, in doing so, we obtain a value of the statistic that would occure rarely when the hypothesis is true, we would have reason to reject the hypothesis.

With this procedure it is customary to reject the hypothesis tested when our statistic has a value that is among those that, theoretically, would be expected to occure no more than 5 out of every 100 times that a random sample (of the same size) is drawn from the population in question when the hypothesis is, in fact, true. Much of the text here is devoted to explanations of exactly how this kind of theoretical expectation is developed.

Finally, it is noteworthy that the appropriate conduct of any statistical test invariably requires many thoughtful decisions. It is, for example, always necessary to decide what statistic to use, what sample size to employ and what criteria to establish for rejection of the hypothesis tested.

Statistics Numerical values summarizing sample data.

Stem and leaf plot A method of organizing data for the purpose of comparison where the leaf is the number in the smallest place value and the stem includes the numbers in the larger place values.

Stem Vertical axis of a stem-and-leaf display containing the leading digits.

Stem-and-leaf display Graphical display presenting original data arranged into a histogram.

Step function A function whose graph resembles a staircase.

Stepwise procedures A set of rules for deriving a regression equation by adding or subtracting one variable at a time from the regression equation.

Stepwise regression model A method in multiple regression studies aimed to find the best model. This method seeks a model that balances a relatively small number of variables with a good fit to the data by seeking a model with high R^2_a (the most parsimonious model with the highest percentage accounted for). The stepwise regression can be started from a null or a full model and can go forward or backward, respectively. At any step in the procedure, the statistically most important variable will be the one that produces the greatest change in the log-likelihood relative to a model lacking the variable. This would be the variable, which would result in the largest likelihood ratio statistics, G (a high percentage accounted for gives an indication that the model fits well). See also multiple regression correlation coefficient – R^2.

Straight angle An angle of 180° (in other words, a straight line).

Straight line A straight line in an axiomatic geometry is an undefined basic term. In analytic geometry, a straight line is a set of points satisfying a linear equation.

Stratification The partitioning of subjects into subgroups that are matched on important variables.

Stratum (plural strata) When data are stratified according to its characteristics, each subgroup is a stratum.

Structural model A theoretical model assumed to underlie the data that expresses the relationship between the dependent variable and the independent variables.

Structured English Modified form of the English language used to specify the logic of information system processes. Although there is no single standard, Structured English typically relies on action verbs and noun phrases and contains no adjectives or adverbs.

Stub testing A technique used in testing modules, especially where modules are written and tested in a top-down fashion, where a few lines of code are used to substitute for subordinate modules.

Studentized range statistic (q) A test statistic for testing the difference between the largest and smallest means in a set.

Studentized residual A statistic for evaluating the residual for an observation in multiple regression.

Student's t distribution The sampling distribution of the t statistic.

Student's t-test A parametric test for the significance between means (two independent sample t-test or paired samples t-test) or between a mean and a hypothesised value (one-sample t-test). One assumption is that the observations must be normally distributed, and the ratio of variances in two samples should not be more than three. If the assumptions are not met, there are non-parametric equivalents of the t-test to use. It is inappropriate to use the t-test for multiple comparisons as a post hoc test. The t-test for independent samples tests whether or not two means are significantly different from each other but only if they were the only two samples taken.

Subgroup A subset H of a group G is a subgroup (of G) provided it's a group with respect to the group operation of G.

Subgroup analysis Analysis of subgroups of a sample either because of a prior hypothesis (gender or age-specific effect/association) or as a fishing expedition / data dredging. This practice increases type I error rates.

Subjective probability Definition of probability in terms of personal subjective belief in the likelihood of an outcome.

Subsequent state A subsequent state is a state which follows a transition. A state may be a prior state of some transitions and a subsequent state of others.

Subsequent state conjunction A transition may turn on more than one state when it finishes. The set of subsequent states which are turned on is a subsequent state conjunction.

Subset A set all of whose elements are elements in a larger set.

Subset A subset of a given set is a collection of things that belong to the original set. For example, A={a,b} could include, a, b, a and b, or the null set (neither).

Substitution Given MPFNs A,B,C, it may be possible to form a new MPFN D by noting the prime powers that change in going from the factorization of A to the factorization of B, and substituting these changes in the factorization of C. In order to produce a MPFN, the following conditions must hold: every prime power in A which does not exactly divide B must exactly divide C, every prime in B that does not divide A must not divide C, and, if index(A) != index(B), then index(C)*(index(B)/index(A)) must be an integer. If these conditions hold, D = C*(B/A) and index(D) = index(C)*(index(B)/index(A)).

A common substitution in small MPFNs involves replacing 19^2 127 by 19^4 151 911; the index is unchanged.

Success/Failure An arbitrary designation of the two possible outcomes in Bernoulli trials.

Successive adjustment The successive adjustment method attempts to find MPFNs as follows:

Step 1. Set N equal to a seed.

Step 2. Compute S = sigma(N)/N.

Step 3. If S is an integer, output N.

Step 4. Let p be the largest prime occurring in S. Multiply N by p^k, where k is the multiplicity with which p appears in S (if p appears in the denominator of S, k is negative).

Step 5. Go to step 2.

The process is ended when the values of N enter a cycle.

Sufficiency The variance of a random variable is a measure of its statistical dispersion, indicating how far from the expected value its values typically are. The variance of random variable

X is typically designated as var $(X), \sigma_X^2$, or simply σ^2.

Sufficient statistic A statistic that uses all of the information in a sample.

Sum identities for tangent Identities involving the tangents of sums of angles.

Sum identity for cosine One of the trigonometric addition identities.

Sum identity for sine One of the trigonometric addition identities.

Sum of squares In an ANOVA the term Sum of Squares refers to an interim quantity used in the calculation of an estimate of the population variance.

In a one-way ANOVA the Within Sum of Squares is a quantity used in the calculation of the Within Mean Square. More specifically the Within Sum of Squares divided by its Degrees of Freedom is equal to the Within Mean Square. Under the Null Hypothesis, the Within Mean Square is an unbiased estimate of the population variance.

Similarly, the Between Sum of Squares is a quantity used in the calculation of the Between Mean Square. More specifically the Between Mean Square is equal to the Between Sum of Squares divided by its Degrees of Freedom. Under the Null Hypothesis, the Between Mean Square is also an unbiased estimate of the population variance.

Sum product identities Useful in writing the sum and difference of trig functions as the product of trig functions.

Supercentenarian The noun and adjective supercentenarian means a person one hundred ten (110) years of age or more.

Superscript In mathematics, superscripts are numbers or letters written above and to the right of other numbers or letters or symbols indicating how many times the latter is to be used as a factor. When typing, one can represent a superscript by using the ^ symbol to indicate raising the number. For example, x^3 is the same as x^3, which equals x * x * x .

Supplementary angles Two angles are called supplementary angles if the sum of their degree measurements equals 180 degrees. One of the supplementary angles is said to be the supplement of the other.

Example: These two angles are supplementary.

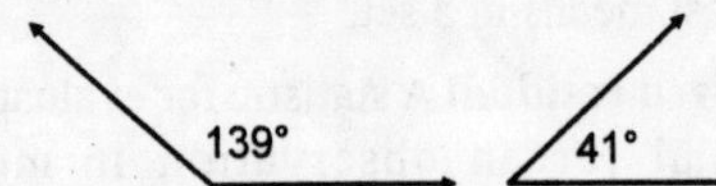

Note that these two angles can be pasted together to form a straight line.

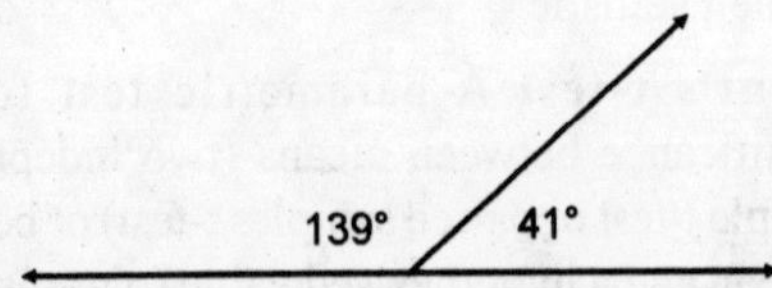

Support Providing ongoing educational and problem-solving assistance to information system users. Support material and jobs must be designed along with the associated information system.

Suppressor variable A variable whose correlation with the criterion is opposite in sign from its regression coefficient.

Surface area The surface area of a space figure is the total area of all the faces of the figure.

Example:

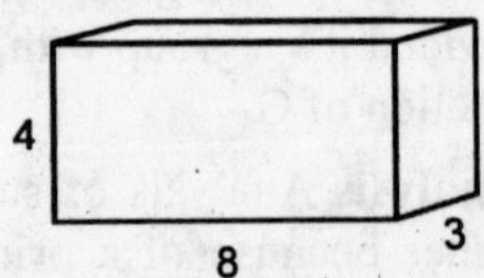

What is the surface area of a box whose length is 8, width is 3, and height is 4? This box has 6 faces: two rectangular faces are 8 by 4, two rectangular faces are 4 by 3, and two rectangular faces are 8 by 3. Adding the areas of all these faces, we get the surface area of the box: $8\times4+8\times4+4\times3+4\times3+8\times3+8\times3=32+32+12+12+24+24=$ 136.

Survival function A time to failure function that gives the probability that an individual survives past a time point (does not experience an event like death, metastasis, conception etc)). Where the event is death, the value of the survival function at time T is the probability that a subject will die at some time greater than T. The survival function always has a value between 0 and 1 and is nonincreasing.

Symmetric figure A figure that can be folded flat along a line so that the two halves match perfectly is a symmetric figure; such a line is called a line of symmetry.

Examples: The triangle below is a symmetric figure. The dotted line is the line of symmetry.

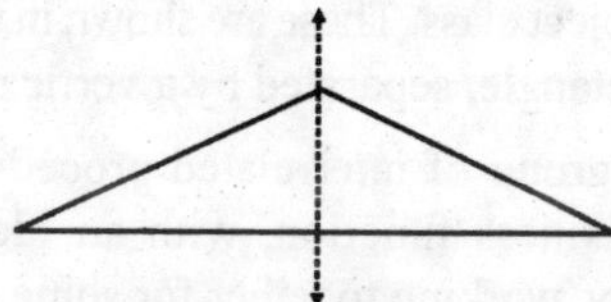

The square below is a symmetric figure. It has four different lines of symmetry shown below.

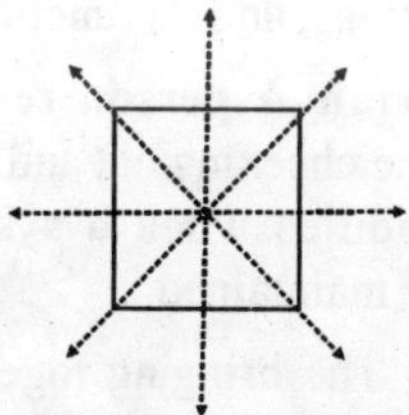

The rectangle below is a symmetric figure. It has two different lines of symmetry shown below.

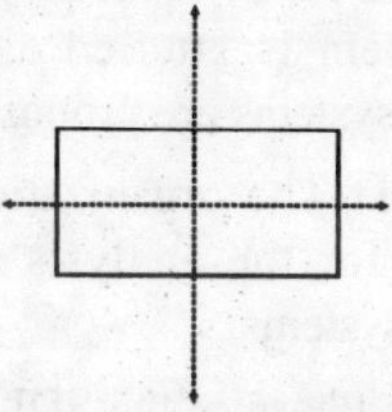

The regular pentagon below is a symmetric figure. It has five different lines of symmetry shown below.

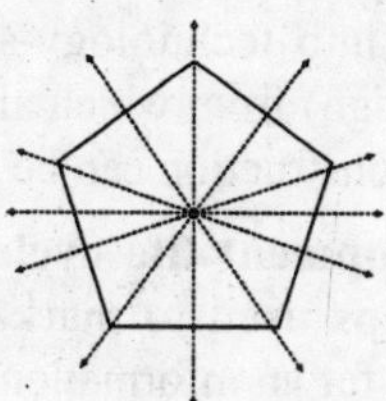

The circle below is a symmetric figure. Any line that passes through its center is a line of symmetry!

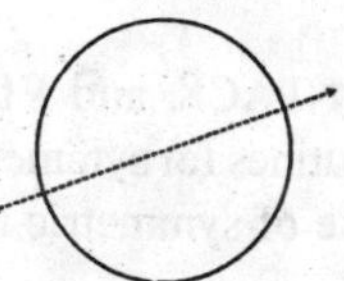

The figures shown below are not symmetric.

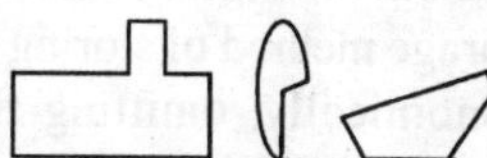

Symmetric Having the same shape on both sides of the center.

Symmetric matrix A symmetric matrix A is equal to its transpose, that is,

A = A'

or, for every pair of indices I and J:

$A_{i,j} = A_{j,i}$

Every matrix A can be decomposed into the sum of an antisymmetric and a symmetric matrix:

A = B + C = (1/2) * ((A – A') + (A + A'))

Here is an example of a symmetric matrix:

1 2 0 4

2 9 4 8

0 4 5 3

4 8 3 7

Simple facts about a symmetric matrix A:

A is normal;

the inverse of A is symmetric;

The eigenvalues of A are real;

Eigenvectors associated with distinct eigenvalues are orthogonal vectors;

A has a complete set of eigenvectors, which can be made into an orthonormal set;

A is orthogonally similar to a diagonal matrix.

A has an LDL Factorization.

The eigenvalues of successive members of the sequence of principal minors of a symmetric matrix have the Sturm sequence property, a strict interlacing relationship. The k+1 eigenvalues of the principal minor of order k+1 are strictly separated by the k eigenvalues of the minor of order k.

LAPACK, LINPACK and EISPACK include specialized routines for symmetric matrices, and include the use of symmetric matrix storage to save space.

Symmetric matrix storage Symmetric storage is a matrix storage method of storing a symmetric matrix economically, omitting the repeated elements.

The strict lower triangle of a symmetric or Hermitian matrix is redundant. A symmetric storage scheme packs the upper triangle of the matrix into a linear vector of length (N * (N + 1)) / 2. The data is organized by columns, with each column starting in row 1 of the original matrix, and proceeding down to the diagonal.

If A was the matrix:

11 12 13 14 15

12 22 23 24 25

13 23 33 34 35

14 24 34 44 45

15 25 35 45 55

then A could be symmetrically stored as:

1 2 3 4 5

(11, 12, 22, 13, 23, 33, 14, 24, 34, 44, 15, 25, 35, 45, 55).

LAPACK, LINPACK and EISPACK include routines which can operate on data stored in this format.

Symmetric relationships Log-linear models in which all variables are treated as dependent variables.

Symmetry The correspondence in size, form, or arrangement of parts on a plane or line. In line symmetry, each point on one side of the line has a corresponding point on the opposite side of the line (picture a butterfly, with wings that are identical on either tessellation.

A tessellation is a repeated geometric design that covers a plane without gaps or overlaps .

Synchronous message A type of message in which the caller has to wait for the receiving object to finish executing the called operation before it can resume execution itself.

Synergism A joint effect of two treatments being greater than the sum of their effects when administered separately (positive synergism) or the opposite (negative synergism).

Synonym Synonyms are multiple names for a single object class. These are shown in the object class rectangle, separated by a vertical bar (|).

System A group of interrelated procedures used for a business function, with an identifiable boundary, working together for some purpose.

System documentation Detailed information about a system's design specifications, its internal workings, and its functionality.

System librarian A person responsible for controlling the checking out and checking in of baseline modules when a system is being developed or maintained.

System testing The bringing together of all the programs that a system comprises for testing purposes. Programs are typically integrated in a top-down, incremental fashion.

Systems analysis Phase of the SDLC in which the current system is studied and alternative replacement systems are proposed.

Systems analyst The organizational role most responsible for the analysis and design of information systems.

Systems design Phase of the SDLC in which the system chosen for development in systems analysis is first described independent of any computer platform (logical design) and is then transformed into technology-specific details (physical design) from which all programming and system construction can be accomplished.

Systems development life cycle (SDLC) The series of steps used to mark the phases of development for an information system.

Systems development methodology A standard process followed in an organization to conduct all the steps necessary to analyze, design, implement, and maintain information systems.

Systems implementation and operation Final phase of the SDLC, in which the information system is coded, tested, and installed in the organization, and in which the information system is systematically repaired and improved.

Systems planning and selection The first phase of the SDLC, in which an organization's total information system needs are analyzed and arranged, and in which a potential information systems project is identified and an argument for continuing or not continuing with the project is presented.

Syto The nonstandard prefix syto- denotes a multiple of ten to the negative twenty-seventh power (10^{-27} = 10 raised to power –27 = 0.000 000 000 000 000 000 000 000 001).

T

T The metric prefix symbol T- (metric prefix: tera-) denotes a multiple of ten to the twelfth power (10^{12} = 1,000,000,000,000) as defined by Le Système International d'Unités (SI).

T scores A set of scores with a mean of 50 and a standard deviation of 10.

t Test The t test employs the statistic (t) to test a given statistical hypothesis about the mean of a population (or about the means of two populations).

Tabled distribution of chi square The table showing the critical values of chi-square for various degrees of freedom and levels of ? when the null hypothesis is true.

Take out formula A formula for finding the take out of an elbow. The take out formula is: Take out = Tan 0/2 x radius of the elbow.

Take out The distance that a fitting extends over the center line of a hypotenuse of an offset triangle.

Tandem dual The compound noun and adjective tandem dual means a pair (2) of dual truck axles mounted in tandem for a total of eight (8) wheels.

Tandem The noun, adjective, and adverb tandem means a pair (2) of objects one behind the other. A tandem bicycle has two seats one behind the other. A tandem carriage is pulled by two horses harnessed one behind the other. Tandem axles are two axles mounted one behind the other.

Tangent A trigonometric ratio equal to the opposite side divided by the adjacent side.

Tangent line A straight line which touches just one point on a circle.

Tangible benefit A benefit derived from the creation of an information system that can be measured in dollars and with certainty.

Tangible cost A cost associated with an information system that can be easily measured in dollars and with certainty.

Tebi The binary prefix tebi- (prefix symbol: Ti-) denotes a multiple of two to the fortieth power (2^{40} = 1,099,511,627,776) as defined by the International Electrotechnical Commission (IEC).

Tebin The adjective and noun tebin means two to the fortieth power (2^{40} = 1,099,511,627,776).

Technical feasibility The process of assessing the development organization's ability to construct a proposed system.

Teen The noun and adjective teen means at least thirteen (13) but less than twenty (20). Teen most commonly refers to a person between thirteen (13) and twenty (20) years of age.

Teenage The noun and adjective teenage means the period from thirteen (13) to twenty (20) years of age.

Template based HTML Templates to display and process common attributes of higher-level, more abstract items.

Ten million The compound noun and adjective ten million (10,000,000) is the standard name for the cardinal number that equals ten to the seventh power (10^7 = 10 raised to power +07 = 10,000,000).

Ten thousand The compound noun and adjective ten thousand (10,000) is the standard name for the cardinal number that equals ten to the fourth power (10^4 = 10 raised to power +04 = 10,000).

Ten-millionth The compound adjective and noun ten-millionth (1/10,000,000 = 0.000 000 100) means divided by ten million (10,000,000).

Tens 1. The plural noun and adjective tens (10s) means an integer multiple of ten (10).

2. The plural noun and adjective tens (10s) means the digit position in the common decimal number notation that represents a multiple of ten (10).

3. The plural noun and adjective tens (10s) means the range of numbers at least ten (10) but less than twenty (20).

4. The plural noun and adjective tens (10s) means a span of ten years in which the tens (10s) digit of the year number is one (1).

Tenth 1. The adjective, adverb, and noun tenth (10th) is the standard name for the ordinal number that follows ninth (9th). Tenth corresponds with the cardinal number ten (10).

2. The adjective and noun tenth (1/10 = 0.100) means divided by ten (10).

Ten-thousandth The compound adjective and noun ten-thousandth (1/10,000 = 0.000 100) means divided by ten thousand (10,000).

Tera The metric prefix tera- (prefix symbol: T-) denotes a multiple of multiple of ten to the twelfth power (10^{12} = 10 raised to power +12 = 1,000,000,000,000) as defined by Le Système International d'Unités (SI).

Terabinary The adjective terabinary means two to the fortieth power (2^{40} = 1,099,511,627,776).

Tercentennial 1. The adjective and noun tercentennial means recurring every three hundred (300) years.

2. The adjective and noun tercentennial means lasting three hundred (300) years.

3. The noun and adjective tercentennial means a three hundred (300) year anniversary.

Terminal point The ending point of a vector.

Terminal side Side of angle where angle measurement ends.

Terms of a sequence The numbers in a sequence.

Tern The noun tern means a set of three (3). In gambling, a tern is a set of three (3) lottery numbers. In sailing; a tern is a schooner with three (3) masts.

Ternary relationship A simultaneous relationship among instances of three entity types.

Terquasquicentenary 1. The noun and adjective terquasquicentenary means a group of one hundred seventy-five (175).

2. The noun and adjective terquasquicentenary means a period of one hundred seventy-five (175) years.

3. The noun and adjective terquasquicentenary means a one hundred seventy-five (175) year anniversary.

4. The adjective and noun terquasquicentenary means one hundred seventy-fifth (175th) in rank or order.

Terquasquicentennial 1. The adjective and noun terquasquicentennial means recurring every one hundred seventy-five (175) years.

2. The adjective and noun terquasquicentennial means lasting one hundred seventy-five (175) years.

3. The noun and adjective terquasquicentennial means a one hundred seventy-five (175) year anniversary.

Terquasquicentennium The noun and adjective terquasquicentennium means a period of one hundred seventy-five (175) years.

Tertiary The adjective and noun tertiary means third (3rd or 3[d]) in order of importance.

Test statistics The results of a statistical test.

Tetra The prefix tetra- means four (4).

Tetrachoric correlation The correlation between two dichotomous variables when underlying normality of each variable is assumed. Rarely used.

Tetraconta The prefix tetraconta- means forty (40).

Tetracontad The noun and adjective tetracontad means a group of forty (40) members.

Tetracontadi The prefix tetracontadi- means forty-two (42). Tetracontadi- is an alternate form of the prefix tetracontakaidi-.

Tetracontaennea The prefix tetracontaennea- means forty-nine (49). Tetracontaennea- is an alternate form of the prefix tetracontakaiennea.

Tetracontagon The noun and adjective tetracontagon means a plain figure with forty (40) straight sides.

Tetracontagram The noun and adjective tetracontagram means an interconnected star figure with forty (40) points formed by stellating a tetracontagon.

Tetracontahedron The noun and adjective tetracontahedron means a solid figure with forty (40) planar faces.

Tetracontahena The prefix tetracontahena- means forty-one (41). Tetracontahena- is an alternate form of the prefix tetracontakaihena-.

Tetracontahepta The prefix tetracontahepta- means forty-seven (47). Tetracontahepta- is an alternate form of the prefix tetracontakaihepta.

Tetracontahexa The prefix tetracontahexa- means forty-six (46). Tetracontahexa- is an alternate form of the prefix tetracontakaihexa-.

Tetracontakaidi The prefix tetracontakaidi- means forty-two (42).

Tetracontakaidigon The noun and adjective tetracontakaidigon means a plain figure with forty-two (42) straight sides.

Tetracontakaidigram The noun and adjective tetracontakaidigram means an interconnected star figure with forty-two (42) points formed by stellating a tetracontakaidigon.

Tetracontakaidihedron The noun and adjective tetracontakaidihedron means a solid figure with forty-two (42) planar faces.

Tetracontakaidyad The noun and adjective tetracontakaidyad means a group of forty-two (42) members.

Tetracontakaiennea The prefix tetracontakaiennea means forty-nine (49).

Tetracontakaiennead The noun and adjective tetracontakaiennead means a group of forty-nine (49) members.

Tetracontakaienneagon The noun and adjective tetracontakaienneagon means a plain figure with forty-nine (49) straight sides.

Tetracontakaienneagram The noun and adjective tetracontakaienneagram means an interconnected star figure with forty-nine (49) points formed by stellating a tetracontakaienneagon.

Tetracontakaienneahedron The noun and adjective tetracontakaienneahedron means a solid figure with forty-nine (49) planar faces.

Tetracontakaihena The prefix tetracontakaihena- means forty-one (41).

Tetracontakaihenad The noun and adjective tetracontakaihenad means a group of forty-one (41) members.

Tetracontakaihenagon The noun and adjective tetracontakaihenagon means a plain figure with forty-one (41) straight sides.

Tetracontakaihenagram The noun and adjective tetracontakaihenagram means an interconnected star figure with forty-one (41) points formed by stellating a tetracontakaihenagon.

Tetracontakaihenahedron The noun and adjective tetracontakaihenahedron means a solid figure with forty-one (41) planar faces.

Tetracontakaihepta The prefix tetracontakaihepta means forty-seven (47).

Tetracontakaiheptad The noun and adjective tetracontakaiheptad means a group of forty-seven (47) members.

Tetracontakaiheptagon The noun and adjective tetracontakaiheptagon means a plain figure with forty-seven (47) straight sides.

Tetracontakaiheptagram The noun and adjective tetracontakaiheptagram means an interconnected star figure with forty-seven (47) points formed by stellating a tetracon-takaiheptagon.

Tetracontakaiheptahedron The noun and adjective tetracontakaiheptahedron means a solid figure with forty-seven (47) planar faces.

Tetracontakaihexa The prefix tetracontakaihexa-means forty-six (46).

Tetracontakaihexad The noun and adjective tetracontakaihexad means a group of forty-six (46) members.

Tetracontakaihexagon The noun and adjective tetracontakaihexagon means a plain figure with forty-six (46) straight sides.

Tetracontakaihexagram The noun and adjective tetracontakaihexagram means an interconnected star figure with forty-six (46) points formed by stellating a tetracontakaihexagon.

Tetracontakaihexahedron The noun and adjective tetracontakaihexahedron means a solid figure with forty-six (46) planar faces.

Tetracontakaiocta The prefix tetracontakaiocta-means forty-eight (48).

Tetracontakaioctad The noun and adjective tetracontakaioctad means a group of forty-eight (48) members.

Tetracontakaioctagon The noun and adjective tetracontakaioctagon means a plain figure with forty-eight (48) straight sides.

Tetracontakaioctagram The noun and adjective tetracontakaioctagram means an interconnected star figure with forty-eight (48) points formed by stellating a tetracontakaioctagon.

Tetracontakaioctahedron The noun and adjective tetracontakaioctahedron means a solid figure with forty-eight (48) planar faces.

Tetracontakaipenta The prefix tetracontakaipenta means forty-five (45).

Tetracontakaipentad The noun and adjective tetracontakaipentad means a group of forty-five (45) members.

Tetracontakaipentagon The noun and adjective tetracontakaipentagon means a plain figure with forty-five (45) straight sides.

Tetracontakaipentagram The noun and adjective tetracontakaipentagram means an interconnected star figure with forty-five (45) points formed by stellating a tetracontakaipentagon.

Tetracontakaipentahedron The noun and adjective tetracontakaipentahedron means a solid figure with forty-five (45) planar faces.

Tetracontakaitetra The prefix tetracontakaitetra-means forty-four (44).

Tetracontakaitetrad The noun and adjective tetracontakaitetrad means a group of forty-four (44) members.

Tetracontakaitetragon The noun and adjective tetracontakaitetragon means a plain figure with forty-four (44) straight sides.

Tetracontakaitetragram The noun and adjective tetracontakaitetragram means an interconnected star figure with forty-four (44) points formed by stellating a tetracontakaitetragon.

Tetracontakaitetrahedron The noun and adjective tetracontakaitetrahedron means a solid figure with forty-four (44) planar faces.

Tetracontakaitri The prefix tetracontakaitri-means forty-three (43).

Tetracontakaitriad The noun and adjective tetracontakaitriad means a group of forty-three (43) members.

Tetracontakaitrigon The noun and adjective tetracontakaitrigon means a plain figure with forty-three (43) straight sides.

Tetracontakaitrigram The noun and adjective tetracontakaitrigram means an interconnected star figure with forty-three (43) points formed by stellating a tetracontakaitrigon.

Tetracontakaitrihedron The noun and adjective tetracontakaitrihedron means a solid figure with forty-three (43) planar faces.

Tetracontaocta The prefix tetracontaocta- means forty-eight (48). Tetracontaocta- is an alternate form of the prefix tetracontakaiocta-.

Tetracontatetra The prefix tetracontatetra- means forty-four (44). Tetracontatetra- is an alternate form of the prefix tetracontakaitetra-.

Tetracontatri The prefix tetracontatri- means forty-three (43). Tetracontatri- is an alternate form of the prefix tetracontakaitri-.

Tetracta The prefix tetracta- means four hundred (400).

Tetrad The noun and adjective tetrad means a group of four (4) members.

Tetradeca The prefix tetradeca- means fourteen (14).

Tetradecagon The noun tetradecagon means a plane figure with fourteen (14) straight sides. Tetragon The noun and adjective tetragon means a plain figure with four (4) straight sides.

Tetragram The noun and adjective tetragram means a word or character sequence consisting of four (4) characters.

Tetrahedron A tetrahedron is a 4-sided space figure. Each face of a tetrahedron is a triangle.

Example: The picture below is a tetrahedron. The grayed lines are edges hidden from view.

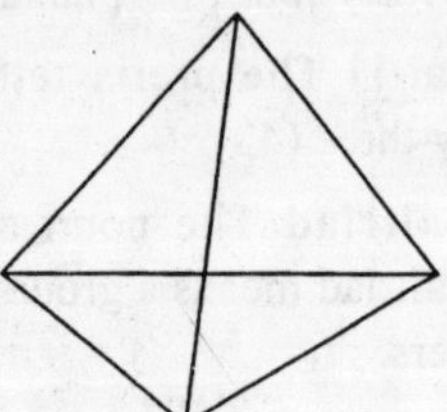

Tetraheptaconta The prefix tetraheptaconta- means seventy-four (74).

Tetrahexaconta The prefix tetrahexaconta- means sixty-four (64).

Tetraicosa The prefix tetraicosa- means twenty-four (24).

Tetrakaideca The prefix tetrakaideca- means fourteen (14).

Tetrakaidecad The noun and adjective tetrakaidecad means a group of fourteen (14) members.

Tetrakaidecagon The noun and adjective tetrakaidecagon means a plain figure with fourteen (14) straight sides.

Tetrakaidecagram The noun and adjective tetrakaidecagram means an interconnected star figure with fourteen (14) points formed by stellating a tetrakaidecagon.

Tetrakaidecahedron The noun and adjective tetrakaidecahedron means a solid figure with fourteen (14) planar faces.

Tetrameter The noun tetrameter means a verse consisting of either four (4) dipodies or four (4) metrical feet.

Tetratriaconta The prefix tetratriaconta- means thirty-four (34).

The Expected value (or expectation) A random variable is the sum of the probability of each possible outcome of the experiment multiplied by its payoff (value). Thus, it represents the average amount one expects to win per bet if bets with identical odds are repeated many times. For example, the expected value of a six-sided die roll is 3.5. The concept is similar to the mean. The expected value of random variable X is typically written E(X) or μ (mu).

The Identity Matrix The identity matrix, usually denoted I, is a square matrix with 1's on the main diagonal and 0's elsewhere. The identity matrix behaves like the number 1 in matrix multiplication. For any matrix A:

A * I = I * A = A.

Here is the 3 by 3 identity matrix:

1 0 0

0 1 0

0 0 1

Simple facts about the identity matrix I:

The norm of I must be 1, for any vector-bound matrix norm

The condition number of I must be 1, for any vector-bound matrix norm;

I' = I^{-1} = I;

The eigenvalues of I are 1, with algebraic and geometric multiplicity N;

Every nonzero vector x is an eigenvector of I. A basis for the eigenspace of I is the set of columns of I.

I is diagonal;

I is idempotent;

I is involutory;

I is symmetric;

I is a projection matrix;

The incomplete cholesky factorization The incomplete Cholesky factorization is an approximate Cholesky factorization, of a matrix A, comprising a lower triangular matrix L for which A is approximately equal to L*L'.

The incomplete Cholesky factorization is a special case of the Incomplete LU Factorization, modified for symmetric positive definite matrices. There are known cases where it will fail.

Typically, the matrix A is large and sparse, (as well as being symmetric and positive definite) and an iterative scheme is being used to solve the linear system A*x=b. The incomplete Cholesky factorization is intended as a preconditioner, which modifies the linear system, improving the convergence rate of the iterative scheme. In particular, the preconditioner matrix is M=L*L', and we are well advised to store L instead of M, since this gives us an easy way of solving linear systems associated with M.

The computation of the incomplete Cholesky factorization is similar to that of the usual Cholesky factor except for two points. First, we assume that no pivoting is required. Secondly, if the original matrix A has a zero entry, then we require that the corresponding entry of the L matrix be zero. This means that there is no fill in (and it is also why L*L' will not equal A). Because there is no fill in, the factor L can actually be stored in a data structure that is the same as that of A.

Because no pivoting is allowed, the algorithm may break down because of negative elements that may appear on the diagonal.

The incomplete LU factorization The incomplete LU factorization is an approximate LU factorization of a matrix A, comprising a unit lower triangular matrix L and an upper triangular matrix U.

Typically, the matrix A is large and sparse, and an iterative scheme is being used to solve the linear system A*x=b. The incomplete LU factorization is intended as a preconditioner, which modifies the linear system, improving the convergence rate of the iterative scheme. In particular, the preconditioner matrix is M=L*U, and we are well advised to store L and U instead of M, since this gives us an easy way of solving linear systems associated with M.

The computation of the incomplete LU factorization is similar to that of the usual PLU factors except for two points. First, we assume that no pivoting is required, hence the P factor can be omitted. Secondly, if the original matrix A has a zero entry, then we require that the corresponding entry of the U matrix be zero. This means that there is no fill in (and it is also why L*U will not equal A). Because there is no fill in, the factors L and U can actually be stored in a data structure that is the same as that of A. The elements of L are stored in the positions devoted to lower triangular elements of A, while U goes in the diagonal and upper triangular locations.

If the matrix A is symmetric and positive definite, then the related Incomplete Cholesky Factorization may be tried.

The Jacobi Algorithm for Linear Equations The Jacobi algorithm for linear equations is an iterative method for solving linear systems of equations A * x = b.

The method is similar to the Gauss Seidel and Successive Overrelation Method (SOR), but has the distinction of being the only simple iterative method that is easy to program in parallel.

The Jacobi iteration is only appropriate for matrices which are strictly diagonally dominant or else symmetric and positive definite.

Each step of the Jacobi iteration begins with an approximate solution x. An improved approximate solution xnew is computed by solving N independent linear equations:

xnew(i) = [b(i)

– A(i,1) * x(1)

– A(i,2) * x(2)

...

– A(i,i–1) * x(i–1)

– 0.0 * x(i) <— note that we are skipping x(i) here

– A(i,i+1) * x(i+1)

...

– A(i,n) * x(n)] / A(i,i)

Once xnew is computed, the residual error is calculated:

r = A * xnew – b

and the solution increment:

dx = xnew – x.

If the norms of these vectors are satisfactorily small, the iteration may be halted. Otherwise, xnew becomes the new starting guess x for the next step of the iteration.

The Jacobi iteration can be considered in terms of its matrix splitting. That is, if we decompose the matrix A into its strictly lower triangular, diagonal, and strictly upper triangular parts:

A = L + D + U

then the method is equivalent to the iteration

D * xnew = b – (L + U) * x.

which means that the convergence of the algorithm can be understood in terms of the behavior of powers of the iteration matrix:

$- D^{-1} * (L + U)$,

which in turn may best be understood by looking at the eigenvalues. Note that if A is symmetric, then so is the iteration matrix.

The jacobi preconditioner The Jacobi preconditioner is a very simple preconditioning matrix for use with an iterative method for linear equations.

For a given system matrix A, the Jacobi preconditioner matrix M is the diagonal matrix whose entries are defined by

M = diag (A)

In effect, using the Jacobi preconditioner amounts to dividing each equation and right hand side entry by the corresponding diagonal entry of the original coefficient matrix.

The lanczos iteration The Lanczos iteration is an iterative method that seeks to approximate some of the eigenvalues of a very large real symmetric (or Hermitian) matrix.

The Lanczos iteration is a version of the Arnoldi iteration that takes advantage of the special properties associated with a real symmetric or Hermitian matrix.

Consider the unitary similarity transform that converts the m by m symmetric or Hermitian matrix A into tridiagonal form:

A = Q T Q*

Let Q_n indicate the m by n matrix formed from the first n columns of Q; similarly, T_n is the n by n submatrix of T. Then we write

$T_n = Q_n * A_n Q_n$

As we pass from n to n+1, we simply add a new column q_{n+1} to Q_n, and to T_n we add a new diagonal value $alpha_{n+1}$ and an offdiagonal value $beta_n$. Thus, the iteration allows us to gradually build up the matrices Q_n and T_n. For reasons not completely understood, even for n much smaller than m, the eigenvalues of T_n will tend to some of the eigenvalues of A, and typically the extreme ones.

The new vector q_{n+1} is a Krylov vector; actually, because of orthogonalization, it represents the new component introduced into the Krylov subspace by the product A^*q_n.

Thus, a typical procedure is to compute T_n and apply a standard eigenvalue procedure for tridiagonal matrices to extract its eigenvalues.

The entries of the matrix T_n are denoted by

alpha$_1$ beta$_1$

beta$_1$ alpha$_2$ beta$_2$

beta$_{n-2}$ alpha$_{n-1}$ beta$_{n-1}$

beta$_{n-1}$ alpha$_n$

An efficient form for the Lanczos iteration is:

beta$_0$ = 0;

q(0) = 0;

Let b be an arbitrary initial vector;

Set q(1) = b / ||b||;

for n = 1, ...

v = A * q(n)

alpha(n) = q(n)' * v

v = v – beta(n–1) * q(n–1) – alpha(n) * q(n)

beta(n) = ||v||

q(n+1) = v / beta(n)

end for

The Mean A random variable is its expected value. The mean (or sample mean of a data set is just the average value.

The MPS storage format The MPS data format is an industry standard for the description of a variety of linear programming problems. Each section begins with a single line containing the name of the section in the Indicator field, and may be followed by a number of data lines, which have a very particular format.

An MPS file may also have comment lines. Such lines may occur anywhere in the file. They begin with an asterisk '*' in column 1. They may contain information intended to explain the problem, or other facts of interest to a human, and of no interest to a program parsing the file.

The data lines of an MPS file are divided into fields of a specific width:

Field 1 Field 2 Field 3 Field 4 Field 5 Field 6

Columns 2-3 5-12 15-22 25-36 40-47 50-61

Contents Indicator Name 1 Name 2 Value 1 Name 3 Value 2

Indicator and Name fields contain names of things; Value fields contain numeric values. The particular use of these fields depends on the section in which the data line occurs.

We now discuss in detail the format of the data lines that occur in each data section.

The neumann series The Neumann series for a square matrix A is the infinite sum

$I + A + A^2 + A^3 + A^4 + \dots$

If the spectral radius of the matrix A satisfies rho(A) < 1, then I-A is nonsingular, and the Neumann series converges to $(I-A)^{-1}$.

Conversely, if the Neumann series converges, then the spectral radius of A must be less than 1.

The statistic (T) This statistic is a measure on a random sample (or pair of samples) in which a mean (or pair of means) appears in the numerator and an estimate of the numerator's standard deviation appears in the denominator. The later estimate is based on the calculated s square or s squares of the samples.

If these calculations yield a value of (t) that is sufficiently different from zero, the test is considered to be statistically significant.

Here is an example of the statistic (t) using 1 sample:

$$t_{df=n-1} = \frac{\bar{x} - \mu_h}{\sqrt{\frac{S_x^2}{n}}}$$

In this equation μ_h is the statistical hypothesis and $\bar{x}$ is the mean of a random sample of n items and S_x^2 is an estimate of the population vanance where $S_x^2 = \dfrac{\sum(x-\bar{x})^2}{n-1}$

The statistic Z This statistic is a measure on a random sample (or pair of samples) in which a mean (or pair of means) appears in the numerator and the numerator's standard deviation appears in the denominator.

If these calculations yield a value of (z) that is sufficiently different from zero, the test is considered to be statistically significant.

$$Z_{\bar{x}} \frac{\bar{x}-\mu_h}{\sqrt{\dfrac{\sigma_x^2}{n}}}$$

This is the formula for the statistic Z when it is used to test the statistical hypothesis that two random samples como from populations with means equal to μ_{x_1} and μ_{x_2}

It is assumed that the values of $\sigma_{x_1}^2$ and $\sigma_{x_2}^2$ are both known.

$$Z_{(\bar{x}_1-\bar{x}_2)} = \frac{(\bar{x}_1-\bar{x}_2)-(\mu_{h1}-\mu_{h2})}{\sqrt{\dfrac{\sigma_{x_1}^2}{n_1}+\dfrac{\sigma_x^2}{n_2}2}}$$

Theorem A statement that has been proven to be true.

Theories of probability A theory of probability is a way of understanding probability statements. That is, a theory of probability connects the mathematics of probability, which is the set of consequences of the axioms of probability, with the real world of observation and experiment. There are several common theories of probability. According to the frequency theory of probability, the probability of an event is the limit of the percentage of times that the event occurs in repeated, independent trials under essentially the same circumstances. According to the subjective theory of probability, probability is a number that measures how strongly we believe an event will occur. The number is on a scale of 0% to 100% (or 0 to 1), with 0% indicating that we are completely sure it won't occur, and 100% indicating that we are completely sure that it will occur.

Theta (θ) A letter of the Greek alphabet that is used as a variable for unknown angles or when referring to any angle.

Third 1. The adjective, adverb, and noun third (3rd or 3d) is the standard name for the ordinal number that follows second (2nd or 2d). Third corresponds with the cardinal number three (3).

2. The adjective and noun third (1/3) means divided by three (3).

Third normal form (3NF) A relation is in third normal form if it is in second normal form and there are no functional (transitive) dependencies between two (or more) nonprimary key attributes.

Third order determinant A determinant with three rows and three columns.

Thirteen The noun, pronoun, and adjective thirteen (13) is the standard name for the cardinal number that follows twelve (12). Thirteen corresponds with the ordinal number thirteenth (13th).

Thirteenth 1. The adjective, adverb, and noun thirteenth (13th) is the standard name for the ordinal number that follows twelfth (12th). Thirteenth corresponds with the cardinal number thirteen (13).

2. The adjective and noun thirteenth (1/13) means divided by thirteen (13).

Thirties 1. The plural noun and adjective thirties (30s) means an integer multiple of thirty (30).

2. The plural noun and adjective thirties (30's) means the range of numbers at least thirty (30) but less than forty (40). When refering to dates, thirties means a span of ten years in which the tens (10s) digit of the year number is three (3).

Thirtieth 1. The adjective, adverb, and noun thirtieth (30th) is the standard name for the ordinal number that follows twenty-ninth (29th). Thirtieth corresponds with the cardinal number thirty (30).

2. The adjective and noun thirtieth (1/30) means divided by thirty (30).

Thirty The noun, pronoun, and adjective thirty (30) is the standard name for the cardinal number that follows twenty-nine (29). Thirty corresponds with the ordinal number thirtieth (30th).

Thirty-eight The noun and adjective thirty-eight (38) is the standard name for the cardinal number that follows thirty-seven (37). Thirty-eight corresponds with the ordinal number thirty-eighth (38th).

Thirty-five The noun and adjective thirty-five (35) is the standard name for the cardinal number that follows thirty-four (34). Thirty-five corresponds with the ordinal number thirty-fifth (35th).

Thirty-four The noun and adjective thirty-four (34) is the standard name for the cardinal number that follows thirty-three (33). Thirty-four corresponds with the ordinal number thirty-fourth (34th).

Thirty-ninth The compound adjective, adverb, and noun thirty-ninth (39th) is the standard name for the ordinal number that follows thirty-eighth (38th). Thirty-ninth corresponds with the cardinal number thirty-nine (39).

Thirty-one The noun and adjective thirty-one (31) is the standard name for the cardinal number that follows thirty (30). Thirty-one corresponds with the ordinal number thirty-first (31st).

Thirty-seven The noun and adjective thirty-seven (37) is the standard name for the cardinal number that follows thirty-six (36). Thirty-seven corresponds with the ordinal number thirty-seventh (37th).

Thirty-sixth 1. The compound adjective, adverb, and noun thirty-sixth (36th) is the standard name for the ordinal number that follows thirty-fifth (35th). Thirty-sixth corresponds with the cardinal number thirty-six (36).

2. The compound adjective and noun thirty-sixth (1/36) means divided by thirty-six (36).

Thirty-Something The compound noun and adjective thirty-something means at least thirty (30) but less than forty (40). Thirty-something most commonly means between thirty (30)) and forty (40) years of age.

Thirty-third The compound adjective, adverb, and noun thirty-third (33rd or 33^{d}) is the standard name for the ordinal number that follows thirty-second (32nd or 32^{d}). Thirty-third corresponds with the cardinal number thirty-three (33).

Thirty-two The noun and adjective thirty-two (32) is the standard name for the cardinal number that follows thirty-one (31). Thirty-two equals two to the fifth power (2^5). Thirty-two corresponds with the ordinal number thirty-second (32nd or 32^{d}).

Thousand The noun and adjective thousand means multiplied by one thousand (1,000).

Thousandth 1. As part of the name of an ordinal number, the noun and adjective thousandth means a multiple of one thousand (1,000).

2. As part of the name of a rational fraction, the noun and adjective thousandth means divided by one thousand (1/1,000 = 0.001).

Thread A thread is a flow of control or execution within an object's behavior.

Three dimensional An object that has length, width, and height.

Three fourths The compound noun, pronoun, and adjective three fourths (3/4 = 0.750) is the standard name for the rational fraction equal to three (3) divided by four (4).

Three quarters The compound noun, pronoun, and adjective three quarters is an alternate name for the rational fraction three fourths (3/4 = 0.750).

Three The noun, pronoun, and adjective three (3) is the standard name for the cardinal number that follows two (2). Three corresponds with the ordinal number third (3rd or 3^{d}).

Threescore And Ten The compound noun and adjective threescore and ten means seventy (70).

Ti The compound binary prefix symbol Ti- (binary prefix: tebi-) denotes a multiple of two to the fortieth power (2^{40} = 1,099,511,627,776) as defined by the International Electrotechnical Commission (IEC).

Time constrained interaction A time-constrained interaction is a time-critical interaction that must take place in a specified finite amount of time.

Time value of money (TVM) The process of comparing present cash outlays to future expected returns.

Tip tail rule A process for doing vector addition.

Tithe 1. The noun tithe means one tenth (1/10 = 0.100).

2. The verb tithe means to give or pay one tenth (1/10 = 0.100) part.

Tolerance One minus the squared correlation of a predictor with all other predictors.

Tolerance Tolerance is the amount of error accepted in a given situation.

Topological space In a topological space every point has a collection of neighborhoods to which it belongs.

Trailing digits (least significant digits) Right-most digits of a number.

Transcendental Number A transcendental number is a real number that is not an algebraic number. Examples include e, e·i, π, e^{π}, and sin(1).

Transformation A rule for moving every point in a plane figure to a new location.

Transformation matrix A matrix used for multiplying a point matrix to give a new point matrix.

Transformation rules In statistics the transformation rules describe the changes in the mean, variance and standard deviation of a distribution when every item in a distribution is either increased or decreased by a constant amount. These rules also describe the changes in the mean, variance and standard deviation of a distribution when every item in the distribution is either multiplied or divided by a constant amount.

Transformation rule (1): Adding a constant to every item in a distribution adds the constant to the mean of the distribution, but it leaves the variance and standard deviation, unchanged.

Transformation rule (2): Multiplying every item in a distribution by a constant multiplies the mean and standard deviation of that distribution by the constant and it multiplies the variance of the distribution by the square of the constant.

$$\mu_{(x+C)} = \frac{\sum(x+C)}{N} = \mu_x + C$$

$$\sigma^2_{(x+C)} = \sigma^2_x \quad \mu_{cx} = C\mu_x \quad \sigma^2_{Cx} = C^2\sigma^2_x$$

Transformations (Ladder of powers) Transformation deals with non-normality of the data points and non-homogeneous variance. The power transformations form the following ladder: ..., x^{-2}, x^{-1}, $x^{-1/2}$, log x, $x^{1/2}$; x^1, x^2, x^3, ... Provided x > 1, powers below 1 (such as $x^{1/2}$ or log x) reduce the high values relative to the low values as in positively skewed data, whereas, powers above 1 (such as x^2) have the opposite effect of stretching out high values relative to low ones, as in negatively skewed data. All power transformations are monotonic when applied to positive data (they are either increasing or decreasing, but not first increasing and then decreasing, or vice versa). The square root transformation often renders Poisson data approximately normal.

Transition A transition is how an object changes state. Transitions consist of a trigger description, an action description, and an optional transition identifier.

Translate In a tessellation, to translate an object means repeating it by sliding it over a certain distance in a certain direction.

Translation A transformation made by substitutions that shift each point of a given graph to a different location without changing the shape of the graph.

Transversal A line or ray that divides other lines or rays.

Trapezoid A four-sided polygon having exactly one pair of parallel sides. The two sides that are parallel are called the bases of the trapezoid. The sum of the angles of a trapezoid is 360 degrees.

Examples:

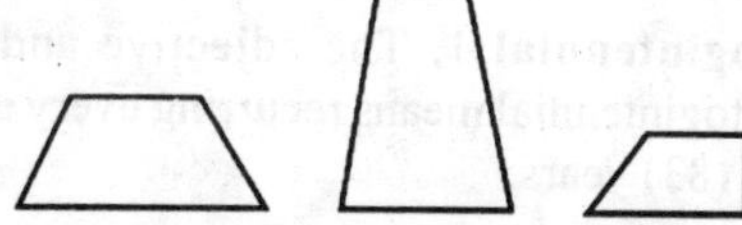

Tre The numeric prefix tre- means three (3). This prefix is derived from the Latin cardinal number tres (III).

Treatment effect The difference between the mean of one treatment (or condition) and the grand mean.

Treatment In experiments, a treatment is what is administered to experimental units (explanatory variables). It does not have to be a medical treatment. Fertilizers in agricultural experiments; different books and multimedia methods in teaching; and chemotherapy of bone marrow transplantation in the treatment of leukemia are examples of treatments in regression analysis.

Tredecennary The noun and adjective tredecennary means a group of thirteen (13) or a period of thirteen (13) years.

Tredecennial 1. The adjective and noun tredecennial means recurring every thirteen (13) years.

2. The adjective and noun tredecennial means lasting thirteen (13) years.

3. The noun and adjective tredecennial means a thirteen (13) year anniversary.

Tredecennium The noun and adjective tredecennium means a period of thirteen (13) years.

Tredecentenary 1. The noun and adjective tredecentenary means a group of one thousand three hundred (1300).

2. The noun and adjective tredecentenary means a period of one thousand three hundred (1300) years.

3. The noun and adjective tredecentenary means a one thousand three hundred (1300) year anniversary.

4. The adjective and noun tredecentenary means one thousand three hundredth (1300th) in rank or order.

Tredecentennial 1. The adjective and noun tredecentennial means recurring every one thousand three hundred (1300) years.

2. The adjective and noun tredecentennial means lasting one thousand three hundred (1300) years.

3. The noun and adjective tredecentennial means a one thousand three hundred (1300) year anniversary.

Tredecentennium The noun and adjective tredecentennium means a period of one thousand three hundred (1300) years.

Tredeci The numeric prefix tredeci- means thirteen (13). This prefix is derived from the Latin cardinal number tredecim (XIII).

Tredecillion The noun and adjective tredecillion means multiplied by 10^{42} in the American number naming system, but means multiplied by 10^{78} in the Chuquet number naming system. .

Tredo The nonstandard prefix tredo- denotes a multiple of ten to the negative thirtieth power (10^{-30} = 10 raised to power −30 = 0.000 000 000 000 000 000 000 000 000 001).

Tree Trees, as special graphs, consist of nodes and edges and are best defined recursively. For every tree one node is singled out and is called the root. One node constitutes a tree and, naturally, is that tree's root. A collection of more than one node is a tree if by removing the root the remaining nodes fall into disjoint trees. Nodes connected to a tree root are called siblings.

Trefoil The noun trefoil means a plant with leaves having three (3) leaflets or a symbolic leaf with three (3) leaflets usually representing the Christian Trinity.

Trend test for counts and proportions A special application of the Chi-squared test (with a different formula) for ordinal data tabulated as a 2xk table. It should be used when the intention is not just to compare the differences between the two groups but to see whether there is a consistent trend towards decrease or increase in the difference between the groups. An example is the association of parental HLA sharing (one-to-four antigens in two loci) with fetal loss in a case-control study (those with recurrent miscarriages and normal fertile couples). A frequent application is the analysis of dose-response relationships. The Chi-squared test for trend has one degree of freedom. The associated P value obtained by the trend test is always smaller than the corresponding P value of an ordinary Chi-squared test. The trend test for counts and proportions is called Cochrane-Armitage trend test. Alternative tests for the analysis of trend are Wilcoxon-Mann-Whitney test or the t-test with use of ordered scores

Trend The general drift, tendency, or direction of a set of data.

Trenonaginta The numeric prefix trenonaginta- means ninety-three (93). This prefix is derived from the Latin cardinal number tres et nonaginta (LXXXXIII).

Trenonagintennary The noun and adjective trenonagintennary means a group of ninety-three (93) or a period of ninety-three (93) years.

Trenonagintennial 1. The adjective and noun trenonagintennial means recurring every ninety-three (93) years.

2. The adjective and noun trenonagintennial means lasting ninety-three (93) years.

3. The noun and adjective trenonagintennial means a ninety-three (93) year anniversary.

Trenonagintennium The noun and adjective trenonagintennium means a period of ninety-three (93) years.

Trenonagintillion The noun and adjective trenonagintillion means multiplied by 10^{282} in the American number naming system, but means multiplied by 10^{558} in the Chuquet number naming system. .

Treoctoginta The numeric prefix treoctoginta- means eighty-three (83). This prefix is derived from the Latin cardinal number tres et octoginta (LXXXIII).

Treoctogintennary The noun and adjective treoctogintennary means a group of eighty-three (83) or a period of eighty-three (83) years.

Treoctogintennial 1. The adjective and noun treoctogintennial means recurring every eighty-three (83) years.

2. The adjective and noun treoctogintennial means lasting eighty-three (83) years.

3. The noun and adjective treoctogintennial means an eighty-three (83) year anniversary.

Treoctogintennium The noun and adjective treoctogintennium means a period of eighty-three (83) years.

Treoctogintillion The noun and adjective treoctogintillion means multiplied by 10^{252} in the American number naming system, but means multiplied by 10^{498} in the Chuquet number naming system. .

Trequadraginta The numeric prefix trequadraginta- means forty-three (43). This prefix is derived from the Latin cardinal number tres et quadraginta (XXXXIII).

Trequadragintennary The noun and adjective trequadragintennary means a group of forty-three (43) or a period of forty-three (43) years.

Trequadragintennial 1. The adjective and noun trequadragintennial means recurring every forty-three (43) years.

2. The adjective and noun trequadragintennial means lasting forty-three (43) years.

3. The noun and adjective trequadragintennial means a forty-three (43) year anniversary.

Trequadragintennium The noun and adjective trequadragintennium means a period of forty-three (43) years.

Trequadragintillion The noun and adjective trequadragintillion means multiplied by 10^{132} in the American number naming system, but means multiplied by 10^{258} in the Chuquet number naming system. .

Trequinquaginta The numeric prefix trequinquaginta- means fifty-three (53). This prefix is derived from the Latin cardinal number tres et quinquaginta (LIII).

Trequinquagintennary The noun and adjective trequinquagintennary means a group of fifty-three (53) or a period of fifty-three (53) years.

Trequinquagintennial 1. The adjective and noun trequinquagintennial means recurring every fifty-three (53) years.

2. The adjective and noun trequinquagintennial means lasting fifty-three (53) years.

3. The noun and adjective trequinquagintennial means a fifty-three (53) year anniversary.

Trequinquagintennium The noun and adjective trequinquagintennium means a period of fifty-three (53) years.

Trequinquagintillion The noun and adjective trequinquagintillion means multiplied by 10^{162} in the American number naming system, but means multiplied by 10^{318} in the Chuquet number naming system. .

Treseptuaginta The numeric prefix treseptuaginta- means seventy-three (73). This prefix is derived from the Latin cardinal number tres et septuaginta (LXXIII).

Treseptuagintennary The noun and adjective treseptuagintennary means a group of seventy-three (73) or a period of seventy-three (73) years.

Treseptuagintennial 1. The adjective and noun treseptuagintennial means recurring every seventy-three (73) years.

2. The adjective and noun treseptuagintennial means lasting seventy-three (73) years.

3. The noun and adjective treseptuagintennial means a seventy-three (73) year anniversary.

Treseptuagintennium The noun and adjective treseptuagintennium means a period of seventy-three (73) years.

Treseptuagintillion The noun and adjective treseptuagintillion means multiplied by 10^{222} in the American number naming system, but means multiplied by 10^{438} in the Chuquet number naming system. .

Tresexaginta The numeric prefix tresexaginta- means sixty-three (63). This prefix is derived from the Latin cardinal number tres et sexaginta (LXIII).

Tresexagintennary The noun and adjective tresexagintennary means a group of sixty-three (63) or a period of sixty-three (63) years.

Tresexagintennial 1. The adjective and noun tresexagintennial means recurring every sixty-three (63) years.

2. The adjective and noun tresexagintennial means lasting sixty-three (63) years.

3. The noun and adjective tresexagintennial means a sixty-three (63) year anniversary.

Tresexagintennium The noun and adjective tresexagintennium means a period of sixty-three (63) years.

Tresexagintillion The noun and adjective tresexagintillion means multiplied by 10^{192} in the American number naming system, but means multiplied by 10^{378} in the Chuquet number naming system. .

Tretriginta The numeric prefix tretriginta- means thirty-three (33). This prefix is derived from the Latin cardinal number tres et triginta (XXXIII).

Tretrigintacentenary 1. The noun and adjective tretrigintacentenary means a group of three thousand three hundred (3300).

2. The noun and adjective tretrigintacentenary means a period of three thousand three hundred (3300) years.

3. The noun and adjective tretrigintacentenary means a three thousand three hundred (3300) year anniversary.

4. The adjective and noun tretrigintacentenary means three thousand three hundredth (3300th) in rank or order.

Tretrigintacentennial 1. The adjective and noun tretrigintacentennial means recurring every three thousand three hundred (3300) years.

2. The adjective and noun tretrigintacentennial means lasting three thousand three hundred (3300) years.

3. The noun and adjective tretrigintacentennial means a three thousand three hundred (3300) year anniversary.

Tretrigintacentennium The noun and adjective tretrigintacentennium means a period of three thousand three hundred (3300) years.

Tretrigintennary The noun and adjective tretrigintennary means a group of thirty-three (33) or a period of thirty-three (33) years.

Tretrigintennial 1. The adjective and noun tretrigintennial means recurring every thirty-three (33) years.

2. The adjective and noun tretrigintennial means lasting thirty-three (33) years.

3. The noun and adjective tretrigintennial means a thirty-three (33) year anniversary.

Tretrigintennium The noun and adjective tretrigintennium means a period of thirty-three (33) years.

Tretrigintillion The noun and adjective tretrigintillion means multiplied by 10^{102} in the American number naming system, but means multiplied by 10^{198} in the Chuquet number naming system. .

Trevigintennary The noun and adjective trevigintennary means a group of twenty-three (23) or a period of twenty-three (23) years.

Trevigintennial 1. The adjective and noun trevigintennial means recurring every twenty-three (23) years.

2. The adjective and noun trevigintennial means lasting twenty-three (23) years.

3. The noun and adjective trevigintennial means a twenty-three (23) year anniversary.

Trevigintennium The noun and adjective trevigintennium means a period of twenty-three (23) years.

Treviginti The numeric prefix treviginti- means twenty-three (23). This prefix is derived from the Latin cardinal number tres et viginti (XXIII).

Trevigínticentenary 1. The noun and adjective trevigínticentenary means a group of two thousand three hundred (2300).

2. The noun and adjective trevigínticentenary means a period of two thousand three hundred (2300) years.

3. The noun and adjective trevigínticentenary means a two thousand three hundred (2300) year anniversary.

4. The adjective and noun trevigínticentenary means two thousand three hundredth (2300th) in rank or order.

Trevigínticentennial 1. The adjective and noun trevigínticentennial means recurring every two thousand three hundred (2300) years.

2. The adjective and noun trevigínticentennial means lasting two thousand three hundred (2300) years.

3. The noun and adjective trevigínticentennial means a two thousand three hundred (2300) year anniversary.

Trevigínticentennium The noun and adjective trevigínticentennium means a period of two thousand three hundred (2300) years.

Trevigintillion The noun and adjective trevigintillion means multiplied by 10^{72} in the American number naming system, but means multiplied by 10^{138} in the Chuquet number naming system. .

Trey The noun trey means a playing card, die, or domino that represents the number three (3).

Tri 1. The Greek-based prefix tri- means three (3).

2. The Latin-based numeric prefix tri- also means three (3). This prefix is derived from the Latin cardinal number tres (III).

Triaconta The prefix triaconta- means thirty (30).

Triacontad The noun and adjective triacontad means a group of thirty (30) members.

Triacontadi The prefix triacontadi- means thirty-two (32). Triacontadi- is an alternate form of the prefix triacontakaidi-.

Triacontaennea The prefix triacontaennea- means thirty-nine (39). Triacontaennea- is an alternate form of the prefix triacontakaiennea-.

Triacontagon The noun and adjective triacontagon means a plain figure with thirty (30) straight sides.

Triacontagram The noun and adjective triacontagram means an interconnected star figure with thirty (30) points formed by stellating a triacontagon.

Triacontahedron The noun and adjective triacontahedron means a solid figure with thirty (30) planar faces.

Triacontahena The prefix triacontahena- means thirty-one (31). Triacontahena- is an alternate form of the prefix triacontakaihena-.

Triacontahepta The prefix triacontahepta- means thirty-seven (37). Triacontahepta- is an alternate form of the prefix triacontakaihepta-.

Triacontahexa The prefix triacontahexa- means thirty-six (36). Triacontahexa- is an alternate form of the prefix triacontakaihexa-.

Triacontakaidi The prefix triacontakaidi- means thirty-two (32).

Triacontakaidigon The noun and adjective triacontakaidigon means a plain figure with thirty-two (32) straight sides.

Triacontakaidigram The noun and adjective triacontakaidigram means an interconnected star figure with thirty-two (32) points formed by stellating a triacontakaidigon.

Triacontakaidihedron The noun and adjective triacontakaidihedron means a solid figure with thirty-two (32) planar faces.

Triacontakaidyad The noun and adjective triacontakaidyad means a group of thirty-two (32) members.

Triacontakaiennea The prefix triacontakaiennea- means thirty-nine (39).

Triacontakaiennead The noun and adjective triacontakaiennead means a group of thirty-nine (39) members.

Triacontakaienneagon The noun and adjective triacontakaienneagon means a plain figure with thirty-nine (39) straight sides.

Triacontakaienneagram The noun and adjective triacontakaienneagram means an interconnected star figure with thirty-nine (39) points formed by stellating a triacontakaienneagon.

Triacontakaienneahedron The noun and adjective triacontakaienneahedron means a solid figure with thirty-nine (39) planar faces.

Triacontakaihena The prefix triacontakaihena- means thirty-one (31).

Triacontakaihenad The noun and adjective triacontakaihenad means a group of thirty-one (31) members.

Triacontakaihenagon The noun and adjective triacontakaihenagon means a plain figure with thirty-one (31) straight sides.

Triacontakaihenagram The noun and adjective triacontakaihenagram means an interconnected star figure with thirty-one (31) points formed by stellating a triacontakaihenagon.

Triacontakaihenahedron The noun and adjective triacontakaihenahedron means a solid figure with thirty-one (31) planar faces.

Triacontakaihepta The prefix triacontakaihepta- means thirty-seven (37).

Triacontakaiheptad The noun and adjective triacontakaiheptad means a group of thirty-seven (37) members.

Triacontakaiheptagon The noun and adjective triacontakaiheptagon means a plain figure with thirty-seven (37) straight sides.

Triacontakaiheptagram The noun and adjective triacontakaiheptagram means an interconnected star figure with thirty-seven (37) points formed by stellating a triacontakaiheptagon.

Triacontakaiheptahedron The noun and adjective triacontakaiheptahedron means a solid figure with thirty-seven (37) planar faces.

Triacontakaihexa The prefix triacontakaihexa- means thirty-six (36).

Triacontakaihexad The noun and adjective triacontakaihexad means a group of thirty-six (36) members.

Triacontakaihexagon The noun and adjective triacontakaihexagon means a plain figure with thirty-six (36) straight sides.

Triacontakaihexagram The noun and adjective triacontakaihexagram means an interconnected star figure with thirty-six (36) points formed by stellating a triacontakaihexagon.

Triacontakaihexahedron The noun and adjective triacontakaihexahedron means a solid figure with thirty-six (36) planar faces.

Triacontakaiocta The prefix triacontakaiocta- means thirty-eight (38).

Triacontakaioctad The noun and adjective triacontakaioctad means a group of thirty-eight (38) members.

Triacontakaioctagon The noun and adjective triacontakaioctagon means a plain figure with thirty-eight (38) straight sides.

Triacontakaioctagram The noun and adjective triacontakaioctagram means an interconnected star figure with thirty-eight (38) points formed by stellating a triacontakaioctagon.

Triacontakaioctahedron The noun and adjective triacontakaioctahedron means a solid figure with thirty-eight (38) planar faces.

Triacontakaipenta The prefix triacontakaipenta- means thirty-five (35).

Triacontakaipentad The noun and adjective triacontakaipentad means a group of thirty-five (35) members.

Triacontakaipentagon The noun and adjective triacontakaipentagon means a plain figure with thirty-five (35) straight sides.

Triacontakaipentagram The noun and adjective triacontakaipentagram means an interconnected star figure with thirty-five (35) points formed by stellating a triacontakaipentagon.

Triacontakaipentahedron The noun and adjective triacontakaipentahedron means a solid figure with thirty-five (35) planar faces.

Triacontakaitetra The prefix triacontakaitetra- means thirty-four (34).

Triacontakaitetrad The noun and adjective triacontakaitetrad means a group of thirty-four (34) members.

Triacontakaitetragon The noun and adjective triacontakaitetragon means a plain figure with thirty-four (34) straight sides.

Triacontakaitetragram The noun and adjective triacontakaitetragram means an interconnected star figure with thirty-four (34) points formed by stellating a triacontakaitetragon.

Triacontakaitetrahedron The noun and adjective triacontakaitetrahedron means a solid figure with thirty-four (34) planar faces.

Triacontakaitri The prefix triacontakaitri- means thirty-three (33).

Triacontakaitriad The noun and adjective triacontakaitriad means a group of thirty-three (33) members.

Triacontakaitrigon The noun and adjective triacontakaitrigon means a plain figure with thirty-three (33) straight sides.

Triacontakaitrigram The noun and adjective triacontakaitrigram means an interconnected star figure with thirty-three (33) points formed by stellating a triacontakaitrigon.

Triacontakaitrihedron The noun and adjective triacontakaitrihedron means a solid figure with thirty-three (33) planar faces.

Triacontaocta The prefix triacontaocta- means thirty-eight (38). Triacontaocta- is an alternate form of the prefix triacontakaiocta-.

Triacontatetra The prefix triacontatetra- means thirty-four (34). Triacontatetra- is an alternate form of the prefix triacontakaitetra-.

Triacontatri The prefix triacontatri- means thirty-three (33). Triacontatri- is an alternate form of the prefix triacontakaitri-.

Triad The noun and adjective triad means a group of three (3) members.

Trial The adjective trial means based on the number three (3).

Triangle of roll A triangle in a rolling offset box that contains the angle of roll. It is usually the first triangle calculated.

Triangle The noun triangle means a plane figure with three (3) vertices.

Triangular matrix An upper triangular matrix is entirely zero below the main diagonal while a lower triangular matrix is zero above the main diagonal.

It the entries of the main diagonal are all equal to 1, the matrix is said to be unit upper triangular or unit lower triangular. For example, you will encounter a unit lower triangular matrix in Gauss elimination.

Simple facts about a triangular matrix A:

The determinant of A is the product of the diagonal entries;

The eigenvalues of A are the diagonal entries;

The inverse of A is also a triangular matrix;

The linear system A * x = y is very easy to solve;

Unless A is actually a diagonal matrix, it is not normal, and hence not unitarily or orthogonally diagonalizable.

Triathlon The noun triathlon means an athletic competition consisting of three (3) separate events.

Tricennial The noun and adjective tricennial means a thirty (30) year anniversary.

Tricentenary 1. The noun and adjective tricentenary means a group of three hundred (300).

2. The noun and adjective tricentenary means a period of three hundred (300) years.

3. The noun and adjective tricentenary means a three hundred (300) year anniversary.

4. The adjective and noun tricentenary means three hundredth (300th) in rank or order.

Tricentennial 1. The adjective and noun tricentennial means recurring every three hundred (300) years.

2. The adjective and noun tricentennial means lasting three hundred (300) years.

3. The noun and adjective tricentennial means a three hundred (300) year anniversary.

Tricentennium The noun and adjective tricentennium means a period of three hundred (300) years.

Tricentisexagesimal The adjective tricentisexagesimal means based on the number three hundred sixty (360).

Tricesimo-secundo The noun tricesimo-secundo (32mo) means a book with leaves printed on one thirty-second (1/32) of the printer sheet.

Tricolor The noun and adjective tricolor means having three (3) distinct colors. The French version of tricolor, tricolore, usually refers to the three-color French flag.

Tricta The prefix tricta- means three hundred (300).

Tricycle The noun, adjective, and verb tricycle means having three (3) wheels.

Trideca The prefix trideca- means thirteen (13).

Trident The noun and adjective trident means having three (3) teeth, processes, or points. Most commonly, trident refers to a three-pronged spear. The trident is the symbol of Poseidon, the Greek god of the sea (known to the Romans as Neptune), and the planet Neptune.

Tridiagonal matrix A tridiagonal matrix is a matrix whose only nonzero entries occur on the main diagonal or on the two diagonals which are immediate neighbors of the main diagonal.

The diagonals immediately below and above the main diagonal are referred to as the (principal) subdiagonal and (principal) superdiagonal respectively.

Here is an example of a tridiagonal matrix which is also positive definite symmetric:

$$\begin{matrix} -2 & 1 & 0 & 0 \\ 1 & -2 & 1 & 0 \\ 0 & 1 & -2 & 1 \\ 0 & 0 & 1 & -2 \end{matrix}$$

Simple facts about a tridiagonal matrix A:

A is a band matrix.

Every matrix is similar to a tridiagonal matrix.

A tridiagonal matrix A is irreducible if and only if every subdiagonal and superdiagonal element is nonzero.

If Gauss elimination can be performed on A without using pivoting, then the L factor is zero except for the diagonal and first subdiagonal, and the U factor is zero except for the diagonal and first superdiagonal, in other words, L and U are lower and upper bidiagonal matrices.

A^{-1} is generally not tridiagonal.

If it is true that none of the subdiagonal elements are zero, and none of the superdiagonal elements are zero, and

$|A_{1,1}| > |A_{1,2}|$

$|A_{n,n}| > |A_{n,n-1}|$

and, for $2 <= i <= n-1$,

$|A_{i,i}| >= |A_{i,i-1}| + |A_{i,i-1}|$

then A is nonsingular, and Gauss elimination can be performed without pivoting. THis is true, for instance, for the –1,2,–1 matrix used to approximate the second derivative, and the 2,1;1,4,1;1,2 matrix used in cubic spline interpolation.

If a (real) matrix A is tridiagonal and irreducible, then its eigenvalues are real and distinct. Moreover, the eigenvalues of the sequence of principal minors of A have a strict interlacing property: the k+1 eigenvalues of the principal minor of order k+1 are strictly separated by the k eigenvalues of the minor of order k. Since the determinant of the minors can be easily computed recursively, and since the sign sequence of these determinants carries information about the number of negative eigenvalues associated with each minor, this suggests how a bisection method can be employed to hunt for the eigenvalues of a tridiagonal matrix.

LAPACK and LINPACK include special routines for efficiently solving linear systems with a tridiagonal coefficient matrix. LAPACK and EISPACK have routines for finding eigenvalues of a tridiagonal matrix.

Cyclic reduction is a method of solving a tridiagonal system of equations which can give a large speedup on vector or parallel processors.

Triennary The noun and adjective triennary means a group of three (3) or a period of three (3) years.

Triennial 1. The adjective and noun triennial means recurring every three (3) years.

2. The adjective and noun triennial means lasting three (3) years.

3. The noun and adjective triennial means a three (3) year anniversary.

Triennium The noun and adjective triennium means a period of three (3) years.

Trigesimal The adjective trigesimal means based on the number thirty (30).

Trigger A trigger is a combination of events and conditions which may activate a state transition.

Triginta The numeric prefix triginta- means thirty (30). This prefix is derived from the Latin cardinal number triginta (XXX).

Trigintennary The noun and adjective trigintennary means a group of thirty (30) or a period of thirty (30) years.

Trigintennial 1. The adjective and noun trigintennial means recurring every thirty (30) years.

2. The adjective and noun trigintennial means lasting thirty (30) years.

3. The noun and adjective trigintennial means a thirty (30) year anniversary.

Trigintennium The noun and adjective trigintennium means a period of thirty (30) years.

Trigintillion The noun and adjective trigintillion means multiplied by 10^{93} in the American number naming system, but means multiplied by 10^{180} in the Chuquet number naming system. .

Trigon The noun and adjective trigon means a plain figure with three (3) straight sides.

Trigonometric addition identities Identities involving the trig functions of sums and differences of angles.

Trigonometric identity An equation made up of trigonometric functions of an angle that is valid for all values of the angle.

Trigonometric ratio A comparison of the measures of the lengths of two sides of a right triangle expressed in fractional or decimal form; there are six trigonometric ratios (sine, cosine, tangent, cotangent, secant, and cosecant) associated with any angle.

Trigonometry The study of right triangle measurements and ratios, useful for calculating indirect measurements.

Trigram The noun and adjective trigram means a word or character sequence consisting of three (3) characters.

Triheptaconta The prefix triheptaconta- means seventy-three (73).

Triicosa The prefix triicosa- means twenty-three (23).

Trilia The prefix trilia- means three thousand (3000).

Trillion The noun and adjective trillion means multiplied by ten to the twelfth power (10^{12} = 10 raised to power +12 = 1,000,000,000,000) in the American number naming system, but means multiplied by ten to the eighteenth power (10^{18} = 10 raised to power +18 = 1,000,000,000,000,000,000) in the Chuquet number naming system. .

Trilogy The noun trilogy means a series of three (3) literary works or musical compositions with a common theme.

Trimeter The noun trimeter means a verse consisting of either three (3) dipodies or three (3) metrical feet.

Trimillennial 1. The adjective and noun trimillennial means recurring every three thousand (3000) years.

2. The adjective and noun trimillennial means lasting three thousand (3000) years.

3. The noun and adjective trimillennial means a three thousand (3000) year anniversary.

Trimillennium The noun and adjective trimillennium means a period of three thousand (3000) years.

Trimmed samples Samples with a percentage of extreme scores removed.

Trimmed statistics Statistics calculated on trimmed samples.

Trimonthly The adjective trimonthly means recurring every three (3) months.

Trinity The noun trinity means the state of having three (3) aspects.

Trinomial A polynomial with three terms.

Trioctaconta The prefix trioctaconta- means eighty-three (83).

Triperfect A number is triperfect if it is multiperfect with index 3.

Only six triperfect numbers are known.

Triple The verb, adjective, adverb, and noun triple means multiply by three (3).

Triplet The noun and adjective triplet means one of three (3) siblings born together or one of three (3) very similar things.

Triplicate The verb, noun, and adjective triplicate means to make two (2) copies of one original, for a total of three (3) objects.

Tris The prefix tris- means three (3).

Trisdecagon The noun trisdecagon means a plane figure with thirteen (13) sides.

Triskaideca The prefix triskaideca- means thirteen (13).

Triskaidecad The noun and adjective triskaidecad means a group of thirteen (13) members.

Triskaidecagon The noun and adjective triskaidecagon means a plain figure with thirteen (13) straight sides.

Triskaidecagram The noun and adjective triskaidecagram means an interconnected star figure with thirteen (13) points formed by stellating a triskaidecagon.

Triskaidecahedron The noun and adjective triskaidecahedron means a solid figure with thirteen (13) planar faces.

Tritetraconta The prefix tritetraconta- means forty-three (43).

Tritriaconta The prefix tritriaconta- means thirty-three (33).

Triumvir The noun triumvir means one (1) of three (3) ruling men. (In Latin, triumvir specifically refers to an adult male.) The plural of triumvir is triumviri.

Triumvirate The noun and adjective triumvirate means a ruling body of three (3) men.

Triyearly 1. The adjective triyearly means recurring every three (2) years.

2. The adjective triyearly can also mean recurring thrice (3) per year.

True The adjective, adverb, and noun TRUE is a logical value. FALSE and TRUE are often used as a binary logic pair.

t-statistics Defined as difference of sample means divided by standard error of difference of sample means.

Tukey's HSD test A multiple comparison procedure for making pairwise comparisons among means while holding the familywise error rate at ?.

Turn of the century The phrase turn of the century means the beginning of a one hundred (100) year period.

Turn of the decade The phrase turn of the decade means the beginning of a ten (10) year period.

Turn of the millennium The phrase turn of the millennium means the beginning of a one thousand (1,000) year period.

Turn To change direction.

Tuvalu The name of the island nation of Tuvalu means the group of eight (8) (islands) in the Tuvaluan language.

Twain The noun, pronoun, and adjective twain is an archaic form of the number two (2).

Tweak Similar to a boost, a tweak is a transformation performed on a candidate MPFN to produce another candidate. Unlike a boost, not all powers are increased. For example, if the candidate contains 43^3 79^2, replacing that with 43^2 79^3 is reasonable because 43^2 provides support for the additional factor of 79, but changing 79^2 to 79^3 removes support for one factor of 43.

Twelfth 1. The adjective, adverb, and noun twelfth (12th) is the standard name for the ordinal number that follows eleventh (11th). Twelfth corresponds with the cardinal number twelve (12).

2. The adjective and noun twelfth (1/12) means divided by twelve (12).

Twelve The noun, pronoun, and adjective twelve (12) is the standard name for the cardinal number that follows eleven (11). Twelve corresponds with the ordinal number twelfth (12th).

Twelvemo The noun twelvemo (12mo) means a book with leaves printed on one twelfth (1/12) of the printer sheet.

Twenties 1. The plural noun and adjective twenties (20s) means an integer multiple of twenty (20).

2. The plural noun and adjective twenties (20's) means the range of numbers at least twenty (20) but less than thirty (30). When refering to dates, twenties means a span of ten years in which the tens (10s) digit of the year number is two (2).

Twentieth 1. The adjective, adverb, and noun twentieth (20th) is the standard name for the ordinal number that follows nineteenth (19th). Twentieth corresponds with the cardinal number twenty (20).

2. The adjective and noun twentieth (1/20) means divided by twenty (20).

Twenty The noun, pronoun, and adjective twenty (20) is the standard name for the cardinal number that follows nineteen (19). Twenty corresponds with the ordinal number twentieth (20th).

Twenty-five The noun and adjective twenty-five (25) is the standard name for the cardinal number that follows twenty-four (24). Twenty-five corresponds with the ordinal number twenty-fifth (25th).

Twenty-Fourmo The noun and adjective twenty-fourmo (24mo) means a book with leaves printed on one twenty-fourth (1/24) of the printer sheet.

Twenty-ninth The compound adjective, adverb, and noun twenty-ninth (29th) is the standard name for the ordinal number that follows twenty-eighth (28th). Twenty-ninth corresponds with the cardinal number twenty-nine (29).

Twenty-one The noun and adjective twenty-one (21) is the standard name for the cardinal number that follows twenty (20). Twenty-one corresponds with the ordinal number twenty-first (21st).

Twenty-second The compound adjective, adverb, and noun twenty-second (22nd or 22^{d}) is the standard name for the ordinal number that follows twenty-first (21st). Twenty-second corresponds with the cardinal number twenty-two (22).

Twenty-seven The noun and adjective twenty-seven (27) is the standard name for the cardinal number that follows twenty-six (26). Twenty-seven corresponds with the ordinal number twenty-seventh (27th).

Twenty-sixth The compound adjective, adverb, and noun twenty-sixth (26th) is the standard name for the ordinal number that follows twenty-fifth (25th). Twenty-sixth corresponds with the cardinal number twenty-six (26).

Twenty-something The compound noun and adjective twenty-something means at least twenty (20) but less than thirty (30). Twenty-something most commonly means between twenty (20) and thirty (30) years of age.

Twenty-three The noun and adjective twenty-three (23) is the standard name for the cardinal number that follows twenty-two (22). Twenty-three corresponds with the ordinal number twenty-third (23rd or 23^{d}).

Twin The noun and adjective twin means one of two (2) siblings born together or one of two (2) very similar things.

Two bits In American use, the compound plural noun two (2) bits means one quarter of a dollar ($0.25). In the early years of the American Republic, one-dollar coins were often split into eight equal parts called bits.

Two tailed test A test that rejects extreme outcomes in either tail of the distribution.

Two tailed test The test of a given statistical hypothesis in which a value of the statistic that is either sufficiently small or sufficiently large will lead to rejection of the hypothesis tested.

The statistical tables for Z and for t provide criticaal values for both one and two tailed tests. That is, they provide the critical values that cut off an entire alpha region at one or the other end of the sampling distribution as well as the critical vfalues that cut off the 1/2 alpha regions at both ends of the sampling distribution.

Two The noun, pronoun, and adjective two (2) is the standard name for the cardinal number that follows one (1). Two corresponds with the ordinal number second (2nd or 2^{d}).

Two thirds The compound noun, pronoun, and adjective two thirds (2/3) is the standard name for the rational fraction equal to two (2) divided by three (3).

Two-bit In American use, the compound adjective two-bit (2-bit) literally means worth one quarter of a dollar ($0.25).

Two-dimensional A figure that has length and width but not height (i.e., a plane figure such as a rectangle or circle).

Two-fisted The compound adjective two-fisted (2-fisted) literally means using both hands. More generally, two-fisted means very vigorous.

Twoscore and ten The compound noun and adjective twoscore and ten means fifty (50).

Two-way anova This method studies the effects of two factors (with several levels) separately (main effect) and, if desired, their effect in combination (interaction).

Type I error You have committed a Type One error if you have rejected the hypothesis tested when it was true.

In a given statistical test, the probability of a type 1 error is equal to the value you have set for alpha

Type II error You have committed a Type II error if you failed to reject the hypothesis tested when a given alternative hypothesis was true.

In a given statistical test, the probability of a type II error is equal to the value calculated for Beta.

U

Ultimate The adjective and adverb ultimate means the last member of a series or the end of an interval.

Umpteen The slang noun umpteen literally means at least thirteen (13) but less than twenty (20). Umpteen more commonly means more than reasonable.

Umpteenth The slang adverb umpteenth literally means at least the thirteenth (13th) but less than the twentieth (20th). Umpteenth more commonly means more often than reasonable.

Una The nonstandard prefix una- denotes a multiple of ten to the thirty-third power (10^{33} = 10 raised to power +33 = 1,000,000,000,000, 000,000,000,000,000, 000,000).

Unary relationship (recursive relationship) A relationship between the instances of one entity type.

Unbiased estimate statistic is said to be an unbiased estimate of a given parameter when the mean of the sampling distribution of that statistic can be shown to be equal to the parameter being estimated.

For example, the mean of a sample is an unbiased estimated of the mean of the population from which the sample was drawn.

s-square calculated on a sample is an unbiased estimate of the variance of the population from which the sample was drawn.

s-square divided by n (the size of the sample) is an unbiased estimate of the variance of the sampling distribution of means for random samples of size n and the square root of this quantity is called the standard error of the mean. It is a commonly used index of the error entailed in estimating a population mean based on the information in a random sample of size n.

We will use the symbol → to indicate that the mean of the sampling distribution of the statistic to the left of the arrow is equal to the quantity indicated to the right of the arrow

For example:

$$\bar{X} \rightarrow \mu_x$$

$$S_x^2 \rightarrow \sigma_x^2$$

Unbiased estimator A statistic whose expected value is equal to the parameter to be estimated.

Unconditional probability The probability of one event ignoring the occurrence or nonoccurrence of some other event.

Uncountable The adjective uncountable means infinite, or more generally, a very large number of individuals or discrete objects.

Uncounted The adjective uncounted means a very large number of individuals or discrete objects.

Undeca The prefix undeca- means eleven (11).

Undecagon The noun undecagon means a plane figure with eleven (11) straight sides.

Undecennary The noun and adjective undecennary means a group of eleven (11) or a period of eleven (11) years.

Undecennial 1. The adjective and noun undecennial means recurring every eleven (11) years.

2. The adjective and noun undecennial means lasting eleven (11) years.

3. The noun and adjective undecennial means an eleven (11) year anniversary.

Undecennium The noun and adjective undecennium means a period of eleven (11) years.

Undecentenary 1. The noun and adjective undecentenary means a group of one thousand one hundred (1100).

2. The noun and adjective undecentenary means a period of one thousand one hundred (1100) years.

3. The noun and adjective undecentenary means a one thousand one hundred (1100) year anniversary.

4. The adjective and noun undecentenary means one thousand one hundredth (1100th) in rank or order.

Undecentennial 1. The adjective and noun undecentennial means recurring every one thousand one hundred (1100) years.

2. The adjective and noun undecentennial means lasting one thousand one hundred (1100) years.

3. The noun and adjective undecentennial means a one thousand one hundred (1100) year anniversary.

Undecentennium The noun and adjective undecentennium means a period of one thousand one hundred (1100) years.

Undeci The numeric prefix undeci- means eleven (11). This prefix is derived from the Latin cardinal number undecim (XI).

Undecillion The noun and adjective undecillion means multiplied by 10^{36} in the American number naming system, but means multiplied by 10^{66} in the Chuquet number naming system. .

Undecimal The adjective undecimal means based on the number eleven (11).

Undenary 1. The adjective and noun undenary means eleventh (11th) in order or importance.

2. The adjective undenary means having eleven (11) parts.

Underdetermined system An underdetermined linear system is a linear system A * x = b in which the values of N variables are sought, but for fewer than N linearly independent equations are given.

Of course, it is possible to set up a linear system where the number of equations M equals or exceeds the number of variables, but for which the number of linearly independent equations is less than N. Thus, you can't assume that M e" N rules out the undetermined case. It is the number of linearly independent equations that sets this. But of course, if you start with fewer than N equations, you are guaranteed an undetermined system.

An underdetermined linear system typically has an infinite number of solutions which constitute an affine linear space. The dimension of the linear space is usually called the number of degrees of freedom.

Uni The numeric prefix uni- means one (1). This prefix is derived from the Latin cardinal number unus (I).

Unicycle The noun, adjective, and verb unicycle means having one (1) wheel.

Unified modeling language (UML) A notation that allows the modeler to specify, visualize, and construct the artifacts of software systems, as well as business models.

Uniform circular motion Circular motion about a point at a uniform linear and angular velocity.

Uniform distribution A distribution in which all possible outcomes have an equal chance of occurring. Also known as a rectangular distribution.

Unimodal A distribution having one distinct peak.

Unimodular matrix A unimodular matrix is a square matrix whose determinant has absolute value 1.

Facts about a unimodular matrix A:

The inverse matrix A^{-1} is unimodular;

If A is an integer matrix, then so is A^{-1};

If B is unimodular, so is A*B;

Examples of unimodular matrices:

the identity matrix;

any permutation matrix;

any orthogonal matrix;

Any diagonal, upper triangular, or lower triangular matrix whose diagonal elements have a product of +1 or –1.

Union constraint The union constraint indicates that every object in a generalization object class must also be a member of at least one of the specialization object classes.

Union of sets The union of two or more sets is the set of all the objects contained by at least one of the sets. The symbol for union is U.

Unique The adjective unique means singular or one (1) of a kind.

Unique is one of the most misused words in contemporary English. Unique is often erroneously used to mean distinctive, uncommon, or unusual. While all individuals are unique, a class can be unique only if it has but one member. Thus a restaurant can be unique only if it is the only restaurant in existence. Since unique is absolute, it has no comparative or superlative forms. Thus more unique and most unique are incorrect.

Unit circle A circle with a radius of one unit.

Unit testing Each module is tested alone in an attempt to discover any errors in its code.

Unit The noun unit means one (1) entity. The plural noun and adjective units (1s) means an integer multiple of one (1).

Unitary matrix A unitary matrix is a complex matrix U whose transpose complex conjugate is equal to its inverse:

$U^{-1} = U^H$

Facts about a unitary matrix U

$U * U^H = U^H * U = I$

U is L2–norm preserving: $||U*x||_2 = ||x||_2$

the columns of U are pairwise orthogonal vectors, and have unit vector L2 norm.

In real arithmetic, the corresponding concept is an orthogonal matrix.

Unitary similarity transformation A unitary transformation is a relationship between two complex matrices A and B, and a unitary matrix V, of the form:

$A = V^{-1} * B * V.$

Units of measurement Terms which describe fixed standard measurements.

Univariate design An experimental design having only one dependent variable.

Unnonaginta The numeric prefix unnonaginta– means ninety–one (91). This prefix is derived from the Latin cardinal number unus et nonaginta (LXXXXI).

Unnonagintennary The noun and adjective unnonagintennary means a group of ninety-one (91) or a period of ninety-one (91) years.

Unnonagintennial 1. The adjective and noun unnonagintennial means recurring every ninety-one (91) years.

2. The adjective and noun unnonagintennial means lasting ninety-one (91) years.

3. The noun and adjective unnonagintennial means a ninety-one (91) year anniversary.

Unnonagintennium The noun and adjective unnonagintennium means a period of ninety-one (91) years.

Unnonagintillion The noun and adjective unnonagintillion means multiplied by 10^{276} in the American number naming system, but means multiplied by 10^{546} in the Chuquet number naming system.

Unnumberable The adjective unnumberable means infinite, or more generally, a very large number of individuals or discrete objects.

Unnumbered The adjective unnumbered means a very large number of individuals or discrete objects.

Unoctoginta The numeric prefix unoctoginta- means eighty-one (81). This prefix is derived from the Latin cardinal number unus et octoginta (LXXXI).

Unoctogintennary The noun and adjective unoctogintennary means a group of eighty-one (81) or a period of eighty-one (81) years.

Unoctogintennial 1. The adjective and noun unoctogintennial means recurring every eighty-one (81) years.

2. The adjective and noun unoctogintennial means lasting eighty-one (81) years.

3. The noun and adjective unoctogintennial means an eighty-one (81) year anniversary.

Unoctogintennium The noun and adjective unoctogintennium means a period of eighty-one (81) years.

Unoctogintillion The noun and adjective unoctogintillion means multiplied by 10^{246} in the American number naming system, but means multiplied by 10^{486} in the Chuquet number naming system.

Unquadraginta The numeric prefix unquadraginta- means forty-one (41). This prefix is derived from the Latin cardinal number unus et quadraginta (XXXXI).

Unquadragintennary The noun and adjective unquadragintennary means a group of forty-one (41) or a period of forty-one (41) years.

Unquadragintennial 1. The adjective and noun unquadragintennial means recurring every forty-one (41) years.

2. The adjective and noun unquadragintennial means lasting forty-one (41) years.

3. The noun and adjective unquadragintennial means a forty-one (41) year anniversary.

Unquadragintennium The noun and adjective unquadragintennium means a period of forty-one (41) years.

Unquadragintillion The noun and adjective unquadragintillion means multiplied by 10^{126} in the American number naming system, but means multiplied by 10^{246} in the Chuquet number naming system. .

Unquinquaginta The numeric prefix unquinquaginta- means fifty-one (51). This prefix is derived from the Latin cardinal number unus et quinquaginta (LI).

Unquinquagintennary The noun and adjective unquinquagintennary means a group of fifty-one (51) or a period of fifty-one (51) years.

Unquinquagintennial 1. The adjective and noun unquinquagintennial means recurring every fifty-one (51) years.

2. The adjective and noun unquinquagintennial means lasting fifty-one (51) years.

3. The noun and adjective unquinquagintennial means a fifty-one (51) year anniversary.

Unquinquagintennium The noun and adjective unquinquagintennium means a period of fifty-one (51) years.

Unquinquagintillion The noun and adjective unquinquagintillion means multiplied by 10^{156} in the American number naming system, but means multiplied by 10^{306} in the Chuquet number naming system.

Unreduced matrix A reduced matrix is a (usually square) matrix A which may be broken up into submatrices

A11 | A12

——+—

0 | A22

where A11 is square, and the submatrix below it is completely zero.

An unreduced matrix is of course a matrix which does not have this form.

An irreducible matrix is one which cannot be put into reduced form by a renumbering of the variables. (The same permutation must be applied to both rows and columns.)

An unreduced tridiagonal matrix is one which has no extra zeroes, that is, the immediate superdiagonal and subdiagonal have no zero entries.

An unreduced upper Hessenberg matrix is one which has no extra zeroes, that is, the immediate subdiagonal has no zero entries.

Facts about a (square) reduced matrix A:

A is singular if and only if one or both of A11 and A22 are singular.

det(A) = det(A11) * det(A22)

the eigenvalues of A are the eigenvalues of A11 plus the eigenvalues of A22.

Unreplicated factorial A single replicate of a 2^k design (where each of k factors of interest has only two levels).

Unseptuaginta The numeric prefix unseptuaginta- means seventy-one (71). This prefix is derived from the Latin cardinal number unus et septuaginta (LXXI).

Unseptuagintennary The noun and adjective unseptuagintennary means a group of seventy-one (71) or a period of seventy-one (71) years.

Unseptuagintennial 1. The adjective and noun unseptuagintennial means recurring every seventy-one (71) years.

2. The adjective and noun unseptuagintennial means lasting seventy-one (71) years.

3. The noun and adjective unseptuagintennial means a seventy-one (71) year anniversary.

Unseptuagintennium The noun and adjective unseptuagintennium means a period of seventy-one (71) years.

Unseptuagintillion The noun and adjective unseptuagintillion means multiplied by 10^{216} in the American number naming system, but means multiplied by 10^{426} in the Chuquet number naming system. .

Unsexaginta The numeric prefix unsexaginta- means sixty-one (61). This prefix is derived from the Latin cardinal number unus et sexaginta (LXI).

Unsexagintennary The noun and adjective unsexagintennary means a group of sixty-one (61) or a period of sixty-one (61) years.

Unsexagintennial 1. The adjective and noun unsexagintennial means recurring every sixty-one (61) years.

2. The adjective and noun unsexagintennial means lasting sixty-one (61) years.

3. The noun and adjective unsexagintennial means a sixty-one (61) year anniversary.

Unsexagintennium The noun and adjective unsexagintennium means a period of sixty-one (61) years.

Unsexagintillion The noun and adjective unsexagintillion means multiplied by 10^{186} in the American number naming system, but means multiplied by 10^{366} in the Chuquet number naming system. .

Untriginta The numeric prefix untriginta- means thirty-one (31). This prefix is derived from the Latin cardinal number unus et triginta (XXXI).

Untrigintacentenary 1. The noun and adjective untrigintacentenary means a group of three thousand one hundred (3100).

2. The noun and adjective untrigintacentenary means a period of three thousand one hundred (3100) years.

3. The noun and adjective untrigintacentenary means a three thousand one hundred (3100) year anniversary.

4. The adjective and noun untrigintacentenary means three thousand one hundredth (3100th) in rank or order.

Untrigintacentennial 1. The adjective and noun untrigintacentennial means recurring every three thousand one hundred (3100) years.

2. The adjective and noun untrigintacentennial means lasting three thousand one hundred (3100) years.

3. The noun and adjective untrigintacentennial means a three thousand one hundred (3100) year anniversary.

Untrigintacentennium The noun and adjective untrigintacentennium means a period of three thousand one hundred (3100) years.

Untrigintennary The noun and adjective untrigintennary means a group of thirty-one (31) or a period of thirty-one (31) years.

Untrigintennial 1. The adjective and noun untrigintennial means recurring every thirty-one (31) years.

2. The adjective and noun untrigintennial means lasting thirty-one (31) years.

3. The noun and adjective untrigintennial means a thirty-one (31) year anniversary.

Untrigintennium The noun and adjective untrigintennium means a period of thirty-one (31) years.

Untrigintillion The noun and adjective untrigintillion means multiplied by 10^{96} in the American number naming system, but means multiplied by 10^{186} in the Chuquet number naming system. .

Unvigintennary The noun and adjective unvigintennary means a group of twenty-one (21) or a period of twenty-one (21) years.

Unvigintennial 1. The adjective and noun unvigintennial means recurring every twenty-one (21) years.

2. The adjective and noun unvigintennial means lasting twenty-one (21) years.

3. The noun and adjective unvigintennial means a twenty-one (21) year anniversary.

Unvigintennium The noun and adjective unvigintennium means a period of twenty-one (21) years.

Unviginti The numeric prefix unviginti- means twenty-one (21). This prefix is derived from the Latin cardinal number unus et viginti (XXI).

Unvigínticentenary 1. The noun and adjective unvigínticentenary means a group of two thousand one hundred (2100).

2. The noun and adjective unvigínticentenary means a period of two thousand one hundred (2100) years.

3. The noun and adjective unvigínticentenary means a two thousand one hundred (2100) year anniversary.

4. The adjective and noun unvigínticentenary means two thousand one hundredth (2100th) in rank or order.

Unvigínticentennial 1. The adjective and noun unvigínticentennial means recurring every two thousand one hundred (2100) years.

2. The adjective and noun unvigínticentennial means lasting two thousand one hundred (2100) years.

3. The noun and adjective unvigínticentennial means a two thousand one hundred (2100) year anniversary.

Unvigínticentennium The noun and adjective unvigínticentennium means a period of two thousand one hundred (2100) years.

Unvigintillion The noun and adjective unvigintillion means multiplied by 10^{66} in the American number naming system, but means multiplied by 10^{126} in the Chuquet number naming system. .

Unweighted means Row or column means based on the average of the cell means in that row or column—without giving greater weight to cells with more observations. Also known as equally weighted means.

Upper case CASE tools designed to support the systems planning and selection, systems analysis, and systems design phases of the systems development life cycle.

Upshift matrix The upshift matrix A circularly shifts all vector entries or matrix rows up 1 position.

Example:

0 1 0 0

0 0 1 0

0 0 0 1

1 0 0 0

Facts about the upshift matrix A:

A is a permutation matrix;

A is an N-th root of the identity matrix;

A is persymmetric;

A is a circulant matrix;

the inverse of the upshift matrix is the downshift matrix.

Use case A complete sequence of related actions initiated by an actor; it represents a specific way to use the system.

Use case diagram A diagram that depicts the use cases and actors for a system.

User documentation Written or other visual information about an application system, how it works, and how to use it.

Valid argument An explicit demonstration or proof that has been shown to be true.

Validate To give evidence that a solution or process is correct.

Validities The correlations of individual predictor variables with the criterion.

Validity The degree to which a variable measures what it is intended to measure.

Value A function f:A->B is defined in A and takes values in B. Thus, for a $\in$ A (a in A), f takes on the value of f(a) $\in$ B.

Value of a series The simplified sum of a series.

Variability Numbers that describe how spread out a set of data is. (e.g., range and quartile).

Variable Some characteristic that varies among experimental units (subjects) or from time to time. A variable may be quantitative or categorical. A quantitative variable is either discrete (assigning meaningful numerical values to observations: number of children, dosage in mg) or continuous (such as height, weight, temperature, blood pressure; also called interval veriable). A categorical variable is either nominal (assigning observations to categories: gender, treatment, disease subtype, groups) or ordinal (ranked variables: low, median, high dosage). Conventionally, a random variable is shown by a capital letter, and the data values it takes by lower case letters.

Variance inflation factor (VIF) The reciprocal of the tolerance—the degree to which the standard error of b_j is increased because of the degree to which X_j is correlated with the other predictors.

Variance ratio Mean square ratio obtained by dividing the mean square (regression) by mean square (residual). The variance ratio is assessed by the F-test using the two degrees of freedom (k-1, N-k).

Variance sum law The rule giving the variance of the sum (or difference) of two or more variables.

$$S_{X\pm Y} = S_X^2 + S_Y^2 \pm 2rs_X s_Y$$

Variance The major measure of variability for a data set. To calculate the variance, all data values, their mean, and the number of data values are required. It is expressed in the squared unit of measurement. Its square root is the standard deviation. It is symbolised by P^2 for a population and S^2 for a sample .

Vector addition Process of combining two vectors.

Vector bound matrix norm A vector-bound matrix norm is a matrix norm that has been (or can be) derived from a vector norm by the following formula:

||A|| = supremum ||A*x|| / ||x||

where the supremum (roughly, the maximum) is taken over all nonzero vectors x.

If such a relationship holds, then expressions involving the matrix norm and vector norm can be mingled to produce useful inequalities, based on the guaranteed compatiblity relationship:

$\|A*x\| <= \|A\| * \|x\|$

Matrix norms which are vector bound with some vector norm include the L1 matrix norm, the L2 matrix norm, and the L Infinity matrix norm.

Vector norm A vector norm is a function $\|*\|$ that measures the size of a vector.

A vector norm must have the following properties:

$\| V \| >= 0$, and $\| V \| = 0$ if and only if V is the zero vector (positivity);

$\| s * V \| = | s | * \| V \|$ for any scalar s (linearity).

$\| V + W \| <= \| V \| + \| W \|$ for any vectors V and W (triangle inequality).

Commonly used vector norms include:

The L1 vector norm;

The L2 vector norm;

The L Infinity vector norm.

Given two points in space, x and y, we can define the distance between the points, d(x,y), in terms of a vector norm operating on the vectors of position coordinates:

$d(x,y) = \| x - y \|$

and this quantity will have the expected properties of a distance function.

For a given vector norm, it is important to know which matrix norms are compatible, so that expressions like

$\|A*x\| <= \|A\| * \|x\|$

may be asserted.

Vector quantity A quantity that has both size and direction.

Vector space A vector is a quantity having magnitude and direction, represented by a directed arrow indicating its orientation in space. Vector space is the three dimensional area where vectors can be plotted.

Vectors Matrices with either one row or one column.

Velocity The rate of change of position over time is velocity, calculated by dividing distance by time.

Velocity This is the rate of change of position (only differing from speed in that it can be negative if you are moving left along the number line.) It is obtained by taking the derivative of your position function.

Velocity vector A vector representing the speed and direction of a moving object.

Venn diagram A diagram where sets are represented as simple geometric figures, with overlapping and similarity of sets represented by intersections and unions of the figures .

Venn diagrams A way of representing shared and unshared variances by a set of overlapping circles.

Vertex (vertices) The points where two line segments come together (corners).

Vertex of a parabola The point on a parabola midway between the focus and the directrix.

Vertex The point at which two straight lines come together to form the angle.

Vertical angles For any two lines that meet, such as in the diagram below, angle AEB and angle DEC are called vertical angles. Vertical angles have the same degree measurement. Angle BEC and angle AED are also vertical angles.

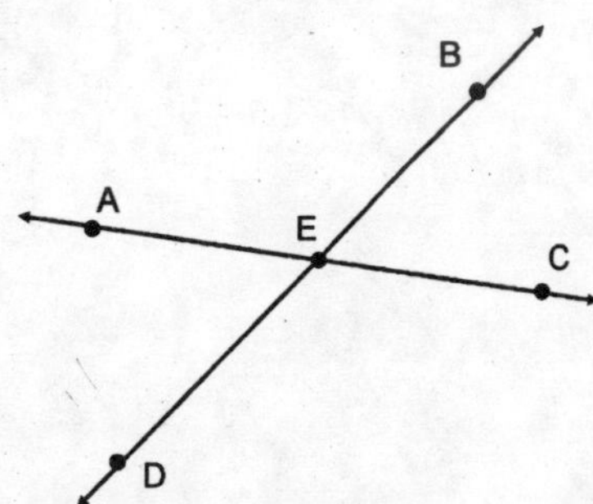

Vertical line test If no vertical line intersects a graph of a relation more than once, then the relation is a function.

Vertical shift The vertical displacement of a function above or below the horizontal axis.

Vertices of a hyperbola The two points in which a hyperbola is intersected by the line through the two focus points.

Vertices of an ellipse The endpoints of the major and minor axes.

Vicesimal The adjective vicesimal means based on the number twenty (20).

Vicesimo-quarto The noun vicesimo-quarto (24mo) means a book with leaves printed on one twenty-fourth (1/24) of the printer sheet.

Vigesimal The adjective vigesimal means based on the number twenty (20).

Vigintennary The noun and adjective vigintennary means a group of twenty (20) or a period of twenty (20) years.

Vigintennial 1. The adjective and noun vigintennial means recurring every twenty (20) years.

2. The adjective and noun vigintennial means lasting twenty (20) years.

3. The noun and adjective vigintennial means a twenty (20) year anniversary.

Vigintennium The noun and adjective vigintennium means a period of twenty (20) years.

Viginti The numeric prefix viginti- means twenty (20). This prefix is derived from the Latin cardinal number viginti (XX).

Vigintillion The noun and adjective vigintillion means multiplied by 10^{63} in the American number naming system, but means multiplied by 10^{120} in the Chuquet number naming system.

Vinculum The bar over the radicand.

Volume Volume is a measure of how much space a space figure takes up. Volume is used to measure a space figure just as area is used to measure a plane figure. The volume of a cube is the cube of the length of one of its sides. The volume of a box is the product of its length, width, and height.

Example: What is the volume of a cube with side-length 6 cm?

The volume of a cube is the cube of its side-length, which is $6^3 = 216$ cubic cm.

Example: What is the volume of a box whose length is 4cm, width is 5 cm, and height is 6 cm?

The volume of a box is the product of its length, width, and height, which is $4 \times 5 \times 6 = 120$ cubic cm.

Walkthrough A peer group review of any product created during the systems development process; also called a structured walkthrough.

Web server A computer that is connected to the Internet and stores files written in HTML (hypertext markup language) that are publicly available through an Internet connection.

Weighted means Row or column means where the cell sizes are used to weight the cell means.

Welch satterthwaite solution A solution to the problem of comparing means with heterogeneous variances—independently arrived at by Welch and Satterthwaite.

Welch-Satterthwaite t-test The Welch-Satterthwaite t-test is an alternative to the pooled-variance t-test, and is used when the assumption that the two populations have equal variances seems unreasonable. It provides a t statistic that asymptotically approaches a t-distribution, allowing for an approximate t-test to be calculated when the population variances are not equal.

Well structured relation (table) A relation that contains a minimum amount of redundancy and allows users to insert, modify, and delete the rows without errors or inconsistencies.

Whiskers Lines drawn in a boxplot from hinges to adjacent values.

Whole number A whole number is a rational number with no fractional part, i.e. an integer.

Wilcoxon matched pairs signed rank T-test A non-parametric significance test analogous to paired t-test. Most suitable for ordinal or interval/ratio variables .It is used to test for the significance of the difference between the distributions of two non-independent samples (repeated measures or matched pairs).

Wilcoxon's matched-pairs signed-ranks test A nonparametric test for comparing the central tendency of two matched (related) samples.

Wilcoxon's rank-sum test A nonparametric test for comparing two independent groups. It is functionally equivalent to the Mann-Whitney U test.

William's correction (for G statistics) This is equivalent to Yates' continuity correction for Chi-squared test but used in likelihood ratio (G) statistics for 2x2 tables .

Winsorized samples Samples in which extreme values have been trimmed and replaced by the most extreme value(s) remaining in the distribution.

Woolf-Haldane analysis A method first described by Woolf and later modified by Haldane for the analysis of 2x2 table and relative incidence (relative risk) calculation. It is the preferred method for relative risk calculation when one of the cells has a zero using the formula: RR = (2a+1)(2d+1) / (2b+1)(2c+1). Since it is a modification of the cross-product ratio, it should be called odds ratio.

Work breakdown structure The process of dividing the project into manageable tasks and logically ordering them to ensure a smooth evolution between tasks.

Wrapping function The function that maps the real numbers onto the points of the unit circle.

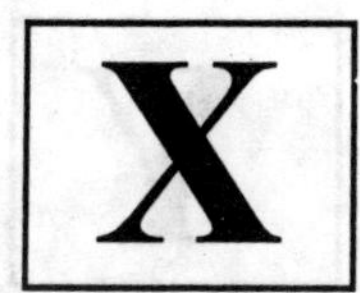

X-intercept The x-coordinate of the point where the line crosses the x-axis.

Y The metric prefix symbol Y- (metric prefix: yotta-) denotes a multiple of ten to the twenty-fourth power (10^{24} = 10 raised to power +24 = 1,000,000,000,000,000,000,000,000) as defined by Le Système International d'Unités (SI).

Yates's correction The approximation of the Chi-square statistic in small 2x2 tables can be improved by reducing the absolute value of differences between expected and observed frequencies by 0.5 before squaring. This correction, which makes the estimation more conservative, is usually applied when the table contains only small observed frequencies (<20). The effect of this correction is to bring the distribution based on discontinuous frequencies nearer to the continuous Chi-squared distribution. This correction is best suited to the contingency tables with fixed marginal totals. Its use in other types of contingency tables (for independence and homogeneity) results in very conservative significance probabilities. This correction is no longer needed since exact tests are available.

Yearly The adjective yearly means recurring once (1) each year.

Yes The adverb and noun yes is a logical value. No and yes are often used as a binary logic pair.

Y-intercept The y-coordinate of the point where the line crosses the y-axis.

Yocto The metric prefix yocto- (prefix symbol: y-) denotes a multiple of ten to the negative twenty-fourth power (10^{-24} = 10 raised to power –24 = 0.000 000 000 000 000 000 000 001) as defined by Le Système International d'Unités (SI).

Yotta The metric prefix yotta- (prefix symbol: Y-) denotes a multiple of ten to the twenty-fourth power (10^{24} = 10 raised to power +24 = 1,000,000,000,000,000,000,000,000) as defined by Le Système International d'Unités (SI).

Z

Z The metric prefix symbol Z- (metric prefix: zetta-) denotes a multiple of ten to the twenty-first power (10^{21} = 10 raised to power +21 = 1,000,000,000,000,000,000,000) as defined by Le Système International d'Unités (SI).

Z matrix A Z matrix is a real (square) matrix with nonpositive off-diagonal entries.

A Z matrix is a candidate to be an M matrix, and over 50 different conditions are known that will guarantee that a Z matrix is, in fact, an M matrix.

Z score Z scores are a special application of the transformation rules. The z score for an item, indicates how far and in what direction, that item deviates from its distribution's mean, expressed in units of its distribution's standard deviation. The mathematics of the z score transformation are such that if every item in a distribution is converted to its z score, the transformed scores will necessarily have a mean of zero and a standard deviation of one.

Z scores are sometimes called standard scores. The z score transformation is especially useful when seeking to compare the relative standings of items from distributions with different means and/or different standard deviations.

Z scores are especially informative when the distribution to which they refer, is normal. In every normal distribution, the distance between the mean and a given Z score cuts off a fixed proportion of the total area under the curve. Statisticians have provided us with tables indicating the value of these proportions for each possible Z score.

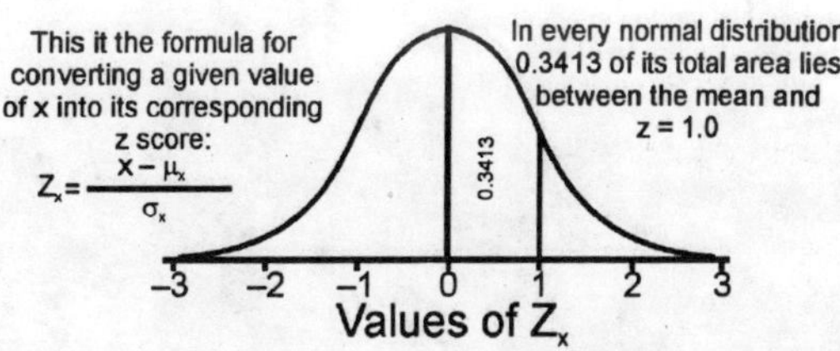

Zeno A Greek philosopher best known for Zeno's Paradox. He pointed out that for a runner to get from A to B, he or she must first traverse half the distance, and then half the remaining distance and then half the remaining distance ad infinitum. Since, clearly, the runner cannot perform infinitely many steps in a finite amount of time, motion is an impossibility, and is therefore an illusion.

Zepto The metric prefix zepto- (prefix symbol: z-) denotes a multiple of ten to the negative twenty-first power (10^{-21} = 10 raised to power −21 = 0.000 000 000 000 000 000 001) as defined by Le Système International d'Unités (SI).

Zero 1. The noun and adjective zero (0) is the standard name for the cardinal number that precedes one (1). Zero may be defined as the cardinal number of the set that contains nothing.

There is no ordinal number that corresponds with the cardinal number zero, although zero is often used to denote a point of origin. For instance, t_0 may indicate the start of a period of observation, t_1 the first event, t_2 the second event, etc.

2. The verb zero means to set to zero (0).

Zero algebraic vector An algebraic vector whose components are both zero.

Zero degree angle An angle with no space between the two lines. In other words, the two lines are occupying the same space. The difference between a straight angle and a zero degree angle is that the vertex is in the middle of the line for a straight angle but at the end of the line for a zero degree angle.

Zero matrix The zero matrix is a matrix all of whose entries are zero.

A zero matrix is sometimes called a trivial matrix. A matrix which has at least one nonzero entry is called a nontrivial matrix or nonzero matrix. (It is not required that all entries be nonzero, just that at least one of them is nonzero)

Zero vector A vector with a magnitude of zero and any direction.

Zeroth The adjective zeroth means pertaining to zero (0). Zeroth is not an ordinal number.

Zetta The metric prefix zetta- (prefix symbol: Z-) denotes a multiple of ten to the twenty-first power (10^{21} = 10 raised to power +21 = 1,000,000,000,000,000,000,000) as defined by Le Système International d'Unités (SI).

Zillion The slang noun and adjective zillion means an extremely large number. Zillion is a parody of the Chuquet number naming system.